Mathematics of Supervised Learning

This book discusses the relevance of probabilistic supervised learning, to the pursuit of automated and reliable prediction of an unknown that is in a state of relationship with another variable. The book provides methods for secured mechanistic learning of the function that represents this relationship between the output and input variables, where said learning is undertaken within the remit of real-world information that can be messy in different ways. For example, the available data may be highly multivariate or be high-dimensional, reflecting the nature of the output variable that could be a vector, or matrix, or even higher in dimension, as is often the case in a real-world application. Additionally, the data is noisy, and often it is small to moderately large in size in multiple applications. Another difficulty that regularly arises is that the training dataset – comprising pairs of values of the input and output –is such, that the sought function cannot be captured by a parametric shape, but is instead underlined by an inhomogeneous correlation structure. These difficulties notwithstanding, we desire a streamlined methodology that allows the learning of the inter-variable relationship – to ultimately permit fast and reliable predictions of the output, at newly recorded values of the input. In fact, occasions arise when one seeks values of the input at which a new output value is recorded, and such a demand is also addressed in the book.

The generic solution to the problem of secured supervised learning amidst real-world messiness, lies in treating the sought inter-variable relation as a (function-valued) random variable, which, being random, is ascribed a probability distribution. Then recalling that distributions on the space of functions are given by stochastic processes, the sought function is proposed to be a sample function of a stochastic process. This process is chosen as one that imposes minimal constraints on the sought function – identified as a Gaussian Process (GP) in the book. Thus, the sought function can be inferred upon, as long as the covariance function of the underlying GP is learnt, given the available training set. The book presents probabilistic techniques to undertake said learning, within the challenges borne by the data, and illustrates such techniques on real data. Learning of a function is always followed by closed-form prediction of the mean and dispersion of the output variable that is realised at a test input.

To help with the background, the book includes reviews on stochastic processes and basic probability theory. This will render the first half of the book useful for students across disciplines, while the latter half will be appreciated by students of numerate subjects at the postgraduate level or higher, including students of computational sciences, statistics and mathematics.

Dr. Dalia Chakrabarty is a Reader in Statistical Data Science in the Department of Mathematics at the University of York. Her D.Phil is from St. Cross College in the University of Oxford, and she works on the development of methods to permit the probabilistic learning of random variables of various kinds, given real world data that is diversely challenging.

Supervised Learning

Mathematical Foundations and Real-world Applications

Dalia Chakrabarty

CRC Press
Taylor & Francis Group
Boca Raton London New York

CRC Press is an imprint of the
Taylor & Francis Group, an **informa** business
A CHAPMAN & HALL BOOK

Designed cover image: Dalia Chakrabarty

First edition published 2025
by CRC Press
2385 NW Executive Center Drive, Suite 320, Boca Raton FL 33431

and by CRC Press
4 Park Square, Milton Park, Abingdon, Oxon, OX14 4RN

CRC Press is an imprint of Taylor & Francis Group, LLC

ISBN: 978-1-032-28330-2 (hbk)
ISBN: 978-1-032-28334-0 (pbk)
ISBN: 978-1-003-29633-1 (ebk)

DOI: 10.1201/9781003296331

Typeset in CMR12 font
by KnowledgeWorks Global Ltd.

Publisher's note: This book has been prepared from camera-ready copy provided by the authors.

prajñapteḥ sanimittatvam anyathā dvayanṣ̄ś̄ataḥ [1]

To my dearest Makhnai - the world's easier since you came.

[1]from the Mandukya Upanishad [40]

Contents

Preface

Success of today's energised pursuit of automation in the real-world, is sensitively predicated upon reliability, and the capacity for reliable predictions to be "quickly" generated. Irrespective of the problem-specificness of such referred quickness, it is true in general, that reliable predictions will stem only from reliability of the learning that underscores the sought predictions. Here reliability harbours reproducibility within itself, such that your attempt at the learning, (followed by your predictions), concurs with similar endeavours undertaken by me – within a given information framework. Then it is imperative that the learning (and the subsequent prediction) is implemented via easy-to-follow protocols, while the all-important demand for reliability drives said protocols to be seeded in mathematically-sound learning methodologies. If invoking the relevant maths allows for fast and reliable predictions at the same time, then the emerging methodology is even better suited for real-world applications. Additionally, the insistence on reliability of the learning and predictions is not independent of the need for objective uncertainties to accompany all reported results. This urges the usage of a fully Bayesian approach that guarantees comprehensive uncertainties on all learning and related predictions. In fact, ease of production and interpretability of uncertainties is maximal within such an approach.

In this book, it is supervised learning that we focus on. So, the book is dedicated to a discussion of the learning of the function that represents the relationship between the input and the target variables. Indeed, the ulterior aim of this function learning is the prediction of outputs that are realised at respective test inputs. So the book covers such a prediction exercise. A crucial challenge is in devising a learning technique that facilitates reliable learning, even when the data – comprising information on the outputs realised at the design points – indicates disparate correlation between a pair of such output variables realised at one pair of design points, as distinguished from another output pair realised elsewhere. The operational word being "reliable", resorting to a parametric model for the sought function is not a possibility – lest the parametric shape of the model is misplaced in some subsets of the available data in a new application, over others. The sensible alternative to such a shortcoming is then addressed only by opting to undertake non-parametric modelling of said function. In other words, we appreciate the need to learn the shape of the sought function, and not impose a parametric shape on this unknown function that we aim to learn given the available data. This need for embracing a non-parametric model is clarified through intuitive arguments in Chapter 1, and the taster of a preliminary attempt at non-parametric modelling is offered in Chapter 3. Such a non-parametric modelling of the sought function, in fact flows from our consideration of the shape of the function to be an unknown. This leads to the function then being considered a random variable within a Bayesian framework, where we seek values of this random variable, given the available data. This in turn justifies our search for a probability distribution for this function variable, culminating in modelling this (random) function as a sample function of a stochastic process. After all, the distribution of a function-valued variable is given by a stochastic process.

The search for a general model that is least constrained – and that which can generalise easily across dimensions – leads to the process being chosen as a Gaussian Process (GP). Another crucial advantage of such GP-based learning is the closed-form prediction of the full distribution (and therefore the expectation and variance) of the output variable that is realised at a test input. Such capacity then links to the imperative requirement of fast (in real time) predictions, as has been mentioned above. Thus, GP-based supervised learning holds the promise of usage in AI implementation, as long as accuracy of the predictions can be ascertained, in the data at hand, i.e. as long as the learning is reliable. Such reliability is assured if the correlation of the data can be accurately mimicked in the construction of the correlation function of the GP that is invoked to generate the sought function. Such competence is developed in Chapters 4 and 6. The protocols for the implementation of the methodologies discussed in the book, are clear streamlined algorithms, rendering GP-based learning reproducible – in addition to being explainable. For completion, we recall from above that predictions are fast. As long as the size of the input-output pairs in the available training set is small to moderately large, (i.e. $\lesssim 10^3$), GP-based supervised learning ticks all the boxes required for automated implementation.

Preliminaries of the GP-based learning exercise are discussed in Chapter 3, and illustrated upon, using a dataset that is used extensively in the book, for the illustration of multiple learning techniques. This is the publicly-available data on energy consumption in a Moroccan city, as associated with multiple metereological parameters. However, what still remains unaddressed, is the aforementioned difficulty arising from heterogeneity – across inputs – in the correlations between output pairs in the data that is used to perform the learning. This then calls for two advancements:

— a clear identification – and parametrisation – of what inhomogeneity in the correlation structure of the data entails;

— and secondly, a reliable duplication of the empirical correlation structure by the correlation structure of the GP that has been invoked to model the sought function.

Early in Chapter 4, is presented a definition of inhomogeneity in the empirical correlation structure, and a new parameter that informs on such inhomogeneity. There, it is also motivated that such inhomogeneities demand that the GP correlation function – that is computed at a pair of input values – be designed to depend on the value of each input in this input pair, not just via their mutual difference. Then imposing a chosen shape on this expressly non-stationary GP correlation function, will not result in a generic model – for similar reasons as stated above when discouraging parametric modelling of the sought function in favour of non-parametric function learning. A parametric model for the GP correlation function might yield good predictions in some data, and compromised results for other datasets. Indeed, accuracy of predictions is of paramount importance if AI implementation of the developed learning technologies is hoped for. Therefore, the need arises for the non-parametric modelling of this GP correlation structure. The simultaneous demand of such non-parametric modelling and of the dependence of the correlation function on the input *pair*, leads to a highly generalised model, which is shown in Chapter 4, to however be unimplementable!

The good news is that the information content of such a model can be replicated by a simpler model, assuming the system to be ergodic. This development is at the core of the 4-th chapter. Crucially, the method is developed while eschewing any unproved assumptions and *ad hoc* imposition of values to relevant parameters. We realise that such a developed method can then learn the originally-sought function non-parametrically, and learn the GP correlation function non-parametrically too. So we refer to this model as fully non-parametric. This model is explicitly non-stationary and generalises easily across dimensions. Chapter 5 is dedicated to an empirical illustration of the learning of a real-world high-dimensional function using the fully non-parametric modelling strategy developed in Chapter 4. The 4-th chapter also includes illustrations of this learning strategy, on the aforementioned energy consumption data – that is first called into Chapter 3. Additionally, Chapter 4 presents details of a multivariate function learning to inform on the sub-surface material density function at any point in the bulk of a lab-grown nano-brick.

Incorporation of measurement noise in the with-uncertainty learning is discussed throughout the book. The extant technique of predicting the mean and variance of each output at a test input is used in all the empirical illustrations, but when discussing the concept of "predicting the input" – at which a test output value is recorded – in Chapter 3, the incorporation of errors in the measurement of the input is also discussed. To be precise, in such exercises, the input is *learnt*, rather than predicted, and in addition to consulting examples of real-world situations that demand this, Chapter 3 includes an empirical illustration of such input learning, (at test output values), as does Chapter 5. In fact, the application in Chapter 5 invokes the learning of a high-dimensional function.

All through the first five chapters, the empirical illustrations employ uninformative priors given the lack of information on parameters, (often of the GP correlation function). However, it is possible to extract an informative prior on the sought parameters, as the learning is progressing. Such a prior – that is referred in the book as "self assembled" – is presented in Chapter 6. Other contexts for usage of a similar prior are discussed in this chapter as well, and the undertaken empirical illustration on the energy consumption data, shows the comparatively higher predictive capacity of the learning strategy that chooses to use such a prior.

As indicated in the blurb on Chapter 6 in the last paragraph, the methods presented in the book sometimes link to the sequential progress in the learning of the relevant parameters. Thus, the methods link to the inference that is being carried out to realise the undertaken learning exercise – to be specific, undertaken within the Bayesian framework. Such inference is via sampling from the posterior of the sought parameters (given the data), i.e. via Markov Chain Monte Carlo (MCMC) based techniques, in the illustrations presented here. However, undertaking Bayesian inference otherwise is most definitely possible within the discussed methods. It then becomes imperative that we discuss MCMC in the book, and this is undertaken in Chapter 2. Indeed, Chapter 2 expounds on some aspects of the Bayesian framework, and includes a few simple illustrative examples. Chapter 1 on the other hand, includes a primer on the aspects of probability theory that are relevant to the learning discussed Chapter 3 onward. Thus, the first two chapters provide the background material that is relevant to the GP-based learning that is discussed here. In fact, Chapter 3 could also be viewed as comprising preliminary material – this time on GP-based learning. It is the 4-th chapter onward that the book includes new techniques, (sometimes motivated by new proofs). In this sense, the first three chapters of the book are within the grasp of undergraduate students of Probability and Statistics/Data Science, while the latter chapters will require a deeper appreciation of Computational Statistics. I have indeed attempted to retain the book as a tool for a wide readership who may not be/have been students of Mathematics at the undergraduate level. Thus, there is an attempt made here to explain terms before usage, and stay away from a necessarily mathematical presentation of the material, as much as was possible. Learning functions that represent inter-variable relations is ubiquitously demanded, across disciplines. The attempt made here is in keeping with such ubiquity, and in encouragement of usage of GP-based supervised learning as an AI tool. This is the attempt irrespective of whether

— the sought function is univariate, or multivariate, and/or high-dimensional.

— the data at hand bears inhomogeneities in its correlation structure – where parametrisation of said inhomogeneities is introduced – forcing the sought function to bear an inhomogeneous correlation structure, and introducing a mathematically full-proof way for a fully non-parametric learning of such functions with a GP.

— the application is blessed with information to generate informative priors, in lieu of which, a self-assembled informative prior is introduced.

— the output has to be predicted at a test input, or we seek the input at which a fresh value of the output is recorded.

I have thoroughly enjoyed the process of writing this book, and am always open to suggestions, offers of discussions and criticisms from colleagues – across disciplines. This of course is an evolving field, and I consider this book to be a temporally-local point of view from the perspective of a statistical learner. I am indebted to so many colleagues over the years for their inputs; their visions; their stubborn affiliation to one mindset over another; their flexibility of thinking; and ultimately, their passion for the subject. I have been humbled and inspired by their contributions.

Lastly, I would please like to thank my husband and our daughter for their love and unwavering support in putting up with me.

Oxfordshire,
July, 2024

Acknowledgements

I gratefully acknowledge the 5th chapter as joint work with Dr. Kangrui Wang (Alan Turing Institute).

There are multiple figures in Chapter 3 that are produced by Mr. Kane Warrior (University of York, Department of Mathematics). I thank Kane for his kind sharing of these figures with me. The text in Chapter 3 specifies which figures are Kane's work.

I also acknowledge the contribution by Prof. S. Paul (De Montford University, Leicester), in growing a nano-brick in his laboratory and imaging the same with Scanning Electron Microscopes, to then share the image data with me. Learning using data generated with such image data is presented in Chapter 4.

I am indebted to Ms. Niharika C. Paul (Max Planck Institute, Mathematics in the Sciences) for her careful reading of the draft and valuable suggestions.

1

Inter-variable relationships

1.1 Introduction

The identification of the relationship between a pair of mutually-associated variables, is of ubiquitous interest across disciplines, where the ulterior motivation behind identifying such inter-variable relation, is to predict values of one variable, at newly recorded values of the other. It could be that interest lies in the learning of the dependence of stock prices on temporally-evolving market parameters; of the progression of a disease on the parametrisation of patient lifestyle, co-morbidities, treatment, etc.; of quality of material grown in a laboratory, on growth conditions; of what parents cook for their child, on the latter's preferences and what the parents judge to be nutritionally optimal; etc. Ultimately, knowledge of the nature of the dependence of the output on the articulated input, is the prime requisite for the desired prediction. The need to identify/understand/quantify/estimate/learn the (functional) relationship between two random variables (r.v.s), manifests in multiple spheres of day-to-day life including: in social sciences [66, 61]; in laboratory-based experimental studies [90, 116, 75, 76]; in applications of finance and in economics [86, 36]; from the pursuit of biological phenomena and medicine [2, 103, 33, 34, 24]; the universe sciences [117, 26]; and elsewhere too [6, 119].

As hinted by the above examples, accessing *training data* is imperative, such that an inter-variable relationship is identified as a function of the input variable, where this function yields values of the output variable. Here, training data refers to the set of input-output pairs, where we know values of the output, realised at a chosen value of the input, i.e. at a *design point* [68]. Then

DOI: 10.1201/9781003296331-1

prediction of either variable, at newly-observed values, (or *test data*) on the other, is possible in principle by simply reading off this identified functional relationship between the input and output variables.

Our recollection of science classes in school, might propose the idea of *fitting* parametric curves to the training data, as a means of identifying the input-output functional relation. However, such an idea is fundamentally flawed – for multiple reasons. Firstly, the data on the output variable, at a given design point, is essentially noisy; noise might accompany the identification of a design point as well. We refer to this by stating that the "training data is noisy". So if we seek to identify the function that links the input to the output, by *fitting* parametric shapes to such a noisy training set, ambiguity will cloud such intended fitting – which part of the error bar on the output (at a design point) do we use to train the parameters of this chosen parametric curve? It would be desirable to achieve the estimation/learning of the sought function in a way that acknowledges measurement errors organically and objectively.

Secondly – and more importantly – we ask if information pre-exists, to constrain the nature of the sought function, between consecutive design points. Framed another way, this is a question on how the extant information can constrain the nature of the correlation between values of the sought function, realised at any two values of the input variable, at least one of which is not a design point. Data is limited to values of this function realised at the design points only, and in lieu of other information about the correlation structure underlying this sought function, there is in general, no empirical justification behind imposing a chosen parametrised form at those values of the input variable, that are not included within the training set. Essentially, one wonders why adding a local maxima – or for that matter, a localised trough – between the i-th and $i+1$-th design points, would not be allowed in the data, for any i where $i \in \{1, \ldots, n\}$, with n data points in the training set.

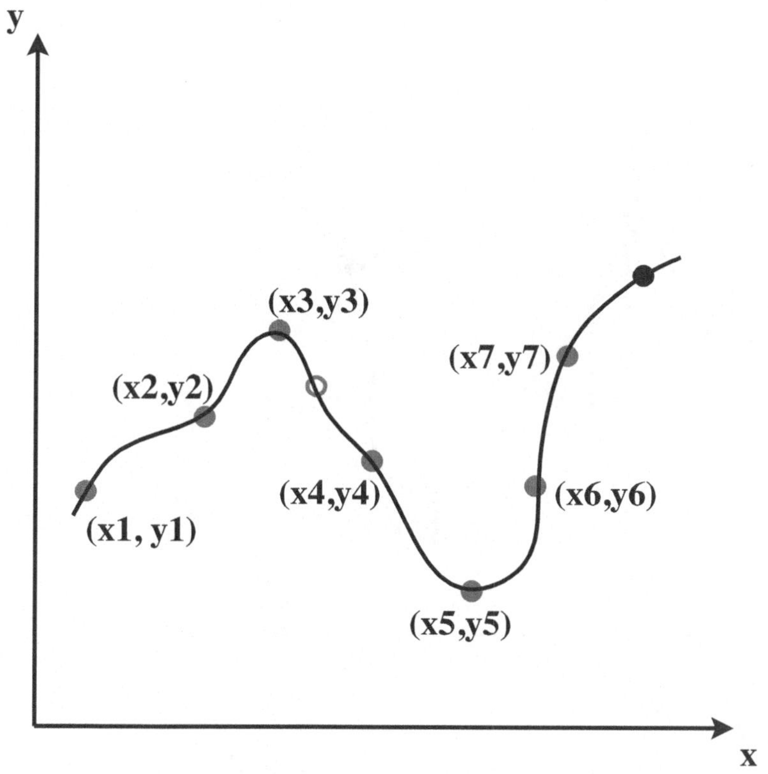

FIGURE 1.1
Cartoon to schematically depict the functional relationship between a random variable (r.v.) X and another r.v. Y that is sought in supervised learning, given the training set {(x1,y1), (x2,y2), ..., (x7,y7)}. A candidate for the sought function is presented as the black curve, while the training data points are depicted in red, (or gray in the monochromatic version). The ultimate aim of learning the functional relation between X and Y is to predict either variable, given test data on the other; in this figure, we seek the prediction of the output of the test point depicted in open circle, or forecast the output depicted in the black filled circle.

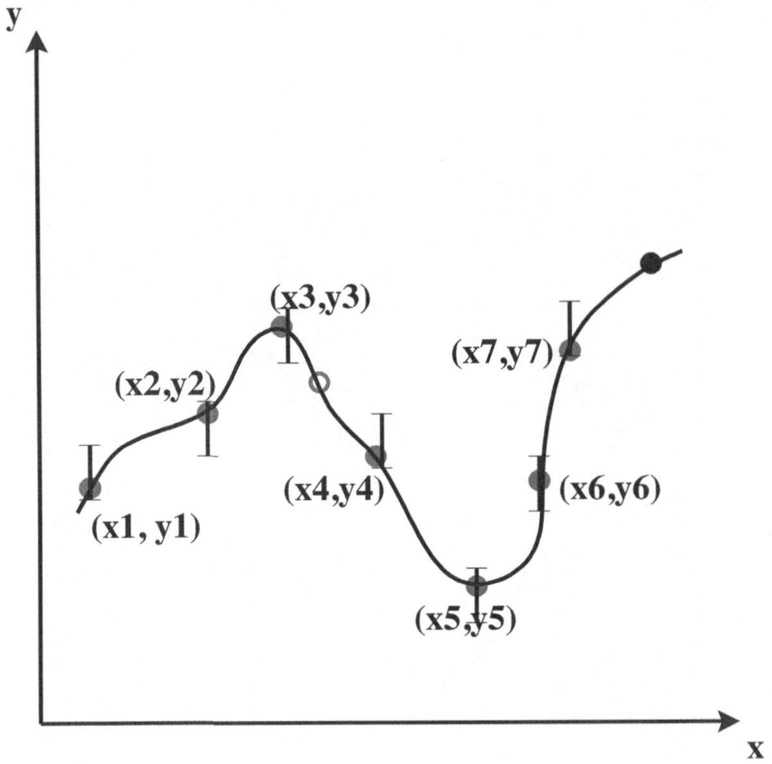

FIGURE 1.2
Cartoon representing the attempt at supervised learning of the functional relation between r.v.s X and Y, as depicted in Figure 1.1, except that here the learning is depicted as rendered difficult when one or both r.v.s admit noisy measurements – noise in the measurement of Y is depicted as error bars.

If the method of identifying the function could somehow learn the correlation structure of the sought inter-variable relationship, given the correlation structure that underlines the data, that would be a desirable property of the method. This is of critical importance, and though we will appear to address it in the third chapter – while advancing a methodology for such correlation learning – we will return to it even later, to expose

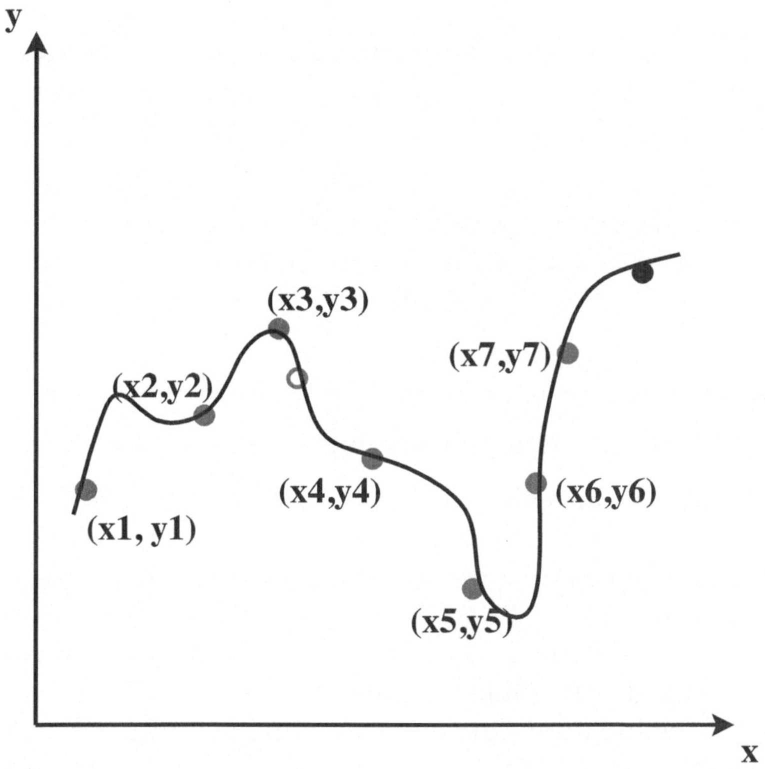

FIGURE 1.3
The learning of the correlation structure of the sought functional relation between r.v.s X and Y is desired within supervised learning. Fitting of parametric forms to the available training data, cannot capture correlation structure of the available data. If fitting parametric forms to the data, one can ask if there exists information to impede the addition of a bump, or a discontinuity between a considered pair of consecutive design points? What stops us from making the bump between x_1 and x_2 in Figure 1.2 more pronounced, or from adding a local minima between x_5 and x_6? In lieu of any information that impedes such arbitrary reshaping of the function, the same have been imposed, as depicted in this figure. Note that prediction performed with the function in the previous figure, will then differ from that done with the function fitted to the data here.

the flaw in that suggestion, and in the same breath, motivate a general model, (Chapter 4).

The drawback of fitting parametric forms to the available data, in an effort to estimate the sought function, is exacerbated when this inter-variable functional relation is high-dimensional.

1.2 Sought function is generally high-dimensional

Indeed, it is possible in real-world situations, that the output is not a scalar, but a vector, or a matrix-valued r.v., or even higher-dimensional; we will consider the output variable to be a *tensor-valued* r.v. in general.

As far as the input variable is concerned, we will see in the third chapter that the higher the dimensionality of the input, the harder the learning of the input-output functional form becomes. We will give the reason for this later, but following on from this suggestion, we now have a reason for choosing the higher-dimensional between the two variables to be the output. In this book, the learning methodology will be motivated, with the input considered a vector-valued r.v., unless otherwise stated. Thus, the univariate, scalar-valued functional relation that is depicted in the cartoons in Figure 1.1, Figure 1.2 and Figure 1.3 is only a schematic representation of the function that we wish to learn in the generic situation, in which

- the output variable $\boldsymbol{Y}$ is a k-th order tensor of dimensionality $m_1 \times m_2 \times \ldots \times m_k$, i.e. $\boldsymbol{Y} \in \mathbb{R}^{m_1 \times m_2 \times \ldots \times m_k}$; $(k, m_i \in \mathbb{Z}^+;\ i = 1, \ldots, k)$, and
- the input variable $\boldsymbol{X}$ is a vector of dimension d, i.e. $\boldsymbol{X} \in \mathbb{R}^d$, $d \in \mathbb{Z}^+$,
- such that (s.t.) the functional relationship between this output variable called $\boldsymbol{Y}$ and the input variable called

$\boldsymbol{X}$ is a k-th ordered, tensor-valued valued function $\boldsymbol{f} : \mathbb{R}^d \longrightarrow \mathbb{R}^{m_1 \times m_2 \times \ldots \times m_k}$,

- which we aim to learn, given the training set $\mathbf{D} = \{(\boldsymbol{x}_i, \boldsymbol{y}_i)\}_{i=1}^n$, where the data on the output alone is denoted $\mathbf{D}_{\boldsymbol{Y}} := \{\boldsymbol{y}_1, \ldots, \boldsymbol{y}_n\}$, with $\boldsymbol{y}_i$ realised at the i-th design point $\boldsymbol{X} = \boldsymbol{x}_i$, $(i = 1, \ldots, n)$,
- to eventually predict the values of $\boldsymbol{Y}$ that are realised at the test data $\boldsymbol{x}^{(test)}$, or to predict $\boldsymbol{X}$ at which test data $\boldsymbol{y}^{(test)}$ is realised.

In this book, we adhere to the convention that the name of a random variable, is in capital letters of the English alphabet, while realisation of the same is in the corresponding small letter.

Hence, the function $\boldsymbol{f}(\cdot)$ that we seek to learn, is s.t. at a given value of the vector-valued input $\boldsymbol{X}$, it churns out the values of the output $\boldsymbol{Y}$ – which, by virtue of being tensor-valued, renders the sought function $\boldsymbol{f}(\cdot)$ tensor-valued too, with the same dimensions as the output $\boldsymbol{Y}$, [57]. This k-th ordered tensor-valued function $\boldsymbol{f}(\cdot)$ is composed of multiple, scalar-valued component functions $f_{q_1,\ldots,q_k}(\cdot)$, $\forall q_i \in \{1, 2, \ldots, m_i\}$; $i = 1, \ldots, k$, any of which is a function of the input variable $\boldsymbol{X}$, s.t. a pair of component functions is correlated in the same way as the corresponding components of the observations on the output variable $\boldsymbol{Y}$.

The very form of $\boldsymbol{f}(\cdot)$ is unknown, and following appreciation of the shortcomings of attempts at fitting parametric forms to the training data, we anticipate the modelling of the unknown functional relation between $\boldsymbol{X}$ and $\boldsymbol{Y}$ as a random function, as a possible way to learn it.

1.3 Imperative that Stochastic Processes be invoked

To appreciate the difficulty in building a high-dimensional function, component-by-component, let us consider the example of a matrix-valued output $\boldsymbol{Y}$, (that is in a state of relationship with the input vector $\boldsymbol{X}$). In our example, $\boldsymbol{Y}$ is an $m_1 \times m_2$-dimensional matrix-valued random variable, (and $\boldsymbol{X}$ is a d-dimensional vector-valued random variable). Then by virtue of the model equation $\boldsymbol{Y} = \boldsymbol{f}(\boldsymbol{X})$, the function $\boldsymbol{f}(\cdot)$ is also a matrix-valued function, of dimensions $m_1 \times m_2$, i.e. learning $\boldsymbol{f}(\cdot)$ is equivalent to the learning of $m_1 m_2$ number of component functions $\{f_{ij}(\cdot)\}_{i=1;j=1}^{m_1,m_2}$.

Before proceeding further, we would like to clarify doubts that may linger amongst readers, about the model equation that we have stated in the last paragraph; absence of an error term may be questioned in the equation $\boldsymbol{Y} = \boldsymbol{f}(\boldsymbol{X})$. Such a question, if relevant, would of course be relevant for an output, (and therefore the function $\boldsymbol{f}(\cdot)$), of any dimensionality. Let us address this question now.

In this model equation, the function $\boldsymbol{f}(\cdot)$ is an unknown function, which we will soon treat as a random function, i.e. a function that attains a certain form with a probability, given the data on $\boldsymbol{Y}$, and the designed $\boldsymbol{X}$ values. In other words, there are uncertainties relevant to the form attained by this function, given a training set, and we will embark on uncertainty-included learning of this random function given the noisy data at hand. When it comes to realisations of the output (or input) variable, the prediction (or learning) of values of the output variable at test inputs, (or of missing values of the input at known output), will be shown to be undertaken, with uncertainties. In our Bayesian approach, all uncertainties will be presented as Highest Probability Density credible regions, [94], but we do not want to get ahead of ourselves at this stage. Here it suffices to say that uncertainties will be very carefully and objectively addressed in all the methods that we will present in this book.

So, to return to the model equation $\boldsymbol{Y} = \boldsymbol{f}(\boldsymbol{X})$, $\boldsymbol{f}(\cdot)$ presents the relationship between the output variable called "$\boldsymbol{Y}$", and the input called "$\boldsymbol{X}$", where $\boldsymbol{f}(\cdot)$, which is a random function that is to-be learnt. Unlike in a linear regression setting, the error in the prediction of the output is *not additive* in general. In fact, unlike in the regression setting, $\mathbb{E}(\boldsymbol{Y}|\boldsymbol{x}^{(test)})$ may not be closed form, for the test input $\boldsymbol{x}^{(test)}$). (We have not discussed prediction yet, but knowing that here we are not linearly regressing $\boldsymbol{Y}$ against $\boldsymbol{X}$, there is no guarantee that the expectation of the output realised at a test input will be closed-form – we shall see). So we will state the $\boldsymbol{X} - \boldsymbol{Y}$ relationship as $\boldsymbol{Y} = \boldsymbol{f}(\boldsymbol{X})$, with the aim of undertaking uncertainty-included learning of the random function $\boldsymbol{f}(\cdot)$ given data $\mathbf{D}$. Thereafter, as indicated in the last paragraph, the uncertainty-included prediction of realisations of $\boldsymbol{Y}$ at $\boldsymbol{X} = \boldsymbol{x}^{(test)}$ will follow in general, by sampling from the probability distribution of $\boldsymbol{Y}$, conditional on this test input and the model of $\boldsymbol{f}(\cdot)$ that we would have learnt given data $\mathbf{D}$. (If said distribution is s.t. such sampling can be undertaken in a closed-form way, then that will simplify the prediction exercise). To summarise, expressing the state of relationship between $\boldsymbol{X}$ and $\boldsymbol{Y}$ as $\boldsymbol{Y} = \boldsymbol{f}(\boldsymbol{X})$ does not in any way suppress the learning of uncertainties.

Now that the validity of the model equation has been established, we proceed to illustrate the difficulty/impossibility in building the matrix-valued $\boldsymbol{f}(\cdot)$ in our considered example, by choosing "adequate" forms for the component functions $f_{ij}(\cdot)$, $(i = 1, \ldots, m_1;\ j = 1, \ldots, m_2)$.

In Figure 1.4 we depict the matrix-valued output $\boldsymbol{Y}$ that is realised at each of n design points $\boldsymbol{x}_1, \ldots, \boldsymbol{x}_n$. The n realisations of $\boldsymbol{Y}$ are stacked up on the left side of this figure, while the values of $\boldsymbol{f}(\cdot)$ computed at these n design points are shown on the right side of the figure. To clarify, in Figure 1.4, realisations of the (matrix-valued) output $\boldsymbol{Y}$ – at each of the design points that reside in the relevant training set – are collated into an array that is depicted to the left of the figure, while values of the function $\boldsymbol{f}(\cdot)$ computed at respective design points are collated into an array that is presented on the right. This figure presents corresponding elements of a realisation of the matrix-valued output $\boldsymbol{Y}$, (shown as a small gray square; a triangle; a circle; and a black filled upturned triangle – drawn on the sheets depicted on the left of the figure), and the components of the matrix-valued function $\boldsymbol{f}(\cdot)$, (shown as a white filled rectangle; a gray triangle; and a black filled rectangle – drawn on the sheets depicted on the right).

Thus, if we are interested in building the random function $\boldsymbol{f}(\cdot)$ component-by-component, we realise the following.

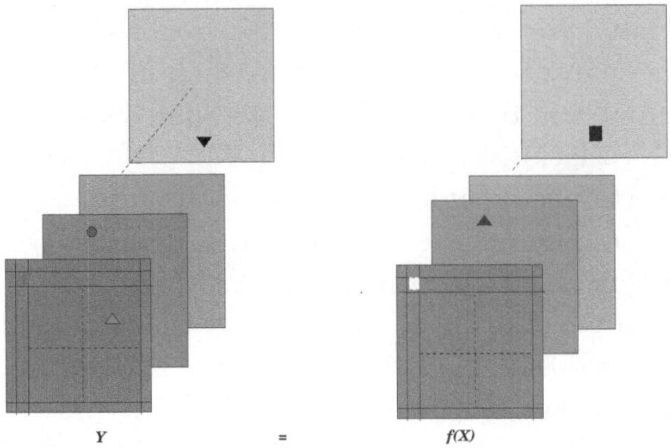

FIGURE 1.4
Cartoon of example model equation for matrix-valued output $\boldsymbol{Y}$. The n values of $\boldsymbol{Y}$ are stacked up, as depicted on the left (in shades of blue in the coloured version), s.t. the i-th rectangle on the left, depicts $\boldsymbol{y}_i$ that is generated at the i-th design point $\boldsymbol{x}_i$. Each matrix value of $\boldsymbol{Y}$ is depicted by a rectangle that comprises m_1 rows and m_2 columns that are shown using straight lines in black, on the front-most rectangle only – in the stack of rectangles on the left. Four token elements of these matrices are depicted: as a small square in red, (or gray in the monochromatic version); as a (green) triangle; as a (purple) circle; and (black) upturned triangle. Values of $\boldsymbol{f}(\cdot)$ computed at the n design points, $(\boldsymbol{x}_1, \ldots, \boldsymbol{x}_n)$ are stacked up in order of the index on the design point, and depicted as rectangles on the right, (in shades of brown in the coloured version). Some of the scalar-valued component functions of the $m_1 \times m_2$-dimensional matrix-valued function $\boldsymbol{f}(\cdot)$, are shown – as a white square; as a (dark blue) triangle embedded on the second-from-front rectangle on the right; and as a black filled square on the last rectangle drawn on the right. The cartoon suggests that the component of the output value $\boldsymbol{y}_1$, depicted as the gray (or red) small square on the front-most left rectangle, is at the same row and column, as is the component function computed at $\boldsymbol{x}_1$, shown in white on the front-most rectangle on the right. Again, the component of $\boldsymbol{y}_2$ shown as a circle (in purple) on the second-from-front rectangle on the left, is the equally-indexed component function of the function $\boldsymbol{f}(\cdot)$ computed at $\boldsymbol{X} = \boldsymbol{x}_2$, where the computed value of this component function is shown as a triangle (in dark blue) on the second-from-from rectangle on the right. Again, corresponding components of $\boldsymbol{y}_n$ and of the function $\boldsymbol{f}(\boldsymbol{x}_n)$, are depicted respectively as the filled upturned black triangle and the filled small black square.

— The correlation ($c_{small-rectangle;\ triangle}$) between components of output (drawn inside the depiction of $\boldsymbol{y}_1$), will constrain the form of the component function computed at $\boldsymbol{X} = \boldsymbol{x}_1$, (drawn as a small white rectangle on the front sheet on the right). Such is affected by imposing $c_{small-rectangle;\ triangle}$ as the correlation between this component function shown in white, with another component function that is also computed at $\boldsymbol{x}_1$, where the latter function bears the same row and column indices as the component of $\boldsymbol{y}_1$ depicted as a triangle.

— Additionally, the component function in white – computed at $\boldsymbol{x}_1$ – will have to be formulated s.t. it is correlated with the component function in the dark gray triangle drawn on the second (from the front) sheet on the right side, (i.e. computed at $\boldsymbol{x}_2$), in the same way, as the rectangular component of $\boldsymbol{y}_1$ correlates with the circular component of $\boldsymbol{y}_2$, drawn on the second (from the front) sheet on the left.

— The component functions shown on the right sheets, as white rectangle and gray triangle will have to be designed s.t. their mutual correlation is dictated by the corresponding elements of $\boldsymbol{Y}$ as stated above, while the same component functions will also need to abide by the constraints that their respective correlations with the function shown as the black rectangle – computed at $\boldsymbol{X} = \boldsymbol{x}_n$ – should be as per the correlation between the small gray rectangle and the circle components of $\boldsymbol{Y}$, with the black upturned triangle component drawn on the left sheets.

— The above are only a few of the inter-component correlations that the sought component functions will have to abide by. Each component function that we hope to choose/design in this approach, computed at distinct design points, will have to abide by a myriad inter-dependent constraints, in the form of their mutual correlations, as imposed via correlations between different components of the output variables.

Said constraints are indeed probabilistic by nature, since the correlation between a pair of random scalar-valued component functions is of relevance here. That too, if the set of design points in the training data changes, the structure of these "probabilistic constraints" will change, and we will need to redesign the function $\boldsymbol{f}(\cdot)$. We appreciate that the complexity of the building of the function in this approach will increase with dimensionality and size of the output. All these problems are further compounded, if noise in values of the output variable is invoked. So such an approach for the learning of a high-dimensional function is infeasible.

The difficulty we faced in this approach stems from our attempt at conveying the inter-component correlation in the data, to the components of the sought function.

1. This inspires us to seek a methodology that permits an easy and streamlined conveying of information on the correlation structure of the data on the output, to that amongst the different components of the sought function.
2. In particular, we would want to incorporate noise in the measured values of the output, and wish to learn the components of the sought function, with uncertainties.
3. Importantly, the method should be such that it can be applicable across dimensionalities.

One wishes that there were a bag of candidate functions, s.t.

— all such candidate functions in this bag were of the desired dimensionality, and

— it is a property of the bag that the correlation structure of the data, be translated into the correlation structure of this sought function that maps the space of the input variable to that of the output.

Then we would sample functions from this bag. What we are hinting at, by invoking such a construct, is that there is a *probability distribution of functions*, and this probability distribution is such that it can accept information about the correlation structure of the data – perhaps via one of its parameters. So the parametrisation of this "bag of functions", *aka* "probability distribution of functions" allows for information about the correlation in the data to permeate down to the correlation structure of the function that is sampled from this bag/distribution. We have not yet discussed probability distributions; nor distributions on function spaces. Indeed, a probability distribution on the space of functions is a Stochastic Process [38]. We review Stochastic Processes later in this chapter, so that in the third chapter, we can recall the ideas broached here, to suggest a formalism for sunch functional learning.

With this taster, we end our current discussion, and move to building the relevant background in probability and Stochastic Processes that will enable appreciation of details of the kind of learning that we have spoken about, so far.

1.4 Reviewing probability

A deterministic experiment is one, the outcome of a trial of which is known, before the trial is conducted - without there being any uncertainties about such knowledge. Before a trial of a *probabilistic experiment* has been undertaken, the outcome from the trial is not known with certainty, i.e. the outcome can only be known with uncertainties.

Let all possible outcomes of a probabilistic experiment live in the space $\boldsymbol{\Omega}$. Let any such outcome be ω. Then $\omega \in \boldsymbol{\Omega}$. Here $\boldsymbol{\Omega}$ is called the *sample space.*

If an event happens, it implies that some *combination* of some of the possible outcomes that live in space $\boldsymbol{\Omega}$, is the outcome of an undertaken trial of this experiment. So, in general, an event is stated to occur if the outcome of a trial is a subset of $\boldsymbol{\Omega}$, i.e.

an event is defined as a subset of $\mathbf{\Omega}$. Now, subsets of $\mathbf{\Omega}$ live in the power set of $\mathbf{\Omega}$, i.e. in $\mathcal{P}(\mathbf{\Omega})$. Then, an event – which is a subset of Ω – lives in $\mathcal{P}(\mathbf{\Omega})$. However, only those subsets of $\mathbf{\Omega}$ that admit *meaningful* combinations of some of the outcomes of this experiment, are defined as events.

As any reader can see from the above, understanding events appears conditional on understanding the qualifications of "meaningful" and "combination". As an event is a set, "combination" above refers to the set operations of union, intersection and complementation, on subsets of $\mathbf{\Omega}$. However, events are not any subset of $\mathbf{\Omega}$, but – to reiterate what we say above – are the meaningful subsets of $\mathbf{\Omega}$. We will soon see that by demanding such meaningfulness, we facilitate quantification of the *size* of the subset of $\mathbf{\Omega}$ that is the event at hand, where the concept of such a *size* will have to of course abide by certain rules. For example, we will require that any parametrisation of the size of any subset of $\mathbf{\Omega}$ - that we identify as an event – be s.t. the size of the union of disjoint subsets be the sum of sizes of such subsets; also that size of the null set be 0; congruent sets bear the same size; and other rules could be possibly motivated within the axiomatic approach one uses.

With the aim of achieving unambiguous assignment of a parametrisation of "size" we can define events as elements of a σ-algebra that is generated by the sample space $\mathbf{\Omega}$, where a σ-algebra $\mathcal{S}$ generated by the set $\boldsymbol{X}$, is next defined.

Definition 1.1 *The σ-algebra $\mathcal{S}$ generated by the set $\boldsymbol{X}$, is the collection of subsets of $\boldsymbol{X}$ s.t. $\mathcal{S}$ is closed under complementation; closed under countably infinite unions; and contains the null set ϕ, i.e.*

(i) $x \in \mathcal{S} \implies x^C \in \mathcal{S}, \forall x \subseteq \boldsymbol{X}$;

(ii) $x_1, x_2, \ldots \in \mathcal{S} \implies \bigcup\limits_{i=1}^{\infty} x_i \in \mathcal{S}$, *where* $x_i \cap x_j = \phi, \forall x_j \neq x_i \subseteq \boldsymbol{X}$.

(iii) $\phi \in \mathcal{S}$.

So, we have just recognised events as elements in the σ-algebra $\mathcal{F}$ that is generated by the sample space $\boldsymbol{\Omega}$.

The above then appears to indicate that there are sample spaces for which, not all elements of the power set $\mathcal{P}(\boldsymbol{\Omega})$ can be assigned a size, but that can be done for subsets of $\boldsymbol{\Omega}$ that constitute the σ-algebra $\mathcal{F}$. (Such subsets are what we had in mind, when we spoke above, of "meaningful" subsets of $\boldsymbol{\Omega}$).

1.4.1 "Size" inaccessible – a digression

We need to develop a better intuition for why any subset of the sample space, is not an event. Can there in fact be some subsets of $\boldsymbol{\Omega}$ – i.e. some elements of the power set $\mathcal{P}(\boldsymbol{\Omega})$ – that violate a fundamental hallmark of an event?

Recalling the idea of "size" of a set, we wonder if there can be $\boldsymbol{A} \in \mathcal{P}(\boldsymbol{\Omega})$, that does not admit a size. For example, if we come to know about

— subset $\boldsymbol{B} \in \mathcal{P}(\boldsymbol{\Omega})$ that is congruent to $\boldsymbol{A}$, that is but shown to bear (ambiguously) distinct size, compared to $\boldsymbol{A}$;

— or we come across multiple replicates of subset $\boldsymbol{A}$ of $\boldsymbol{\Omega}$, that are composed of "linearly" transformed partitions of $\boldsymbol{A}$;

— or $\boldsymbol{A}$ is such a subset of $\boldsymbol{\Omega}$ that is anomalous in the sense that its structure disallows a size to be assigned to itself,

then such a subset $\boldsymbol{A}$ would violate what we intuitively demand from events, with respect to the considered concept of "size", and we may decide to discard such anomalous subsets from the perceived set of events. Alternatively, we can adopt the stance that even subsets (of the sample space) such as $\boldsymbol{A}$ are meaningful, but we need to switch to a distinct formalisation of size of a set, when we consider such anomalous subsets of $\boldsymbol{\Omega}$. The former option is saner.

Indeed in the first of the three points enumerated in the last paragraph, the intuition of size suffers, since congruent events are expected to share a unique size. In the second case, we are being told that "size" of an event is such a concept, that sum of sizes of a pair of disjoint replicates of the original set, equals the size of the original set. These examples then violate the salient properties of "size" of a set that we anticipate formalising. Yet, such examples have been proved to hold. At this point, readers who might prefer to skip further details regarding examples of such violation of the intuition of "size of sets", can proceed to the next subsection, without concerns about loss of continuity. This subsection is indeed a digression, which is but highly interesting, since it discusses the seemingly strange anomalies that are enumerated in the previous paragraph. Secondly, the current subsection helps to clarify the need for constructing the σ-algebra of subsets of the sample space, as the receptacle of events.

1.4.1.1 Banach-Tarski paradox

The Banach-Tarski paradox is a thought experiment that results in the duplication of a hypersphere in $\mathbb{R}^d$, $(d > 2)$, into a pair of congruent hyperspheres – leading indeed to possible further replication – each of which is also congruent to the original set [123, 126].

At `https://www.quantamagazine.org/how-a-mathematical-paradox-allows-infinite-cloning-20210826/`, in the blog titled "Banach-Tarski and the Paradox of Infinite Cloning" authored by Mark Levy, an excellent visualisation is offered in $\mathbb{R}^3$, of this highly interesting result, that appears to be suggesting that size of a set in $\mathbb{R}^d$, for $d \geq 3$ equals the size of a pair of replicates of that original set. In $\mathbb{R}^3$, the paradox invokes a sphere, (that is not hollow). The thought experiment initiates with rotations – having started from any chosen "starting point" on the sphere – through a polar angle, and an azimuthal angle, of irrational degrees each, (which ensures that the result of any number of rotations, is never a point on the sphere that has been visited before). The rotated coordinates of any original "starting point" is our new starting point, and similar rotations are undertaken again, an infinite number of times, to result in countably infinite configurations. This comprises the first step. These configurations, that result from such rotations, are partitioned into 6 disjoint sets (or groups), comprising infinite points each, s.t. under union, these sets will result in the original sphere. Each group is marked by an underlying commonality; for example, one of these groups is the set of all "starting point"s on the original sphere; other groups are characterised by the sense of last rotation undertaken in the first step. To duplicate that original sphere, rotations are undertaken of each starting point in a group, along a sense that is opposite to the nature of the rotations that characterise this group. Such action, followed by translations, produces 2 copies of each of the 6 disjoint groups, that union to produce 2 copies of the original sphere.

Thus, this thought experiment appears to have generated 2 disjoint copies of the sphere, simply via the operations of rotations and translations. Indeed, this contradicts the concept of size of the original set, (of points represented by the original sphere), as sum of sizes of the resulting copies of the original set. However, since the σ-algebra of events is closed to unions, a chosen identification of size will be relevant for all events. So, assuming that the original sphere discussed in this thought experiment is an event, it will imply that it can be replaced by the union of its disjoint copies, which is also an event in the same σ-algebra, s.t. a pre-defined parametrisation of size of the original sphere, equals the similarly-quantified size of the union of its two disjoint copies. This is a contradiction. To escape the difficulty that this though experiment has evoked, we simply rule out the original sphere as an event. In this way, we can relieve ourselves of complications brought in by the Banach-Tarski paradox, when working with events.

1.4.1.2 Hausdorff paradox

The Banach-Starski paradox is a development on the earlier work done by Giuseppe Vitali and Felix Hausdorff [121, 20, 44]. Hausdorff showed that the 2-dimensional sphere $\mathbb{S}^2$ in $\mathbb{R}^3$ can be decomposed into 4 sets A, B, C, D, s.t. A, B, C and $B \cup C$ are congruent, while D is countable. Then if the size of the set $\mathbb{S}^2 - D$ is s, it would imply that

— size of $B \cup C$ is $1/3$, in units of s, since $B \cup C$ has the same size as the congruent sets B and C, that each have a size of 1/3, given that disjoint and congruent sets A, B and C union to a set of size of 1.

— size of $B \cup C$ is 2/3, in units of s, since the disjoint sets A, B and C union to a set of size of 1, with size of each of these 3 sets the same, namely 1/3, given that they are congruent. So B and C union to a set of size 1-1/3 = 2/3, which then is also the size of $B \cup C$.

— size of $B \cup C$ is 1/2, in units of s, since size of union of A and $B \cup C$ is 1 – where each of the congruent sets B and C, has the same size as $B \cup C$ (that B and C each is congruent to, and each is same size as set A), s.t. size of $A \cup (B \cup C)$ is 1.

Then this violates the intuition that congruent sets have equal sizes, where we conceptualise size as additive.

1.4.1.3 Vitali sets

One of Guiseppe Vitali's legacies is the eponymous set that cannot be assigned a size. A Vitali set V is defined as $V \subset [0,1]$, where $\exists$ a unique $v \in V$ for any $x \in \mathbb{R}$, s.t. $v - x \in \mathbb{Q}$, i.e. is a rational number [11]. Thus, there is a one-to-one mapping between the space of irrationals and V. To show that V does evade assignment of a size, we define the set

$$V_k := \{v + q_k | v \in V; q_k \in \mathbb{Q}, q_k \in [-1,1]\}, \text{ where } k = 1, 2, \ldots.$$

Thus, the set of rationals in interval [-1,1] – which is a countable set – is $\{q_1, q_2, \ldots\}$.
Then the sequence $(V_1, V_2, \ldots)$ is countable. Also, the sets V_k and $V_{k'}$ are disjoint $\forall k, k' \in \{1, 2, \ldots\}$; $k \neq k'$.
Let us assume that the size of V can be assigned.

— Now, the union of the disjoint sets $V_1, V_2, \ldots$ is $\bigcup_{k=1} V_k$, which contains [0,1], since $V \subset [0,1]$.

— At the same time, $q_k \in [-1,1]$.

— This in concert with $V \subset [0,1]$, implies that the interval [-1,2] contains $\bigcup_{k=1} V_k$.

— So $[0,1] \subseteq \bigcup_{k=1} V_k \subseteq [-1,2]$.

Now, our intuitive idea about size of a set – which we shall soon examine in detail – lends to the result that size of the union of disjoint sets $V_1, V_2, \ldots$, is the sum of sizes of $V_1, V_2, \ldots$.
In other words, taking size of the last result, we get that the size of [0,1] $\leq \sum_{k=1}$ size of $V_k \leq$ size of [-1,2]. Then using the Lebesgue measure of $[a, b]$ for any $a \leq b$; $a, b \in \mathbb{R}$, we get $1 \leq \sum_{k=1}$ size of $V_k \leq 3$.
However, translating each element of a set by a give constant does not alter the size of the set. So size of V_k is the size of V $\forall k = 1, 2, \ldots$. Then the last result states that $1 \leq \sum_{k=1}$ size of $V \leq 3$.
However, the size of the V – which we assumed exists – is a non-negative number. Then adding an infinite number of this size will yield either 0, or will approach ∞. However, the last result states that the result of this addition lies between 1 and 3. This is a contradiction. Then the original assumption – that size of V exists – must be false.
Therefore, size of V does not exist.

1.4.1.4 Pulling these examples together

So the summary of this subsection is that not all sets can be assigned a size, where size is such that it abides by the intuition that: size of a given set is unique; the size of union of disjoint sets is the sum of the sizes of the individual sets; that congruent sets are equally-sized. Indeed, some special sets can be designed, that cannot be assigned a size. We take the step to avoid such "special sets", and agglomerate those subsets of the sample space into a structure that ensures that assignment of size on any element of this structure is possible, with size broadly interpreted to abide by the above intuition, in addition to other properties (e.g. size of the null set is 0). Different types of size assignment are possible, as we shall see below. The "structure" alluded to above, is a σ-algebra.

1.4.2 Borel σ-algebra

For the case of the sample space $\boldsymbol{\Omega}$ being the space $\mathbb{R}$ of reals, events are elements of the σ-algebra $\mathcal{B}(\mathbb{R})$ that is generated by the collection of open intervals in $\mathbb{R}$. This amounts to $\mathcal{B}(\mathbb{R})$ comprising all possible open intervals in $\mathbb{R}$; their complements; their unions – as well as the null set and $\mathbb{R}$. Then $\mathcal{B}(\mathbb{R})$ is called a Borel σ-algebra on this sample space $\mathbb{R}$.

A smallest Borel σ-algebra exists, that can generate all elements of any collection of subsets of the sample space, via unions, intersections and complementation.

1.4.3 Size of an event

We still have not formalised the concept of "size" of an event. We undertake this discussion now.

The sample space $\boldsymbol{\Omega}$ and the σ-algebra $\mathcal{F}$ defined on it, form the measurable space $(\boldsymbol{\Omega}, \mathcal{F})$.

Definition 1.2 *The function $\mu : \mathcal{F} \longrightarrow [0, \infty)$ is called a measure if and only if*

(i) $\mu(\phi) = 0$;

(ii) μ admits "σ-additivity", i.e. for mutually disjoint events $E_1, E_2, \ldots, \in \mathcal{F}$,

$$\mu\left(\bigcup_{i=1} E_i\right) = \sum_{i=1} \mu(E_i).$$

So, this parametrisation of the size of a set that we refer to as *measure*, abides by the intuition-abiding constraints that

— size is a non-negative number;

— that size of the null set is 0, and

— that size of a countable union of disjoint sets is the sum of sizes of these sets.

The set $(\boldsymbol{\Omega}, \mathcal{F}, \mu)$ is referred to as a measure space.

We appreciate the following.

Definition 1.3 *A measure μ defined on the measurable space $(\boldsymbol{\Omega}, \mathcal{F})$ has to abide by the property that for events $A_1, A_2, \ldots \in \mathcal{F}$ – not all of which are necessarily disjoint -*

$$\mu(\bigcup_{i=1} A_i) \leq \sum_{i=1} \mu(A_i).$$

Also,

$$if\ A_i \subseteq A_j, \mu(A_i) \leq \mu(A_j).$$

Lastly, if there exists a sequence $(A_1, A_2, \ldots)$ s.t. $A_i \in \mathcal{F}$, with $A_1 \subset A_2 \subset A_3 \subset \ldots$, then

$$\mu(\lim_{n \longrightarrow \infty} A_n) = \lim_{n \longrightarrow \infty} \mu(A_n).$$

Using this formalisation of measure, we can now motivate a useful concept of *almost everywhere.* If there is any property that is violated at certain points that are elements of set $\boldsymbol{S}$, but holds everywhere else, then the property is stated to hold almost everywhere under measure μ, if $\mu(\boldsymbol{S}) = 0$.

When the sample space is the space of reals, the *Lebesgue measure* μ on the measurable set $(\mathbb{R}, \mathcal{B}(\mathbb{R}))$. is defined as

$$\mu([a, b]) := b - a, \ \ \forall [a, b] \in \mathcal{B}.$$

Then for the measurable space $([0, 1], \mathcal{B}([0, 1]))$, the Lebesgue measure is the probability measure, but we are getting ahead of ourselves; let us review probability measure first.

1.4.4 Probability measure

The axioms on the probability measure were advanced by Kolmogorov. These state that on the measurable space $(\boldsymbol{\Omega}, \mathcal{F})$, a probability measure Pr is the mapping $\mathcal{F} \longrightarrow [0,1]$ s.t.

(i) $\Pr(\boldsymbol{\Omega}) = 1$;

(ii) $0 \leq \Pr(A_i) \leq 1 \; \forall A_i \in \mathcal{F}$;

(iii) $\Pr(A_1 \cup A_2 \cup \ldots \cup A_n) = \sum_{i=1}^{n} \Pr(A_i) \forall A_i, A_j \in \mathcal{F}$, where $A_i \cap A_j = \phi$.

Then for a probability measure Pr that is defined on the measurable space $(\boldsymbol{\Omega}, \mathcal{F})$, we define the probability space $(\boldsymbol{\Omega}, \mathcal{F}, \Pr)$.

Now that we have reviewed the tools to study probability theory, we are keen to define random variables, which the rest of the book will invoke. In particular, we are interested in clarifying "randomness" of a variable and this will tie in with the concept of a probability distribution. We will proceed to such discussions soon, but to set the groundwork for the same, we first have to explore the concept of a measure on a high-dimensional sample space.

1.4.5 High-dimensional sample spaces

Notwithstanding the earlier discussion, events can definitely be defined in a high-dimensional space as well. We now address the question of defining a measure on any event that is an element of the σ-algebra that is generated by the product of σ-algebras that are respectively generated by the relevant sample spaces. Thus, for the sample space $\boldsymbol{\Omega}$ defined as the product space $\boldsymbol{\Omega}_1 \times \ldots \times \boldsymbol{\Omega}_n$ – where the σ-algebra $\mathcal{F}_i$ is generated on $\boldsymbol{\Omega}_i$, $(i = 1, \ldots, n)$ - we define the space $\boldsymbol{\Psi}$ of the n-tuples as:

$$\boldsymbol{\Psi} := \{(A_1, A_2, \ldots, A_n) | A_i \in \mathcal{F}_i, i = 1, \ldots, n\}.$$

Then, the σ-algebra $\mathcal{G}$ generated by $\boldsymbol{\Psi}$, on the product space $\boldsymbol{\Omega}$ is referred to as the *product σ-field*, i.e. an algebra defined on the power set of the product space $\boldsymbol{\Omega}$. Thus, $(\boldsymbol{\Omega}, \mathcal{G})$ is a measureable space.

Definition 1.4 *A unique measure μ – referred to as the product measure – can be defined on the measurable space $(\boldsymbol{\Omega}, \mathcal{G})$, as long as μ_i is σ-finite $\forall i = 1, \ldots, n$, i.e.*

$$\mu(A_1 \times A_2 \times \ldots \times A_n) := \mu_1(A_1) \ldots \mu_N(A_n), \text{ for } A_i \in \mathcal{F}_i,$$

if μ_i is σ-finite on the measureable space $(\boldsymbol{\Omega}_i, \mathcal{F}_i)$, $\forall i = 1, \ldots, n$. Here, by "σ-finite" we imply that

$$\text{for any } i \in \{1, \ldots, n\}, \exists (A_{1i}, A_{2i} \ldots) \in \mathcal{F}_i,$$

$$\text{s.t.} \bigcup_{j=1} A_{ji} = \boldsymbol{\Omega}_i, \text{ with } \mu_i(A_{ji}) \text{ finite}, \forall j = 1, 2 \ldots.$$

1.4.6 Random variables

Having discussed the generalisation into high-dimensions, we now embark upon the conceptualisation of a random variable. In order to undertake the same, we will need to appreciate what a *measurable function* is. We begin our discourse with measurable functions, from which we will proceed to discuss random variables.

Consider a measure space $(\boldsymbol{\Omega}, \mathcal{F}, \mu)$, and a mapping f from the sample space $\boldsymbol{\Omega}$ to some other space $\boldsymbol{R}$, s.t. the inverse of f on any $A \subset \boldsymbol{R}$ is defined as

$$f^{-1}(A) = \{\omega \in \boldsymbol{\Omega} | f(\omega) \in A\}.$$

Definition 1.5 *f is defined as a measurable function from the measurable space $(\boldsymbol{\Omega}, \mathcal{F})$ to the measurable space $(\boldsymbol{R}, \mathcal{B}_{\boldsymbol{R}})$, if and only if*

$$f^{-1}(\mathcal{B}_{\boldsymbol{R}}) \subseteq \mathcal{F}.$$

Often we use the concept of measurable functions as mappings from a sample space $\boldsymbol{\Omega}$ and its generated σ-algebra $\mathcal{F}$, to $\mathbb{R}^d$ and the Borel σ-algebra $\mathcal{B}_d$ generated on it.

If f is a measurable function that maps $(\boldsymbol{\Omega}, \mathcal{F})$ to $(\mathbb{R}^d, \mathcal{B}_d)$, it is referred to as a *random variable* – indeed, here we refer to real-valued random variables only. All through the book we will use the abbreviation that r.v. stands for "random variable", and we will imply a real-valued r.v. whenever a r.v. is invoked, unless otherwise specified. Thus, we view a r.v. as a real-valued mapping of an outcome of a probabilistic experiment, irrespective of the nature of such an outcome.

Definition 1.6 *Let the r.v. X be a measurable function from $(\boldsymbol{\Omega}, \mathcal{F})$ – on which probability measure* Pr *holds – to $(\mathbb{R}^d, \mathcal{B}_d)$. The measure* $\Pr(X^{-1}(\cdot))$ *is induced on the σ-algebra $\mathcal{B}_d$, s.t.*

$$\Pr(X^{-1}(E)) \equiv \Pr(X \in E) \text{ for any } E \subseteq \mathcal{B}_d.$$

Then $\Pr(X^{-1}(\cdot))$ *is referred to as the distribution of X.*

This distribution is alternatively denoted $\Pr_X(\cdot)$, and is identified as the probability distribution of the r.v. X.

Thus, if the Borel σ-algebra on the sample space $\mathbb{R}^d$ – which is a product space – is $\mathcal{B}_d$, and Pr is the probability measure on the measureable space $(\mathbb{R}^d, \mathcal{B}_d)$. Then the joint *cumulative density function* or (*cdf*) of the real-valued r.v.s $X_1, \ldots, X_n$, computed at $x_1, \ldots, x_n \in \mathbb{R}$, is

$$F_{X_1,\ldots,X_n}(x_1, \ldots, x_n) := \Pr((-\infty, x_1] \times \ldots \times (-\infty, x_n]).$$

Then the *marginal cdf* of r.v. X_i, computed at $x_i \in \mathbb{R}$ is obtained from the joint as

$$F_{X_i}(x_i) = \lim_{x_1 \longrightarrow \infty \ldots x_{i-1} \longrightarrow \infty x_{i+1} \longrightarrow \infty \ldots x_n \longrightarrow \infty} F_{X_1,\ldots,X_n}(x_1, \ldots, x_n).$$

The right hand side (RHS) of the last-but-one equation, computed at $x_j \longrightarrow \infty$, $\forall j \in \{1, 2, \ldots, i-1, i+1, \ldots, n\}$, will result in $\Pr((-\infty, x_i])$, for $x_j \in (-\infty, \infty) = 1, \forall x_j$, given the concept of product measure seen above.

1.4.7 Density functions and mass functions

Let r.v. X be a measurable function in $(\boldsymbol{\Omega}, \mathcal{F})$ – on which probability measure Pr holds – to the measureable space $(\mathcal{X} \subseteq \mathbb{R}, \mathcal{B})$; here $\mathcal{B}$ is the Borel σ-algebra generated by $\mathcal{X}$. Then from above, the measure $\Pr(X^{-1}(\cdot))$ induced on this Borel σ-algebra $\mathcal{B}$, is also denoted $\Pr_X(\cdot)$, and is identified as the probability distribution of the r.v. X. For this r.v. X, the *support* of its probability distribution is the smallest closed set $\mathcal{X} \in \mathbb{R}$ s.t. $\Pr(X \in \mathcal{X}) = 1$.

Now, an r.v. can be either discrete or continuous, i.e. its realisations are either in a countable or an uncountable set respectively. For the continuous r.v. $X \in \mathcal{X} \subseteq \mathbb{R}$ that we invoked in the last paragraph, the probability of X attaining a value in an interval $[a, b] \subseteq \mathcal{X}$ is defined as the integral of the (probability) density function $f_X(\cdot)$ from a to b:

$$\Pr(X \in [a, b]) = \int_a^b f_X(s) ds.$$

Here $f_X(s)$ is the value of the probability density function or *pdf* of X, computed at $s \in \mathbb{R}$. For $s \notin \mathcal{X}$, $f_X(s) = 0$. The cumulative distribution function (or *cdf*) of r.v. X, computed at $X = x$, from what we have seen above, is then $F_X(x) \equiv \Pr(X \in (-\infty, x]) = \int_{-\infty}^{x} f_X(s) ds$. Equivalently, the *pdf* of X is

$$f_X(x) = \frac{dF_X(x)}{dx}.$$

If on the other hand, the r.v. X is discrete, then there is non-zero probability mass at a point in the support $\mathcal{X}$. So the question: "what is $\Pr(X = x)$?", is a meaningful question, with the answer not necessarily 0. We refer to this probability as the value of the probability mass function or *pmf*, computed at x. Then for this discrete r.v. X, its *cdf* computed at x is

$$F_X(x) = \sum_{s \longrightarrow -\infty}^{x} \Pr(X = s).$$

1.4.8 Expectation et al.

For the continuous r.v. $X \in \mathcal{X} \subseteq \mathbb{R}$ that is distributed according to the *cdf* $F_X(\cdot)$ and the *pdf* $f_X(\cdot)$, its expected value is

$$\mathbb{E}(X) = \int_{-\infty}^{\infty} x f_X(x) dx.$$

Then for a discrete r.v. X, its expectation is given as

$$\mathbb{E}(X) = \sum_{x} x \Pr(X = x).$$

While expectation informs on the central tendency of a density, dispersion about this central tendency is informed upon by the variance:

$$Var(X) := \mathbb{E}[(X - \mathbb{E}(X))^2] = \int_{-\infty}^{\infty} (x - \mathbb{E}(X))^2 f_X(x) dx.$$

Expressions for discrete r.v.s are not provided any further.

The influence of one r.v. Y, on the values attained by another r.v. X is informed upon via their covariance:

$$Cov(X, Y) := \mathbb{E}[(X - \mathbb{E}(X))(Y - \mathbb{E}(Y))] =$$

$$\int_{-\infty}^{\infty} \int_{-\infty}^{\infty} (x - \mathbb{E}(X))(y - \mathbb{E}(Y)) f_{X,Y}(x, y) dx dy.$$

The density $f_{X,Y}(x, y)$ is indeed the joint probability density function of the two r.v.s in question. We have discussed joint and marginal distributions of r.v.s above, and we will pick up the discussion of joint, marginal and conditional densities soon. In the current subsection, the salient concept of correlation remains to be covered.

The correlation between the r.v.s X and Y is the covariance, normalised by the product of the standard deviations of the two r.v.s, i.e.

$$Corr(X, Y) = \frac{Cov(X, Y)}{\sqrt{Var(X) Var(Y)}}.$$

So correlation provides a scale-free quantification of the influence of one r.v. on another; then by definition $-1 \leq Corr(X,Y) \leq 1$.

1.4.9 Joint, marginal and conditional densities

The joint *pdf* of r.v.s $X_1, X_2, \ldots, X_n$ is defined using the joint *cdf* defined above, as:

$$f_{X_1,\ldots,X_n}(x_1, x_2, \ldots, x_n) = \partial_{x_1} \ldots \partial_{x_n} F_{X_1,\ldots,X_n}(x_1, x_2, \ldots, x_n).$$

The marginal *pdf* of X_i can be computed by integrating out $X_1, \ldots, X_{i-1}, X_{i+1}, \ldots, X_n$ from the joint *pdf* above i.e.:

$$f_{X_i}(x) = \int \cdots \int f_{X_1,\ldots,X_{i-1},X_{i+1},\ldots,X_n}(x_1, x_2, \ldots, x_{i-1}, x, x_{i+1}, \ldots, x_n) \, dx_1 \ldots dx_{i-1} dx_{i+1} \ldots dx_n. \tag{1.1}$$

While such marginalisation of the joint over all values of $X_1, \ldots, X_{i-1}, X_{i+1}, \ldots, X_n$, can be viewed as a projection of the joint *pdf* onto the space of X_i, it is also possible to compute the value of the joint density at fixed values of the r.v.s $X_1, \ldots, X_{i-1}, X_{i+1}, \ldots, X_n$ – say $X_1 = x_1, \ldots, X_{i-1} = x_{i-1}, X_{i+1} = x_{i+1}, \ldots, X_n = x_n$.

Then the density of X_i, "conditional" on these fixed values of $X_1, \ldots, X_{i-1}, X_{i+1}, \ldots, X_n$ suggests the conditional density of X_i, given $X_1 = x_1, \ldots, X_{i-1} = x_{i-1}, X_{i+1} = x_{i+1}, \ldots, X_n = x_n$. However, the *pdf* computed at this chosen slice in the space of $X_1, \ldots, X_{i-1}, X_i, X_{i+1}, \ldots, X_n$ does not appear to integrate to 1. Then recognising that area under the slice is $f_{X_1,\ldots,X_{i-1},X_{i+1},\ldots,X_n}(x_1, \ldots, x_{i-1}, x_{i+1}, \ldots, x_n)$, we normalise the joint computed at $X_1 = x_1, \ldots, X_{i-1} = x_{i-1}, X_i = x_i, X_{i+1} = x_{i+1}, \ldots, X_n = x_n$, by this area, to get the conditional density of X_i, given the fixed values assigned to the other $n-1$ r.v.s under consideration. Thus, this conditional density is

$$f_{X_i|X_1,\ldots,X_{i-1},X_{i+1},\ldots,X_n}(x|x_1, \ldots, x_{i-1}, x_{i+1}, \ldots, x_n) =$$

$$\frac{f_{X_1,\ldots,X_{i-1},X_i,X_{i+1},\ldots,X_n}(x_1,\ldots,x_{i-1},x_i,x_{i+1},\ldots,x_n)}{f_{X_1,\ldots,X_{i-1},X_{i+1},\ldots,X_n}(x_1,\ldots,x_{i-1},x_{i+1},\ldots,x_n)}.$$

To simplify the notation, we consider the case of only 2 r.v.s, X and Y, s.t. the conditional *pdf* of X, given $Y = y_0$ is

$$f_{X|Y}(x|y_0) = \frac{f_{X,Y}(x, y_0)}{f_Y(y_0)}.$$

Then in light of information that Y has attained the value y_0 with probability density $f_Y(y_0)$, the probability density of X, computed at x is given the RHS of this equation.

Considerations of conditional density have given rise to the whole new discipline of Bayesian statistics that this book is a token reflection of. We discuss Bayesianism in the next chapter.

Excellent texts exists on the material covered above and on the upcoming discussion on Stochastic Processes, [123, 47, 112, among others].

1.5 Reviewing Stochastic Processes

We wrap up this chapter by recalling tenets of Stochastic Processes. All of the methods that we will be discussing in the context of supervised learning in real-world applications, will invoke Stochastic Processes. Indeed we have alluded to the usage of such processes above, when we spoke about the difficulty of building a high-dimensional function, one component function after another, while ensuring that every such component is correlated with every other component via the correlation structure manifest in the data on the output. We spoke about how the easier option would be to sample a function from a "bag of functions" where it is the property of the bag, to convey the intra-data correlation structure to inter-component correlation in any of the resident functions. We had recognised this "bag" as the distribution of the random function variable, and we had anticipated this distribution as given by a Stochastic Process. So here, we reconcile such an intuition with formalisation of a Stochastic Process.

A Stochastic Process is a collection of r.v.s that share a commonality.

Definition 1.7 *A Stochastic Process is a collection of r.v.s* $\{X_t\}_{t \in \mathcal{T}}$*, where all such r.v.s are defined on* $\{\boldsymbol{\Omega}, \mathcal{F}, \Pr\}$*, and are indexed by elements of an index set* $\mathcal{T}$*, s.t. each such r.v. is* $\mathcal{R}$*-valued, with space* $\mathcal{R}$ *measurable on an* σ*-algebra, (*$\mathcal{S}$*).*

This Stochastic Process could also be denoted $\{X(t, \omega) : t \in \mathcal{T}\}$, where $\omega \in \boldsymbol{\Omega}$.

Here, the *state space* of this Stochastic Process is the space $\mathcal{R}$.

The process can be discrete-valued if $\mathcal{R}$ is $\mathbb{Z}$ for example, and continuous-valued if $\mathcal{R}$ is $\mathbb{R}$ (for example). The process is *discrete-timed* if the index set is a set of integers, i.e. $\mathcal{T}$ is $\mathbb{N}$ for example. The process is *continuous-timed* if t is continuous, such as when $\mathcal{T}$ is $\mathbb{R}_{\geq 0}$. Setting the state space as $\mathbb{R}^d$ renders the process vector-valued; higher dimensional generalisations are also possible. If $\mathcal{T} = \mathbb{N}$ and $\mathcal{R} = \mathbb{R}^d$ (for example), the sequence of r.v.s is rendered *one-sided* real-valued, as distinguished from the *two-sided* real-valued one (that follows, if $\mathcal{T} = \mathbb{Z}$). If on the other hand, $\mathcal{T} = \mathbb{Z}^m$ and $\mathcal{R} = \mathbb{R}^d$, the process is an m-dimensional *spatially-discrete random field.*

For this process $\{X(t, \omega) : t \in \mathcal{T}\}$, the function $X(\cdot, \omega)$ is a sample from it, i.e. a *sample function* from this Stochastic Process is

$$X(\cdot, \omega) : \mathcal{T} \longrightarrow \mathcal{R}.$$

When the variable T is time, we will refer to the sample function as a *sample path* as well.

Thus, a realisation of the process at a given ω, is a sample function $X(\cdot, \omega)$ of this process. At a different ω, a different sample function is realised. This motivates a different view of a Stochastic Process – different from the treatment of the process as a collection of random variables. We can now see that the stochastic proocess is in fact, a random variable that is function-valued, and any realisation of this *random function* is a sample function. This random function is simply a r.v. that is a function, and the intuition that a Stochastic Process is a random function is useful in the sense that in this treatment, we think of the process as one single random structure, as distinguished from the treatment motivated above, in which the process is a collection of random variables. So we view the process as the mapping $\boldsymbol{\Omega} \longrightarrow \mathcal{R}^{\mathcal{T}}$, where $\mathcal{R}^{\mathcal{T}}$ is the space of all $\mathcal{R}$-valued functions of $T \in \mathcal{T}$.

We understand that the pocess invokes two arguments; so instead of fixing ω, if we fix the index parameter to $T = t$, the process yields the r.v. $X_t(\omega)$.

1.5.1 Probability distribution of a Stochastic Process

We recall that each of the r.v.s that constitutes the Stochastic Process defined in Definition 1.7, is a mapping from the probability space $(\boldsymbol{\Omega}, \mathcal{F}, \Pr)$ to the space that hosts all $\mathcal{R}$-valued functions of the index parameter $T \in \mathcal{T}$. What is the probability distribution, or "law" of this process? This is what we probe in this subsection.

Effectively, we are trying to understand how the probability measure on any constituent r.v. X_i of the process, can be conveyed to formalise the probability measure that is defined on any collection $X^{\mathcal{U}} := \{X_i : i \in \mathcal{U} \subset \mathcal{T}\}$ of such r.v.s that are indexed by index parameter $T \in \mathcal{U}$, where $\mathcal{U} \subset \mathcal{T}$. So, we need to understand how to convey a probability measure on the measurable space $(\mathcal{R}^{\mathcal{U}}, \mathcal{S}^{\mathcal{U}})$ that any such collection $X^{\mathcal{U}}$ maps into, from the σ-algebra $\mathcal{F}$ generated on $\boldsymbol{\Omega}$.

Above, we should have defined $\mathcal{S}^{\mathcal{U}}$ as a σ-algebra that is generated by $\mathcal{R}^{\mathcal{U}}$. Then, $\mathcal{S}^{\mathcal{U}}$ should contain enough sets $\forall\, X_i \in X^{\mathcal{U}}$ to be measurable. Such an $\mathcal{S}^{\mathcal{U}}$ is then intuitively said to be a "coarse" σ-algebra. We can construct such a σ-algebra, if for every open set $R_i \in \mathcal{S}$ we include the set $X_i^{-1}(R_i)$, $\forall i \in \mathcal{U}$. In fact, we will need to refine such a choice of sets to ensure that the host structure is a σ-algebra. This can be done by taking intersection of the sets $X_i^{-1}(R_i)$, $\forall i \in \mathcal{U}$ – this action will produce the sought (coarsest) σ-algebra on $\mathcal{R}^{\mathcal{U}}$.

The probability measure on any measurable subset $R \subset \mathcal{R}^{\mathcal{U}}$, is obtained by recalling that $\forall X_i \in X^{\mathcal{U}}$, $X_i^{-1}(\cdot)$ is the inverse of the function $X_i : \mathbf{\Omega} \longrightarrow \mathcal{R}^{\mathcal{U}}$, i.e.

$$X_i^{-1}(R) = \{\omega \in \mathbf{\Omega} : X(\omega) \in \mathcal{R}^{\mathcal{U}}\}.$$

Let the probability distribution of this Stochastic Process defined on the probability space $(\Omega, \mathcal{F}, \Pr)$ be $m(\cdot)$, s.t. for this measurable set $R \subset \mathcal{R}^{\mathcal{U}}$, the probability distribution of the process is

$$m(R) = \Pr(\{\omega \in \mathbf{\Omega} : X(\omega) \in \mathcal{R}^{\mathcal{U}}\}).$$

For the measurable set $X^n = \{X_{t_1}(\omega), \ldots, X_{t_n}(\omega)\}$, (where $X_{t_i}(\omega)$ is random variable for $i = 1, \ldots, n$), the finite dimensional distribution of the process is

$$m_{t_1,\ldots,t_n}(X^n) = \Pr(\{\omega \in \mathbf{\Omega} : (X_{t_1}(\omega), \ldots, X_{t_n}(\omega)) \in X^n\}).$$

Thus, in this subsection we have discussed how a probability distribution can be defined on a function space. In this subsection, we understood the σ-algebra that can be defined on this space of functions, and a probability measure can be defined on this σ-algebra.

1.5.2 Parameters of a Stochastic Process

As we have seen above, at a given value t of the index parameter T, the Stochastic Process reduces to the r.v. $X_t(\omega)$ or X_t. Then expectation of a Stochastic Process is the expectation of this r.v. So at another value of T – say at $T = t^/$ – the expectation of the process is $\mathbb{E}(X_{t^/})$. Following this intuition, we realise why the mean of a Stochastic Process is a function of the index parameter T. Thus, the mean function of the process is $\mu(T) := \mathbb{E}(X_T)$.

Again, at time $T = t$, the Stochastic Process is the r.v. X_t and at $T = t^/$, the process is another r.v. $X_{t^/}$, for $t, t^/ \in \mathcal{T}$. Then, it is possible to define the covariance function $Cov(X_t, X_{t^/})$ of the Stochastic Process, as the covariance between the r.v.s $X_t, X_{t^/}$ for $X_t, X_{t^/} \in \mathcal{R}$. When the index parameter T is physically interpreted

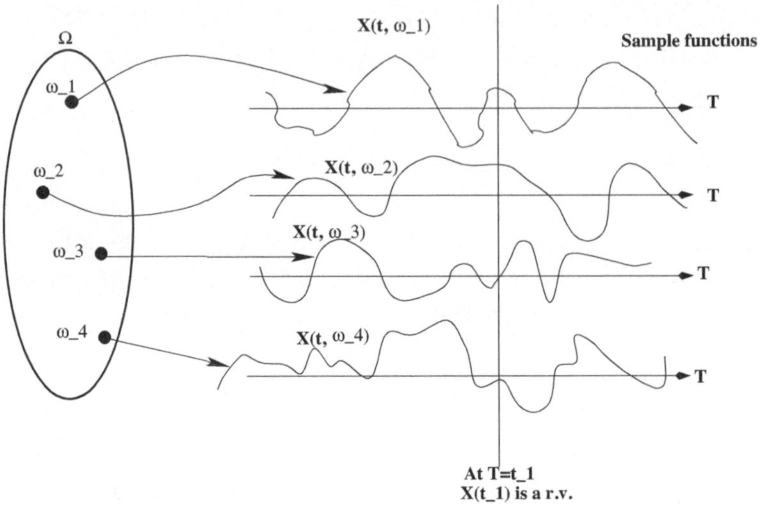

FIGURE 1.5
Figure displays the relation of a stochastic process to realisations of a random function, i.e. sample functions. The figure displays sample functions, as realised at 4 different values $(\omega_1, \omega_2, \omega_3, \omega_4)$ of the outcome ω of a probabilistic experiment. This random function takes the index parameter T as its input.

as time, this covariance function is often referred to as the "autocovariance" function between r.v. X_t and the r.v. $X_{t+\Delta}$ for the lag parameter $\Delta \in \mathbb{R}$, for any value t of time T. The autocovariance function of this Stochastic Process is also denoted as $Cov_X(t, \Delta)$. The correlation function of a process is denoted $R_X(t, t^/) \equiv \mathbb{E}(X_t, X_{t^/})$, so that for time-indexed Stochastic Processes, we refer to the "autocorrelation" function as $R_X(t, \Delta) \equiv \mathbb{E}(X_t, X_{t+\Delta})$.

1.5.3 Stationarity of Stochastic Processes

Definition 1.8 *A Stochastic Process is strictly stationary if all the moment distributions of the process are invariant to a translation in* T*, (i.e. shift-invariant to time for time-indexed Stochastic Processes). This requires that the joint density of all* m *r.v.s* $X(t_1), \ldots, X(t_m)$ *that result from computing the sample function at values* $t_1, \ldots, t_m$ *of the index parameter* T *respectively, is the* *same as the joint density of the sample function computed at the shifted values* $t_1 + \Delta, \ldots, t_m + \Delta$ *of* T *respectively, for all* $m \in \mathbb{N}$. *In other words, the joint is shift-invariant to changes in* T, *for all* m, *i.e.*

$$f_{X(t_1),\ldots,X(t_m)}(x_1, \ldots, x_m) = f_{X(t_1+\Delta),\ldots,X(t_m+\Delta)}(x_1, \ldots, x_m).$$

Weak stationarity imposes such shift-invariance – to changes in values of T, only on the mean and covariance functions of the Stochastic Process. This would require that the mean is independent of the index parameter T, and the covariance function computed at two values of this variable T is dependent only on the Euclidean distance between these two values, i.e.

$$\mathbb{E}[X(t)] = \mu \text{ and } Cov_X[t, t^/] = K(\sqrt{(t - t^/)^2}).$$

Hence, for a weakly stationary, time-indexed Stochastic Process, $Cov_X[t, \Delta] = K(\Delta)$.

1.5.4 Integration and differentiation of Stochastic Processes

Discussions on the integration and differentiation of Stochastic Processes - preceded by discussion on continuity of processes – is wonderfully articulated by [62]. In this subsection, we consider a real-valued Stochastic Process $\{X(t)\}_{t \in \mathcal{T}}$ in this subsection s.t. the mean function ($\mu(\cdot)$) and covariance function ($Cov(\cdot, \cdot)$) of this process are continuous. We assume that with the r.v. $X(t) \in \mathbb{R}^d$, $\forall t \in \mathcal{T}$, a sample function $X(t)$ of this process – at the outcome ω of the relevant probabilistic experiment – if discontiuous at any point within an arbitrary interval $[a, b] \in \mathcal{T}$, is so discontinuous at a finite number of values of the index parameter T.

Process obtained by integrating $X(t)$

Then, the integral of the sample path $X(t)$ of the considered process, is defined as

$$Y(t) = \int_a^t X(u)du,$$

at any $t \in \mathcal{T}$. Here a is a constant, and $\{Y(t)\}_{t\in\mathcal{T}}$ is another continuous sample Stochastic Process, with continuous mean and covariance functions respectively given as:

$$\mu_Y(t) = \int_a^t \mu(u)du,$$

$$Cov(s,t) = \int_a^s dv \int_a^t Cov(u,v)du,$$

for $s, t \in \mathcal{T}$.

Process obtained by differentiating $X(t)$

If the process $\{X(t)\}_{t\in\mathcal{T}}$ has continuous sample functions, then it is differentiable, i.e. we can define a process $\{X^{/}(t)\}_{t\in\mathcal{T}}$ s.t.

$$X(t) - X(a) = \int_a^t X^{/}(u)du.$$

In other words,

$$X^{/}(t) = \frac{dX(t)}{dt},$$

$\forall t \in \mathcal{T}$ at which $X^{/}(t)$ is not discontinuous. Then at such $t \in \mathcal{T}$, the process $\{X^{/}(t)\}_{t\in\mathcal{T}}$ has a mean function

$$\mu_{X^{/}}(t) = \frac{d\mu(t)}{dt},$$

and covariance function of process $\{X^{/}(t)\}_{t\in\mathcal{T}}$, computed at $T = t$ and $T = s$ is

$$Cov(X^{/}(t), X^{/}(s)) = \frac{\partial^2 Cov(X(t), X(s))}{\partial t \partial s}.$$

Proofs of these results are in the 5th chapter of the book by [62].

1.5.5 Examples of Stochastic Processes

1.5.5.1 Markov Processes

An example Stochastic Process that we will encounter often in this book is a Markov Process, [122, 110, among others]. A Markov Process is a Stochastic Process with the property that probability of the current event depends on the probability only of the previous event, but not on other earlier events. Then the realisation from a discrete-timed Markov Process at the i-th time point, is defined as

$$\Pr(X_i = x_i | X_{i-1} = x_{i-1}, X_{i-2} = x_{i-2}, \ldots, X_0 = x_0) = \Pr(X_i = x_i | X_{i-1} = x_{i-1}) \quad (1.2)$$

Thus, the parametrisation of the probability distribution of the sample function at the i-th time, is dependent on the deterministic index parameter T. Here $i = 1, 2, \ldots$.

1.5.5.2 Gaussian Processes

Another Stochastic Process that we will use extensively in this book is Gaussian Process ($\mathcal{GP}$), [105]. A $\mathcal{GP}$ is an example of a continuous-time, continuous-valued Stochastic Process. The index parameter T is indeed more general than physical time in many applications. A $\mathcal{GP}$ offers a distribution on the space of functions s.t. this distribution – on a function $Y = f(X)$ that can be vectorised into an infinite-dimensional vector $(f(x_1), f(x_2), \ldots,)^T$ – is multivariate Normal.

We have seen above, how the underlying probability space that an $\mathcal{R}$-valued Stochastic Process $\{X_t(\omega)\}_t$ is defined on, can convey measurability to an identified σ-algebra defined on a relevant subset of $\mathcal{R}$. Indeed, we have defined above the finite dimensional distribution of a Stochastic Process as $m_{t_1,\ldots,t_n}(X^n) = \Pr(\{\omega \in \boldsymbol{\Omega} : (X_{t_1}(\omega), \ldots, X_{t_n}(\omega)) \in X^n\})$, for the measurable set $X^n = \{X_{t_1}(\omega), \ldots, X_{t_n}(\omega)\}$. Then for a $\mathcal{GP}$, the vector $\boldsymbol{X}_{t_1,\ldots,t_n} := (X_{t_1}, \ldots, X_{t_n})^T$ has a multivariate Normal dstribution for $t_1, \ldots t_n \in \mathcal{T}$. Ultimately, we abbreviate $\boldsymbol{X}_{t_1,\ldots,t_n}$ as $\mathbf{X}$, and note that it is distributed as

$$\frac{1}{(2\pi)^{\frac{n}{2}}\sqrt{|\mathbf{S}|}} \exp\left[-\frac{(\mathbf{x}-\boldsymbol{\mu})^T\mathbf{S}^{-1}(\mathbf{x}-\boldsymbol{\mu})}{2}\right] \tag{1.3}$$

where the mean vector $\boldsymbol{\mu} = \mathbb{E}(\mathbf{X})$, and covariance matrix $\mathbf{S} = [s_{ij}] = \mathbb{E}((x_i - \mu_i)(x_j - \mu_j))$, $i, j = 1, \ldots, n$. So, we can interpret a $\mathcal{GP}$ to define a distribution over a space of functions s.t. any finite set of sample functions realised from this $\mathcal{GP}$, are distributed jointly as a multivariate Normal, with given mean and covariance.

1.5.5.3 Poisson Process

A Poisson Process is a continuous time example of an SP [69], such that number of arrivals during the time interval $(t_1, t_2]$ is $X_{(t_1,t_2]} \in \mathbb{Z}_{\geq 0}$, where for the given time interval $(t_1, t_2]$, $X_{(\cdot,\cdot]}$ is distributed as a Poisson distribution with parameter $\lambda(t_2 - t_1)$. This probability distribution is

$$Pr(X_{(t_1,t_2]} = n) = \frac{\exp\left[-\lambda(t_2 - t_1)\right]\left[\lambda(t_2 - t_1)\right]^n}{n!},$$

where λ is called the rate or intensity parameter of the process. Indeed, as we can already see, this *pmf* depends only on the width $(t_2 - t_1)$ of the time interval, and not the location of the initial and starting time points (i.e. t_1 and t_2 respectively) on the T-axis. This is the "stationarity" property (of a Poisson Process) that we have mentioned above.

So, in this process, the probability distribution of the sample function at time t after the start (at $T = 0$), can be thought of as parametrised by t. In Poisson Processes:

- number of arrivals at time=0 is 0.
- number of arrivals in non-overlapping time intervals is independent and "memoryless" i.e. depends only the length of the interval, not location of interval)
- the inter-arrival times are distributed as an exponential distribution w.p. that is the rate parameter, where

1. the rate parameter is independent of time interval $[t_1, t_2]$ for homogeneous Poisson Process.
2. the rate parameter is a function $(\lambda(t))$ of time interval, for inhomogeneous Poisson Process.

1.6 Summary

In this chapter we came across some of the technical material that constitute the background to the learning of the function that represents the relation between a pair of variables. Such learning is at the heart of this book, and will mainly concern Stochastic Processes, since we will treat the sought function as if it is a sample function of a chosen process. To appreciate the background of Stochastic Processes, we discussed above some basics of Probability theory. In the next chapter we will embark upon a detailed discussion about the Bayesian framework, which is relevant since the learning of the sought function will be undertaken within such a framework in this book.

2

Bayesianism

2.1 Introduction

In the Bayesian world, anything that is an unknown, is random, which equivalently implies that any realisation of an unknown – treated as a random structure ($\boldsymbol{X}$) – occurs with a probability. This probability is underlined by the probability distribution of this random structure.

Indeed, the unknown referred to above - that we seek to learn given the data – may not necessarily be a scalar, vector or (a k-th ordered) tensor valued variable. In other words, this unknown $\boldsymbol{X}$ may not necessarily be a mapping from the measurable space $\{\boldsymbol{\Omega}, \mathcal{F}\}$ to the measurable space comprising the high-dimensional subset $\mathcal{X}$ of $\mathbb{R}^{m_1 \times m_2 \times \ldots \times m_k}$, and the σ-algebra generated on $\mathcal{X}$. Instead, $\boldsymbol{X}$ could be a random function; or could be a random geometric object abiding by certain rules of construction; or could be a random graph – and therefore it is, in general, a mapping onto the appropriate space. We aim to learn such a generic unknown, with uncertainties, given data that is typically noisy and information – that such intended learning is conditional upon – often incomplete. Noise incorporation, and uncertainty learning are both possible within the Bayesian paradigm. Crucially the Bayesian approach allows for readily computable and objectively interpreted uncertainties on the predictions that follow the undertaken learning exercise [51, 70, 74, 81, 8].

The Bayesian approach resorts to implementation of the eponymous rule, to obtain the joint probability distribution of all sought parameters, conditional on the data. This distribution is referred to as the posterior probability distribution of the sought parameters, given the data. Our ulterior aim is, however, to learn values of

DOI: 10.1201/9781003296331-2

each unknown given the data - learning the joint of all unknowns, conditional on the data, is a step in that direction. Proceeding to learn values of individual parameters given the data, using the joint posterior, is Bayesian inference.

If we are able to identify the marginal distribution of each sought parameter, given the data, (using the posterior of all unknowns given the data), we are in a position to subsequently identify a *relevant* interval of values of the parameter, given the data at hand, where criteria qualifying such relevance can be invoked. It then follows that it would be crucially helpful to learn the marginal distribution of each unknown parameter, from the joint posterior of all unknowns given the data. This is possible via techniques that perform Bayesian inference via sampling from the posterior [42, 49, 39, 16, 101, 99, 7, 128, 67, among others].

An important advantage of the approach towards Bayesian inference via sampling from the joint posterior of all unknowns, is that it can potentially facilitate easy and organic computation of the marginal probability distribution of each scalar parameter that learning of any random structure ultimately reduces to, (given the data).

As we indicate in this shaded box above, the aim of the exercise of learning any random structure $\boldsymbol{X}$ – irrespective of what this random structure is – is indeed reducible to an endeavour in learning individual, scalar-valued parameters, given the data. For example,

— as we shall soon see, when learning an unknown function that informs on how a random variable (r.v.) is related to another, the exercise will culminate in the learning of the scalar parameters of the covariance kernel that parametrises the covariance function of the Stochastic Process, from which, the sought function is sampled;

— when learning a tensor, we of course end up learning its scalar-valued components;

— learning a random network implies learning of the binary edge parameter relevant to a pair of vertices, as well as – perhaps – learning each element of the matrix of inter-variable correlations, where any such variable is associated with a node in the network.

By design, the Bayesian approach is advantageous as it permits sequential updating of the undertaken inference, as new data and/or information become available.

This of course calls for a review of what the "Bayesian approach" entails; we embark on the formal exposition of the same next.

2.1.1 Bayes rule

Bayes rule provides the connection between probability for event A to occur, conditional on the occurrence of event B, and for the probability of B to occur, given that A has. It states:

$$\Pr(A|B)\Pr(B) = \Pr(B|A)\Pr(A),$$

$\forall A, B \in \mathcal{F}$. This follows from the commutativity of the set intersection operation, and the definition of conditional probability.

In spite of its simple-looking form, the scope of Bayes rule is enormous. If A represents the event that the parameters $\boldsymbol{\theta}$ of a model take a certain value, and B the event that n observations of certain observable parameters – collated into the vector $\boldsymbol{Y}$ – comprise the data set $\mathbf{D}$, then by Bayes rule,

$$\Pr(\boldsymbol{\theta}|\mathbf{D}) = \frac{\Pr(\mathbf{D}|\boldsymbol{\theta})\Pr(\boldsymbol{\theta})}{\Pr(\mathbf{D})}.$$

Likelihood

Here the probability of data given a value of the model parameters is $\Pr(\mathbf{D}|\boldsymbol{\theta})$. This is referred to as the *likelihood*, which is a function of model parameters, as computed in a given value of the data variable. This motivates reference to the likelihood function as the "likelihood of the model parameters, in the data", though we will be mindful of what the likelihood is – probability of the data variable to attain a value, conditional on a value assigned to the random variable $\boldsymbol{\theta}$ that represents the model parameters. As the probability of the data given the model parameters, the likelihood is the joint probability of the n observations, conditional on $\boldsymbol{\theta}$, s.t. for independent observations, the likelihood is the product of the probabilities of each observation, given $\boldsymbol{\theta}$.

Prior

The term $\Pr(\boldsymbol{\theta})$, on the other hand, is simply the probability of the model parameters, uninformed by any data. Thus, if the probability of the model parameters is enquired upon, prior to the undertaking of the data collection, it is $\Pr(\boldsymbol{\theta})$. Thus, this is referred to as the *prior* probability on the model parameters.

Probability of data

The denominator $\Pr(\mathbf{D})$ on the RHS of the last equation normalises the product of the likelihood and the prior probabilities; it informs on the probability of the data variable to attain a given value. But the probability for the data variable to attain a given value, conditional on a given value of the model parameters, is different from the probability with which the data variable attains said value, given a different value of $\boldsymbol{\theta}$. So, the probability $\Pr(\mathbf{D})$ has to be computed by acknowledging such conditionality on all possible values of $\boldsymbol{\theta}$. Then we anticipate computation of this probability to be difficult. This computation is made possible via the *Law of Total Probability*.

Consider an event that results from the union of its countable disjoint partitions, where any such partition is the intersection of this event with one of a set of disjoint events. Then the Law of Total Probability is essentially a statement on the equality of the probability measure of the event with that of aforementioned union. Thus, this law states,

$$\Pr(A) = \sum_{i=1} \Pr(A \cap B_i) = \sum_{i=1} \Pr(A|B_i)\Pr(B_i),$$

where $A \in \mathcal{F},\ B_i \in \mathcal{F},\ \forall i = 1, 2, \ldots$.

Posterior

The LHS of Bayes rule is then the *posterior probability* of the model parameters given a value of the data variable. Indeed, we could alternatively seek the posterior probability density function of a vector of model parameters $\boldsymbol{\theta}$, conditional on data $\mathbf{D}$. Then by Bayes rule,

$$\pi(\boldsymbol{\theta}|\mathbf{D}) = \frac{\mathcal{L}(\boldsymbol{\theta}|\mathbf{D})\pi_0(\boldsymbol{\theta})}{f_{\mathcal{D}}(\mathbf{D})},$$

where the posterior probability density of $\boldsymbol{\theta}$ given data $\mathbf{D}$ is $\pi(\boldsymbol{\theta}|\mathbf{D})$ – distinguished from the prior probability density of the parameters, $\pi_0(\boldsymbol{\theta})$. The likelihood function of the parameters given the data is $\mathcal{L}(\boldsymbol{\theta}|\mathbf{D})$. Then for independent observations of the observable $\boldsymbol{Y}$, likelihood is given as the product of the *pdf* of the n observations of $\boldsymbol{Y}$, given $\boldsymbol{\theta}$. The *pdf* of the data variable $\mathcal{D}$ is $f_{\mathcal{D}}(\mathbf{D})$. Then by the Law of Total Probability,

$$f_{\mathcal{D}}(\mathbf{D}) = \int f_{\mathcal{D}|\boldsymbol{\theta}}(\mathbf{D}|\boldsymbol{\theta})\pi_0(\boldsymbol{\theta})d\boldsymbol{\theta}.$$

2.1.2 Implications of Bayes rule

Sequential updating

So, what are the broader implications of Bayes rule? One ready interpretation of this result is that as the quality/quantity of the available information changes, inference on the unknown parameters changes. Thus, the Bayesian approach ensures that sequential updating of the available information, permits successive refreshing of the inferred values of the unknowns – given the data. This is pleasing from a knowledge management point of view; in the three-tiered model of knowledge management [124, 48], data and information feed to constitute knowledge, which is thereby rendered "information on information". In this light, it is useful that Bayes rule ensures that knowledge on the originally unknown parameters cannot contract with increase in information.

Learning model for given data

There is an additional gift of Bayes rule that is most productive for domain experts. In general, we only know how to compute the output of a given "formula", i.e. given a model, we know how to compute its output. This output could comprise the observed variables, where this output is computed at chosen values of the inputs that are the parameters of a given model of the relationship between the input and the output. It is much harder to learn the model parameters – reduced perhaps from the intended learning of the model – given a (data) set of values of this observed output, than to compute values of the output given a model.

Bayes rule helps undertake the harder learning option that is demanded, by proxying it with the simpler computation

Example 2.1 *It is easier to compute the magnitude $F_{Sun.\ Earth}$ of the force of gravitational attraction between the Sun and the Earth, given*

> *— values of the gravitational masses M_{Sun} and M_{Earth} of the bodies, and*
>
> *— their mutual separation $R_{Sun,\ Earth}$,*

all input into the inverse square law $F = GM_{Sun}M_{Earth}/R^2_{Sun,\ Earth}$ - than to learn the power α of the inter-body separation $R_{1,2}$, given

> *— data on the force $F_{1,2} = GM_1M_2/R^{\alpha}_{1,2}$ for two bodies,*
>
> *— the gravitational masses of which (M_1, M_2), and separation between which $(R_{1,2})$, are known.*

Here G is the known Gravitational Constant.

However, it is only the former kind of computation that we know how to undertake. In other words, it is the computation of the output given a model (parameter), that is easy to pursue, given data on the observables.

- *Thus, given data on* $M_1, M_2, R_{1,2}$, *for a given value of the model parameter* α *in the model* $F_{1,2} = GM_1M_2/R_{1,2}^{\alpha}$, *we can easily compute* $F_{1,2}$.
- *However, given data on* $M_i, M_j, R_{i,j}$ *and* $F_{i,j}$ *on the* i*-th and* j*-th bodies, (for* $i = 1, \ldots, n$ *and* $j = 1, \ldots, n^{/}$*), it is harder to learn the model parameter* α *defined in the model:* $F_{i,j} = GM_iM_j/R_{i,j}^{\alpha}$.

The Bayesian approach provides the toolkit to undertake such pursuits efficiently, by replacing the *harder-to-compute* probability that we seek – *aka*, posterior probability of the unknown model parameters (conditional on the data) – with computation of probability of the observable given a model. The latter probability contributes to the likelihood.

To summarise, the Bayesian approach suppresses the difficulty of learning about the model, by adopting the easier computation of the output given a model, though it still provides for what was ordered, namely, probability of the unknown model parameters given the data.

So, this approach avoids the fundamental mistake of frequentist approaches – for example, the maximum likelihood approach that attempts identification of the model, given which, probability of the data variable is maximised. On the contrary, it is the probability of the model, conditional on the data, that we in fact seek.

Upcoming empirical illustrations and MCMC

Below, we present illustrations of these two interpretations of Bayes rule; in the first, we showcase the (harder) learning of model parameters given the data, facilitated instead by the undertaking of the (easier) computation of the observable at given model choices, (supplemented by choice of the prior). The second illustration below, discusses the usage of Bayes rule towards sequential updating of inference, given refreshed information. In both illustrations, Bayesian inference is performed by sampling from the posterior probability density of the (sought) model parameters, given the data; we recall that by "Bayesian inference", we imply the learning of the values of the variables given their posterior probability distributions. In this book, we will undertake such posterior sampling using Markov Chain Monte Carlo (MCMC) methods [84, 58, 32, 18, 107, 35]. So, it makes sense for us to equip ourselves with skills for implementing MCMC. We will undertake this discussion next, and present the illustrations thereafter.

2.2 MCMC

Following on from above discussions on Bayesian inference, we anticipate that such inference via posterior sampling will involve drawing a sample $\mathcal{S}$ of values $\boldsymbol{\theta}_1, \ldots, \boldsymbol{\theta}_n$, of the random variable $\boldsymbol{\Theta} \in \mathbb{R}^d$, from the posterior *pdf* $\pi(\boldsymbol{\theta}|\mathbf{D})$ of this parameter vector of the model, given data $\mathbf{D}$.

Puzzlingly though, the posterior *pdf* that we need to sample from, is not known to us, since its parameters – collated into vector $\boldsymbol{\Theta}$ – are unknown. These are the parameters that we are aiming to infer upon, given data $\mathbf{D}$. So, how can we fulfil the desired aim of sampling from this posterior density, informed by the very parameters that we seek to learn? In the MCMC approach, the trick is to generate a sequence of posterior densities that converge to a posterior *pdf*, (or rather, converge to an identifiable set of posterior *pdfs*). This sequence is generated by computing the posterior density at values of each of sequentially updated random variables $\boldsymbol{\Theta}_1, \boldsymbol{\Theta}_2, \ldots$, where such posterior computation is undertaken given the data $\mathbf{D}$, s.t. probability of an element of this sequence of random variables is conditional only on the value attained by the previous element. This sequence of random variables is a Markov Chain. We sample from the computed posterior of the i-th element of this Markov Chain, to generate a value of the next, i.e. the $i+1$-th element of this Chain, at which the posterior is freshly computed. i.e. this time the posterior of $\mathbf{\Theta}_{i+1}$ is computed, given $\mathbf{D}$. The generic form of the posterior *pdf* of each of these r.v.s is typically the same, except that posterior of $\mathbf{\Theta}_i$ is computed at value $\boldsymbol{\theta}_i$ of this r.v., while posterior of $\mathbf{\Theta}_j$ is computed at value $\boldsymbol{\theta}_j$. Here $i, j \in \mathbb{N}$.

Then we need to understand, how to generate $\mathbf{\Theta}_{i+1}$, given $\mathbf{\Theta}_i$, and how to draw random samples from a generic posterior density. Indeed, we would want to know about sampling from posterior *pdf*s of high-dimensional r.v.s, (i.e. tensor-valued variables) in general.

To begin with, we will say that we need the sought sequence of r.v.s to be a Markov Chain that is ergodic; this implies that this chain has a unique stationary distribution. By a "stationary distribution" of a Markov Chain, is implied a probability distribution that is invariant to the passage of time.

Definition 2.1 *An ergodic Markov Chain is one, for which there exists a time step, indexed as $t_\star$, (where $t_\star > 0$), s.t. probability for the Chain to be in any state at time $t > t_\star$, is > 0, with the chain having started from any other state (at $t = 0$). Ergodicity is ensured if the Chain is positive recurrent and aperiodic.*

A chain is recurrent if the waiting time since the first visit, to visit a state the second time, is finite, for all states in the state space $\mathcal{H}$, where $\mathbf{\Theta}_i \in \mathcal{H}$, for $i = 1, 2, 3, \ldots$. In a positive recurrent chain, such is true of the average waiting time till another visit to the state.

An aperiodic chain is one in which there does not exist a state that is visited after each passage of a given interval of time.

Once the Chain converges to a time-invariant, identifiable set of values for the constituent random variables, it would then imply that the Chain has converged to its stationary distribution. This follows from the result that *this Chain – ergodic by construction – has a unique stationary distribution.*

So, now we restart our pursuit of the two things that we need to know how to do;

— firstly, we need the know-how for generating an ergodic Markov Chain, and
— secondly, we need to know how to sample from a posterior probability density that is not necessarily invertible.

Markov Chain Monte Carlo, (or MCMC) methods achieve this. Let us discuss the methodology for sampling from a generic posterior density first, and embedded within this method, we will identify how ergodicity – and thereby a unique stationary distribution – of the Chain is ensured.

Why is it difficult to sample value x of r.v. X from its probability distribution $F_X(\cdot)$? It so happens that such a sampled value is $F_X^{-1}(u)$ where $U = u$, with $U \sim Uniform[0, 1]$. The reason for this suggestion is then immediately queried. We in fact seek that transformation $T(\cdot)$, the sampled value of which, is $T(u) = x$. We will see that $T(\cdot) = F_X^{-1}(\cdot)$.

Theorem 2.1 *For $X \in \mathcal{X} \subseteq \mathbb{R}$, a sample from the cdf $F_X(\cdot)$ of X, is $F_X^{-1}(u) = T(u)$, where $X = T(U)$, with $U \sim Uniform[0, 1]$, and $U = u$; here $T : [0, 1] \longrightarrow \mathbb{R}$ is monotonically non-decreasing and bijective.*

Proof *We express $U \sim Uniform[0, 1]$ equivalently as $U \sim F_U(\cdot)$. Then*

$$F_U(u) = \begin{cases} 0 \; if \; u < 1 \\ u \; if \; u \in [0, 1] \\ 1 \; if \; u > 1. \end{cases} \tag{2.1}$$

Now, $X = T(U)$ is distributed according to cdf $F_X(\cdot)$. So,

$$F_X(u) = \Pr(T(U) \leq u).$$

But

$$T(U) \leq u \implies U \leq T^{-1}(u),$$

since T is monotonic non-decreasing, and its inverse exists, (T is bijective). Therefore

$$F_X(u) = \Pr(U \leq T^{-1}(u)) \equiv F_U(T^{-1}(u)).$$

But a property of r.v. $U \sim Uniform[0,1]$ is that its cdf computed at a value $u \in [0,1]$, is that value, i.e. $F_U(u) = u$. It then follows:

$$F_X(u) = \Pr(U \leq T^{-1}(u)) \equiv F_U(T^{-1}(u)) = T^{-1}(u).$$

But operating F_X^{-1} on both sides of $F_X(u) = T^{-1}(u)$ from above, we get

$$u = F_X^{-1}(T^{-1}(u)) = T^{-1}(F_X^{-1}(u)),$$

where the last equality stems from the independence of $F_X^{-1}(\cdot)$ and $T^{-1}(\cdot)$ in general. It follows that

$$T(u) = F_X^{-1}(u),$$

and therefore,

$$T(\cdot) = F_X^{-1}(\cdot),$$

in general.

Then the sample from $F_X(\cdot)$ is a value x of X, which as we have just proved, is equivalently, the value of $F_X^{-1}(\cdot)$ computed at u. But any generic *cdf* is not invertible in a closed-form way, Also, the inversion task becomes increasingly more difficult as the dimensionality of X increases.

von Neumann suggested an ingenious way to sample from $F_X(x)$ that is non-invertible/difficult-to-invert [46, 88, 21], by suggesting that we sample from a closed-form invertible function $g(x)$ instead, but decide to retain only those samples that abide by such a constraint, that renders the probability density of the sample variable, the *pdf* $f_X(\cdot)$. This sampling technique is referred to as *Rejection Sampling*. Here $f_X(x)$ is the *target density*, i.e. the density that we target drawing samples from. $g(x)$ is the *proposal density*, since we will propose a sample from this density – as a potential sample from $f_X(\cdot)$, where such a sample will be examined under the *acceptance criterion* to check if the proposed sample is indeed one from the target density, or not.

The idea behind Rejection Sampling is expertly simple. With $M > 0$ chosen s.t. the function $Mg(x) \geq f_X(x) \forall x \in \mathbb{R}$, any sample $x^\star$ from the proposal density is accepted as a sample from the target density, if $f(x^\star)/Mg(x^\star) \geq u$, where $U = u$, with $U \sim Uniform[0, 1]$.

Let us see why obeying this criterion implies that the *pdf* of the sample is $f_X(\cdot)$.

Theorem 2.2 *If the random variable $X^\star$ that is sampled from proposal density $g(\cdot)$, is s.t. its value $x^\star$ is accepted as a sample from the target density $f_X(\cdot)$ when $U \leq f_X(x^\star)/Mg(x^\star)$, then it implies that such an $X^\star$ is distributed according to the pdf $f_X(\cdot)$; here, $U \sim Uniform[0, 1]$, $M > 0$ is s.t. $f_X(x^\star) < Mg(x^\star) \forall x^\star$ where $X^\star = x^\star$ and $X^\star \in \mathcal{X} \subseteq \mathbb{R}$.*

Proof *For $X^\star \sim g(\cdot)$, if we can show that the probability of $X^\star$ to be $\in [a, b] \subseteq \mathcal{X}$ – conditional on $x^\star$ accepted as a sample from $f_X(\cdot)$ – is $\int_a^b f_X(s)ds \ \forall [a, b] \subseteq \mathcal{X}$, then we would have shown that the method of accepting the sample r.v. $X^\star$ is correct.*

Now, probability of sampled $X^\star = x^\star$ to be $\in [a, b] \subseteq \mathcal{X}$, given that it is accepted as a sample from $f_X(\cdot)$ is

$$\Pr(x^\star \in [a, b] | x^\star \text{ is accepted}) = \frac{\Pr(x^\star \in [a, b], x^\star \text{ is accepted})}{\Pr(x^\star \text{ is accepted})}. \tag{2.2}$$

From the theorem statement we learn that the random sample $x^\star$ (from $g(\cdot)$) is accepted when $U \leq f_X(x^\star)/Mg(x^\star)$. Therefore,

$$\Pr(x^\star \text{ is accepted}) = \Pr(U \leq f_X(x^\star)/Mg(x^\star)).$$

But $X^\star$ is a random variable, distributed according to density $g(\cdot)$. So, the probability of acceptance, as expressed in the RHS of the last equation is in fact incorrect, as this probability will need to be averaged over all possible values of $X^\star$, i.e. the correct probability of acceptance is

$$\mathbb{E}(\Pr(U \leq f_X(X^\star)/Mg(X^\star)) = \int_{\mathcal{X}} g(x) \Pr(U \leq f_X(x)/Mg(x)) dx.$$

Recalling that for $U \sim Uniform[0,1]$, $\Pr(U \leq s) = F_U(s) = s, \forall s \in [0,1]$, *we simplify the probability of acceptance to be*

$$\begin{aligned}
\mathbb{E}_X(\Pr(U \leq f_X(X^\star)/Mg(X^\star)) &= \int\limits_{\mathcal{X}} g(x) f_X(x)/(Mg(x))\, dx \\
&= \frac{1}{M} \int\limits_{\mathcal{X}} f_X(x) dx \\
&= \frac{1}{M}. \qquad (2.3)
\end{aligned}$$

So, in Equation 2.2, probability of sampled $X^\star = x^\star$ *to be* $\in [a,b] \subseteq \mathcal{X}$, *given that it is accepted as a sample from* $f_X(\cdot)$, *is*

$$M\mathbb{E}(\Pr(X^\star \in [a,b], X^\star \text{ is accepted})), \qquad (2.4)$$

where the probability in the numerator of the RHS of Equation 2.2 is replaced by the expected probability, for the same reason as above.

Now,

$$\mathbb{E}(\Pr(X^\star \in [a,b], X^\star \text{ is accepted})) = \int\limits_a^b g(x) \Pr(U \leq f_X(x)/Mg(x)) dx$$

$$= \int\limits_a^b g(x)(f_X(x)/Mg(x)) dx = \frac{\int_a^b f_X(x)}{M},$$

so that in Equation 2.2, probability of sampled $X^\star$ *to be* $\in [a,b] \subseteq \mathcal{X}$, *given that it is accepted as a sample from* $f_X(\cdot)$, *is*

$$M \frac{\int_a^b f_X(x)}{M} = \int\limits_a^b f_X(x),$$

and this holds $\forall [a,b] \in \mathcal{X}$.

This implies that as long as a sampled value $x^\star$ *of the r.v.* $X^\star$, *from the proposal density* $g(\cdot)$, *is accepted with probability*

$f_X(x^\star)/Mg(x^\star)$, the accepted samples are effectively drawn from the target density $f_X(\cdot)$,

So, we understand how we will randomly sample a value $\boldsymbol{\theta}_{i+1}$ of the r.v. $\boldsymbol{\Theta}_{i+1}$. from the generic posterior $\pi(\boldsymbol{\theta}_i|\mathbf{D})$ of r.v. $\boldsymbol{\Theta}_i$, given data $\mathbf{D}$ – we will use Rejection Sampling for this. We will use the invertible (or approximately-invertible) proposal density $g(\boldsymbol{\theta}_{i+1}|\boldsymbol{\theta}_i)$ to propose a candidate value of $\boldsymbol{\Theta}_{i+1}$ from. This *proposed* value of $\boldsymbol{\Theta}_{i+1}$ will be accepted as a sample from its posterior *pdf* given $\mathbf{D}$, with a probability that is motivated by rejection sampling, s.t. this probability of acceptance – also referred to as "acceptance ratio" $\alpha(\boldsymbol{\theta}_{i+1};\boldsymbol{\theta}_i)$ – is:

$$\alpha(\boldsymbol{\theta}_{i+1};\boldsymbol{\theta}_i) := \min\left(1, \frac{\pi(\boldsymbol{\theta}_{i+1}|\mathbf{D})}{g(\boldsymbol{\theta}_{i+1}|\boldsymbol{\theta}_i)} \Big/ \frac{\pi(\boldsymbol{\theta}_i|\mathbf{D})}{g(\boldsymbol{\theta}_i|\boldsymbol{\theta}_{i+1})}\right).$$

So, the suggestion is that, if $\alpha(\boldsymbol{\theta}_{i+1};\boldsymbol{\theta}_i) \geq u$, we accept the sampled value $\boldsymbol{\theta}_{i+1}$ as a sample from the current posterior density; else reject it. Here u is the value of r.v. $U \sim Uniform[0,1]$, as above.

This form of the acceptance ratio $\alpha(\boldsymbol{\theta}_{i+1};\boldsymbol{\theta}_i)$, emerges out of the need to design a *reversible* Chain. The ergodic Chain $\{\boldsymbol{\theta}_0, \boldsymbol{\theta}_1, \boldsymbol{\theta}_2, \ldots\}$ is reversible, if running the sequence of states backwards gives rise to the same chain. Then in such a reversible chain, if we define a time step index n, s.t. $\boldsymbol{\phi}_i := \boldsymbol{\theta}_{n-i}, \ \forall i = 0, 1, \ldots, n$, then the set of states $\{\boldsymbol{\theta}_i\}_{i=1}^n$ that is generated as time moves forwards, equals the set $\{\boldsymbol{\phi}_i\}_{i=1}^n$ that is generated when time is turned backwards. Then if the stationary distribution of the forward chain is π, (where such stationarity stems from the ergodic nature of the considered Chain), the stationary distribution of the backward chain is also π. That the forward moving and backward moving chains are equal, implies that the probability of moving from the i-th state (at which $\boldsymbol{\theta}_i$ is attained) to the j-th state, equals the probability of returning from the j-th to the i-th state, $\forall i, j = 1, \ldots, n$, i.e.

$$\pi(\boldsymbol{\theta}_j)p(\boldsymbol{\theta}_i|\boldsymbol{\theta}_j) = \pi(\boldsymbol{\theta}_i)p(\boldsymbol{\theta}_j|\boldsymbol{\theta}_i), \quad \forall i, j = 0, \ldots, n,$$

where $p(\boldsymbol{\theta}_i|\boldsymbol{\theta}_j)$ is the probability with which the Chain transits from state $\boldsymbol{\theta}_j$ to $\boldsymbol{\theta}_i$, (for any states $\boldsymbol{\theta}_i, \boldsymbol{\theta}_j$ in the state space). In the above equations, we recall the usage of the same stationary distribution π in the forward and reversed chains. The last set of equations represent "detailed balance". As is evident from these equations, detailed balance implies that the probability of moving from any considered state to another, is retained as the probability of moving from the latter state to the originally considered state.

In the MCMC paradigm, we rephrase the stationary distribution as $\pi(\cdot|\mathbf{D})$, which is the posterior probability distribution that is conditional on data $\mathbf{D}$. In this context, we will need to sample from the proposal density $g(\cdot|\cdot)$ and then check if the sample has been accepted or not, with acceptance probability $a(\cdot;\cdot)$. Thus

$$p(\boldsymbol{\theta}_j;\boldsymbol{\theta}_i) = a(\boldsymbol{\theta}_j;\boldsymbol{\theta}_i)g(\boldsymbol{\theta}_j|\boldsymbol{\theta}_i).$$

Using this in the demand for detailed balance we get

$$\pi(\boldsymbol{\theta}_{i+1}|\mathbf{D})a(\boldsymbol{\theta}_i;\boldsymbol{\theta}_{i+1})g(\boldsymbol{\theta}_i|\boldsymbol{\theta}_{i+1}) = \pi(\boldsymbol{\theta}_i|\mathbf{D})a(\boldsymbol{\theta}_{i+1};\boldsymbol{\theta}_i)g(\boldsymbol{\theta}_{i+1}|\boldsymbol{\theta}_i),$$

or in other words,

$$\frac{a(\boldsymbol{\theta}_{i+1};\boldsymbol{\theta}_i)}{a(\boldsymbol{\theta}_i;\boldsymbol{\theta}_{i+1})} = \frac{\pi(\boldsymbol{\theta}_{i+1}|\mathbf{D})}{g(\boldsymbol{\theta}_{i+1}|\boldsymbol{\theta}_i)} \div \frac{\pi(\boldsymbol{\theta}_i|\mathbf{D})}{g(\boldsymbol{\theta}_i|\boldsymbol{\theta}_{i+1})}.$$

Then this ratio of acceptance probabilities allows for the definition of one "acceptance ratio" – the comparison of which with u allows us to accept (or reject) the candidate $\boldsymbol{\theta}_{i+1}$: in flavours of one type of MCMC algorithms, (called the Metropolis Hasting algorithm), the acceptance ratio is:

$$\alpha(\boldsymbol{\theta}_{i+1};\boldsymbol{\theta}_i) := \min\left(1, \frac{a(\boldsymbol{\theta}_{i+1};\boldsymbol{\theta}_i)}{a(\boldsymbol{\theta}_i;\boldsymbol{\theta}_{i+1})}\right).$$

Thus, the Chain is sequentially updated using this acceptance probability that is implied by detailed balance, which (sufficiently, but not necessarily) ensures that this Chain has a stationary distribution. We also require this stationary distribution to be unique. This is rendered true by requiring the Chain to be ergodic, i.e. positive recurrent and aperiodic.

Above, we have qualified detailed balance to imply stationarity, but stationarity itself is a weaker constraint on the distribution across the states, than detailed balance is. Stationarity will be ensured if the sum of probabilities of leaving a given state to move to all the other states, is the same as the sum of the probabilities of entering this state from all the other states. But detailed balance demands this at a much finer resolution, local to the given state and every one of the other remaining states. Thus, while detailed balanced implies stationarity, stationarity does not imply detailed balance.

We notice that the acceptance ratio is s.t. any normalisation constant that is independent of the sought model parameters – that are being updated in the MCMC Chain – cancels, when the ratio of the posterior probability densities is computed. Therefore, there is no need to compute any such normalisation, from the inference point of view. This is an important adavantage of using MCMC techniques towards Bayesian inference.

So, we have an understanding now, of how to generate the sequence $(\boldsymbol{\theta}_1, \boldsymbol{\theta}_2, \ldots,)$ and sample from the posterior *pdf* computed at any $\boldsymbol{\theta}_i$ (given the data). Firstly, this sample is the result of accepting a proposal from the proposal density, the parameters of which are driven by $\boldsymbol{\theta}_i$. Then, this accepted sample is the next element in the sequence.

2.2.0.1 Random Walk Metropolis Hastings

One concept that we have evaded discussions about so far, is that $\boldsymbol{\Theta}_i \in \mathbb{R}^d$ is vector-valued in our exposition of the MCMC ideology, while our discussion of Rejection Sampling was made using a scalar X. How do we handle the high-dimensional problem? One possibility is to sample each of the d components of $\boldsymbol{\Theta}_{i+1} = (\Theta_{i+1}^{(1)}, \ldots, \Theta_{i+1}^{(d)})^T$, at the same time, from respective proposal densities. So with the posterior of $\boldsymbol{\Theta}_i$ computed given the data, we draw d number of random samples from each of these proposal densities, using: $\Theta_{i+1}^{(k)} \sim g_k(\cdot|\theta_i^{(k)}), \forall k = 1, \ldots, d$. Then the sampled values $\{\theta_{i+1}^{(k\star)}\}_{k=1}^d$ of scalar-valued r.v.s $\{\Theta_{i+1}^{(k)}\}_{k=1}^d$ are checked for acceptance, using

$$\alpha(\theta_{i+1}^{(1\star)}, \ldots, \theta_{i+1}^{(d\star)}; \theta_i^{(1)}, \ldots, \theta_i^{(d)}) =$$

$$\frac{\pi(\theta_{i+1}^{(1\star)}, \ldots, \theta_{i+1}^{(d\star)}|\mathbf{D})/(g_1(\theta_{i+1}^{(1\star)}|\theta_i^{(1)}) \ldots g_d(\theta_{i+1}^{(d\star)}|\theta_i^{(d)}))}{\pi(\theta_{i+1}^{(1)}, \ldots, \theta_i^{(d)}|\mathbf{D})/(g_1(\theta_i^{(1)}|\theta_{i+1}^{(1\star)}) \ldots g_d(\theta_i^{(d)}|\theta_{i+1}^{(d\star)}))}.$$

This is the basic mode of implementation for the Metropolis Hastings (MH) algorithms. One type of MH algorithm can vary from another, insofar as the structure of the proposal density is concered.

Thus, the Random Walk MH (RWMH) algorithm [107, 118] resorts to the usage of a Normal proposal density, in which the mean of the density that variable Θ_{i+1} is proposed from, is the current value of Θ_i – say θ_i. Then we can immediately see what the ratio $g(\theta_{i+1}^\star|\theta_i)/g(\theta_i|\theta_{i+1}^\star)$ reduces to the ratio of a Normal density in $\theta_{i+1}^\star$ with mean θ_i to a Normal density in θ_i with mean $\theta_{i+1}^\star$, where both Normals are parametrised with the same variance. Thus the ratio of the proposal densities then cancels to yield 1.

2.2.0.2 Independent Sampler

Again, it is possible to define a proposal density that Θ_{i+1} is proposed from, is independent of Θ_i, i.e. the ratio of the proposal densities is then $g(\theta_{i+1}^\star)/g(\theta_i)$. This implementation is referred to as the Independent Sampler MH algorithm [107].

2.2.0.3 Gibbs Sampling

Distinguished from the MH approach, it is possible that the d number of scalar-valued unknowns $\Theta_{i+1}^{(1)}, \ldots, \Theta_{i+1}^{(d)}$ are not updated all at the same time, but in a step-wise fashion. Thus, $\Theta_{i+1}^{(1)}$ is first updated; then at the next step, at this updated $\Theta_{i+1}^{(1)}$, $\Theta_{i+1}^{(2)}$ is updated; at the 3rd step, at the updated $\Theta_{i+1}^{(1)}$ and $\Theta_{i+1}^{(2)}$, $\Theta_{i+1}^{(3)}$ is then updated; and we proceed in this way in each iteration, till the last of the d unknowns are updated. This approach is referred to as Gibbs sampling [107, 50, 97].

2.2.0.4 Metropolis-within-Gibbs

Ultimately, it is possible that the full state space vector $\boldsymbol{\Theta}$ that comprises the unknowns that we are interested in making inference upon, using data $\mathbf{D}$, is s.t. its partition $\boldsymbol{\Theta}_A$ comprises scalar-valued r.v.s that have a commonality of feature in their mutual interaction, while partition $\mathbf{\Theta}_B$ comprises the remaining r.v.s that we want to learn, or comprises another set of variables that share a commonality in their inter-correlations, [109, 54]. Thus $\mathbf{\Theta}^T = (\mathbf{\Theta}_A^T, \mathbf{\Theta}_B^T)$; of course it is possible that the full state space vector is identified to readily partition into q number of such "blocks", or partitions, instead of the 2 that we have used in the motivating example above. In such a situation, at each iteration, we will update all components of a partition or block at one go; then at the next step, at the updated elements of the first block, update the components of the second partition – as in Gibbs sampling; proceed to the subsequent block if applicable, and so on. So, it is as if we are exploiting information about the commonalities in the correlations amongst the different variables that we are seeking to make inference upon, to perform the partitioning of the full state space vector, and then updating variables comprising individual partitions using MH, while updating the partitions sequentially as in Gibbs. There are multiple other implementation of MCMC and related sampling techniques [72, 108, 73, for example], that we do not included here.

2.3 Priors

We are all set to undertake a couple of toy empirical illustrations of the Bayesian modelling and inferential strategies, but it appears that we have not embarked on the discussion of an important aspect of Bayesianism – priors. So, we will discuss priors in this section, and proceed to the empirical illustrations in the following section.

As we have noted during the above discussion on Bayes rule, the posterior *pdf* of any random variable, given the data that bears influence upon this variable, is informed upon, by the probability of the data given a choice for the value of this variable, (i.e. the likelihood), as well the *pdf* of this variable that is formulated prior to data collection, i.e. the prior *pdf*. If information is available about the variable attaining a given value with an identified probability, not using the same towards inference on the unknowns, is a wastage of such information – and Bayes rule allows for the usage of such information, while ensuring that data is capacitated within this approach, to swamp effects of such prior assignment if need be, lest inference becomes prior-dominated, and thereby the sought solutions are rendered arbitrary, (driven by the practitioner's inputs).

One may not have the exact value of such prior probability, but one could be aware of whether a given value will be attained by a variable, with a probability that we can identify as "high", as distinguished from "low", or "moderately high". In other words, our probabilistic knowledge of the variable could be diversely uncertain – a limit of this is of course to have minimal prior information, i.e. to consider any value of the variable as attainable, with equal prior probability. There are multiple real-world applications where such a Uniform prior *pdf* might become relevant. In particular, when new methodology is developed, such prior choices are likely to be preferred, lest preliminary computations using the draft method are rendered biased towards one set of solutions over another.

Ultimately, we need to demand from any strategy that addresses problems within the Bayesian paradigm, that

- the model accommodates prior beliefs on the unknown, as might be provided by the practitioner, where the prior *pdf* is formulated s.t. with increasing uncertainty on values attained by each uknown, the chosen prior approaches a Uniform density with support given by an identified interval in the space of the unknown;
- the solution to the problem is not driven by the choice of the prior, s.t. the solution does not change within the learnt uncertainties if the prior is changed, given "sufficiently" large data sets.

(We have not discussed learning uncertainties on sought values of each unknown in the problem, given the data; will do so soon).

But it is difficult to quantify "sufficient" largeness of a data set before the learning of the unknowns has been attempted, as this optimal size depends on details of the problem and model; on the available prior information, etc. In any case, if on increasing the size of the data set, prior influence on the learning appears unchanged, it may indicate existence of a flaw in the model and/or inference.

As with any other theory of knowledge, in Bayesianism too, the approach can be an *objectivist* one, as distinguished from a *subjectivist* one [78, 65]. In the objectivist viewpoint, information on the variables relevant to the model is independent of the practitioners' – in our context, the Bayesian learners' - priors. We might have appeared to be encouraging this perspective so far in this section, but any such conclusion is incorrect, and our stance requires clarification. While it is fine to consider one set of solutions as likeli as another, in lieu of information – such as when developing new methodology – in real world applications, if the practitioner has access to opinions of domain experts, it is advisable to acknowledge their inputs towards solutions, since such inputs are likely to improve the quality of inference, e.g. reduce the width of the "interval representing uncertainties" learnt on any variable that we seek to perform such inference upon.

Let us now talk about an example of the help that domain expertise fetches, as experienced by the author when faced with the task of predicting the *risk score* of a potentially-fatal complication of Bone Marrow Transplants that patients of certain cancers sometimes are recommended to undergo, [24]. While such prediction was undertaken for independent patients, for one patient, one of the comorbidities was entirely different, compared to the comprehensive list of comorbidities that were noted to have been manifest in those patients who comprised the training set. Then how were we to make an informed prediction for this patient, on the basis of the extant learning made using said training set? We requested the oncologists for their opinion on how susceptible to the complication this patient was, given their particular comorbidity; additionally, we requested inputs on how strongly they believed their answer to be correct. To the question of anticipated severity of the complication, as well as to the strength of their belief, we requested the oncologists to state their opinion as in the categories: "high"; "moderately high/low"; "low". Their answer to the question regarding severity of the complication in this patient, was parametrised and assigned an appropriate numerical value; we modelled the prior *pdf* on the *risk score* to be a Normal density, centred at this value. Their answer to the question regarding strength of conviction in their belief was parametrised s.t. this parameter is modelled as the variance of the Normal prior *pdf*. When we explored values of the prior mean to correspond to the "high" to "moderate" assessment of severity of the complication anticipated by the physicians, and explored the prior variance to correspond to "low" to "moderate" convictions that were suggested by the doctors, all combinations produced recurrent score predictions, (within the predicted uncertainties). Indeed, the doctors' beliefs about their patient is a useful piece of information that should be employed within the Bayesian approach. Thus, we encourage the subjectivist approach to Bayesianism when possible – in which, truth about the unknowns is not considered independent of the practitioner, but is relative to the practiotioner's experience.

At the end of the day, when attempting a Bayesian solution to a problem, one is not going to split hair on which mode of conduct is suitable, but one would attempt formulating a verifiable model with reproducible results that are not going to become "siginificantly" different if the priors change, where by "significant" differences, we imply differences outside the learnt uncertainty interval. Models and results that are not prior-sensitive need to be sought. But then is it wrong to choose *conjugate priors*?

To answer this, let us first recall what conjugate priors are. In order to attain closed-form, parametric posterior probability distributions, one idea offered within Statsitics is to choose a prior probability, that is such, that when multiplied by the likelihood, the result is the closed-form posterior that is the same probability distribution as the prior probability distribution. For example, in a problem in which the likelihood is a Binomial distribution, choosing the prior on the model parameter to be a Beta density, allows for the posterior to be a Beta density as well, with identified parameters. This was of course introduced to aid inference, in days when usage of Bayesian inference was seen as numerically challenging. Of course these suggestions are not relevant to generic high-dimensional real-world problems of the type that we will discuss in the book, and so the practice of conjugacy is irrelevant to Machine Learning undertakings within the Bayesian paradigm. While modelling the prior density of a sought r.v. to be a chosen parametric density, as distinguished from any other parametric form, is not wrong – given what a prior is, namely, a guess of the *pdf* of the r.v. independent of data – the conjugacy-driven prior choice may not be typically justifiable. Additionally, if the solutions for the unknown given the data, are identified to be significantly different once the priors are changed, that is definitely cause for concern.

2.4 Quantifying uncertainties of learnt parameters

In this section we discuss the computation of the Highest Probability Density credible regions (HPDs) on each unknown that we learn, given the available data. Identification of the marginal posterior *pdf* of the parameter Θ_i given the data, is crucial to the learning of the HPDs on this parameter, $\forall i = 1, \ldots, d$.

It is clear that each algorithm of an MCMC implementation offers an updated value of each model parameter in every iteration. Thus, within the Chain, the set of the updated values of the i-th such parameter Θ_i is $\{\theta_i^{(1)}, \ldots, \theta_i^{(n)}\}$, $\forall i = 1, \ldots, d$. Then considering only the subset comprising values of Θ_i that are current during each of the post-burnin iterations of the Chain, frequency distribution of that subset offers a ready approximation of the scaled marginal posterior *pdf* of Θ_i, given the data. Thus, the learning of the marginals for all the sought model parameters, is undertaken in an organic fashion, within the MCMC technique.

We want to unambiguously compute the $\alpha\%$ HPD $[\theta_i^{(left)}, \theta_i^{(right)}]$ on the Θ_i updated in the post-burnin part of the Chain, given the data, s.t. the area under the marginal $m_{\Theta_i}(\theta_i|\mathbf{D})$ curve, enclosed within $\Theta_i = \theta_i^{(left)}$ and $\Theta_i = \theta_i^{(right)}$, is $\alpha\%$ of the total area under the marginal, [94, 51]. We have to arrive at a method of this HPD computation, while keeping in mind that the marginal posterior *pdf* of a parameter given a data set, is not necessarily univariate; it would be multivariate in general, and the sought HPD is then anticipated as the union of two or more intervals.

Given this general situation, a robust method for quantifying the uncertainties in the learning of Θ_i given a data set at hand is the following. Consider a line that we view as a variable, s.t. this line is parallel to the $m_{\Theta_i}(\cdot|\cdot) = 0$ line. Basically we consider the equation $m_{\Theta_i}(\cdot|\cdot) = K$, where $K \in \mathbb{R}_{\geq 0}$ is a random variable. Now, we seek the $K = k$ for which the area under $m_{\Theta_i}(\cdot|\cdot)$, enclosed by solutions to the equation:

$$m_{\Theta_i}(\theta_i|\mathbf{D}) = k,$$

is $\alpha\%$ of the total area under $m_{\Theta_i}(\cdot|\cdot)$. For a univariate marginal posterior *pdf*, there are 2 solutions to this equation, as long as k does not exceed the modal marginal density; for a bimodal marginal, there could be 4 solutions or 2 solutions, depending on the shape of the marginal, at a given K; etc.

Typically we will compute the 95% Highest Probability Density credible regions, (or HPDs hereon), by varying the line parallel to the horizontal axis, until area under the marginal *pdf* – enclosed by values of Θ_i at which the marginal intersects this line – is 95% of the total area under the marginal.

2.5 Empirical illustration – Bayes rule and learning model parameters

What is harder – computing area A of a rectangle with length $L = \ell$ and breadth $B = b$, given that the model for area is $A = LB$; or learning the model for area as the function $f(L, B)$, given data on L, B, A of n observed rectangles? Naturally the latter exercise is relatively more challenging. This is what we referred to above as the comparatively higher difficulty in learning model parameters given the data, over churning out values of a datum, given a model. In this section we will try to learn this function $f(L, B)$ that is a model for the area of a rectangle with length $L \geq 0$ and breadth $B \geq 0$, given the data $\mathbf{D} = \{(\ell_i, b_i, a_i)\}_{i=1}^{n}$, where ℓ_i, b_i, a_i are values of the length, breadth and area of the i-th observed rectangle, respectively. We will consider this data set to consist of such information on n toy rectangles, where we will vary n at different stages of the experiment. To begin with, the first 28 lines of the toy data $\mathbf{D}$ that we use, are displayed in Table 2.1; it is these first 28 lines that we will make the learning of the model parameters with.

In fact, we will assign a parametric form to the model $f(L, B)$ of the area, with parameters of this chosen form left as unknowns that we will aim to learn, given data $\mathbf{D}$; we choose to set

$$f(L, B) = L^Q B^R,$$

with the unknowns $Q, R \in \mathcal{A} \subset \mathbb{R}$.

We treat observations of the n distinct observed rectangles to be independent of each other so that we define probability of the data set $\mathbf{D}$, conditional on chosen values of Q and R, to be the product of the *pdf* of the observable $\boldsymbol{Y} := (L, B, A)^T$ for the i-th observed rectangle, given q, r, across all values of $i = 1, \ldots, n$. Here we model the *pdf* of the observable $\boldsymbol{Y}$ given the model parameters Q, R to be a smoothly declining function of the distance between the modelled and observed areas, s.t. when this distance approaches 0, the *pdf* is maximal $\forall q, r$, and for the distance approaching $\pm\infty$, the *pdf* is 0 $\forall q, r$. Also, the *pdf* of $\boldsymbol{Y}$ at $Q = q$ and $R = r$ is the same, irrespective of whether the modelled area exceeds the obseved area by δ, or is less than the observed area by δ, $\forall \delta > 0$. Then a parametric form of any bell-shaped curve, with mode at 0, would satisfy the above constraints.

Does that mean that the likelihood is ambiguous? The answer is yes. The shape of the *pdf* of the observable given the model parameters, is not uniquely constrained by the information available in this problem, but Bayesian inference undertaken with MCMC performed on the model parameters Q and R, by sampling from their ensuing joint posterior *pdf*, will converge in mean, irrespective of the shape of the likelihood.

TABLE 2.1
Observations on length, breadth and area of 28 toy rectangles that were employed to learn the model parameters Q and R that define our used model of area of a rectangle.

Length (units)	Breadth (units)	Area (unit squared)
2	2	3.99
2	3	6.01
3	3	8.98
3	4	11.99
4	4	16.1
4	5	19.99
5	5	25.01
2	4	8.02
5	2	10.02
3	2	6.01
3	5	14.99
5	6	29.99
6	2	12.03
3	6	17.97
3	4	12.01
9	4	36.01
6	2	11.99
3	9	26.98
2	2	3.99
9	7	62.98
8	6	47.98
3	6	18.02
1	2	1.98
4	1	4.01
8	7	56.01
2	9	17.98
7	8	56.02
1	5	4.98

We choose the *pdf* of $\boldsymbol{Y} = \boldsymbol{y}_i = (\ell_i, b_i, a_i)^T$ given $Q = q$, $R = r$, to be $\propto \exp(-(a_i - \ell_i^q b_i^r)^2/2\sigma^2)$, where σ^2 is a chosen constant.

Then assuming independent data points, the likelihood is

$$\mathcal{L}(q, r; \mathbf{D}) = \prod_{i=1}^{n} K \exp\left(-\frac{(a_i - \ell_i^q b_i^r)^2}{2\sigma^2}\right).$$

Here K is a constant $(=(2\pi\sigma^2)^{-n/2})$

We choose Normal priors on the model parameters Q and R. We will experiment with different prior means (μ_q, μ_r) and prior variances (σ_q^2, σ_r^2) within our learning exercise. So, the prior *pdf* is

$$\pi_0(q, r) = K^{/} \exp(-(q - \mu_q)^2/2\sigma_q^2) \exp(-(r - \mu_r)^2/2\sigma_r^2).$$

where $K^{/}$ is a constant.

Then the joint posterior *pdf* of the model parameters Q, R, given the data $\mathbf{D}$, by Bayes rule is

$$\begin{aligned}&\pi(q, r|\mathbf{D}) \propto \\ &\prod_{i=1}^{n} K \exp\left(-\frac{(a_i - \ell_i^q b_i^r)^2}{2\sigma^2}\right) \\ &\exp(-(q - \mu_q)^2/2\sigma_q^2) \exp(-(r - \mu_r)^2/2\sigma_r^2). \qquad (2.5)\end{aligned}$$

To generate posterior samples using MCMC, we require the posterior upto an unknown scale. This is presented above. So, we are in a position to learn values of Q, R using MCMC, given data $\mathbf{D}$.

In Figure 2.1 we examine the trends in the likelihood, prior and posterior *pdf*, as we vary across values of the model parameters Q and R independently, with the functions computed at a chosen q, r, using the data. We assume that minimal prior information is available on the values of Q and R; this is reflected in our choice of the Normal priors with high variance, on each parameter. We centre the prior *pdf* of Q, as well as the prior of R, at 3; that the variance of each prior density is high, is manifest in the flat shape of the priors on Q, shown in (red) open circles in the upper panel of Figure 2.1 as well in the lower panel. However, in the 2nd run of this exercise, the data size is doubled to 56, and the prior variance is made lower, i.e. the misleading priors centred at $q = 3; r = 3$, are made more peaky. The relatively more peakiness of the prior on Q is noted (in open red circles in the lower panel of this figure); this is achieved via a prior variance of 0.001^2, as distinguished from the 1st run that employed a variance of 0.01^2.

The top panel of this figure displays the likelihood – and thereafter posterior density – computed from the 1st run of this learning exercise, using 28 data points, while the lower panel displays results computed using 56 data points in the 2nd run. Likelihood is depicted in black in both panels; we can see that while the likelihood bears a point of inflection at the correct answer of $q = 1; r = 1$, the rate of change of slope as a function of q (or r) is low at q=1 when the data size is smaller, as distinguished from when the data size doubles (lower panel). Then the likelihood attains a sharper maxima at $q = 1$, (and r=1).

We notice that the posterior density given the data in the 2nd run of this exercise, peaks sharply at about the correct answer of $q = 1$, (and r=1), in spite of the very inappropriate prior used in the 1st run, and the even more inappropriate (than in the 1st run) prior used in this 2nd run.

In fact, in this 2nd run, the prior is even worse than in the 1st run since here the prior centred at a value of 3 is more informative than it was in the 1st run. But despite this erroneous guidance from the prior, the data – now that it has inflated in size – leads to the correct answers, swamping the effect of such badly selected priors.

However, this exercise that we have discussed results of in Figure 2.1, is not one in which we are learning values of Q and R given a chosen data set. Above, we simply computed the joint posterior *pdf* of Q and R given a data set, and then plotted the value of the joint posterior at a chosen value of Q, ignoring the trends with the chosen values of R. Thus, the output of this computation – at

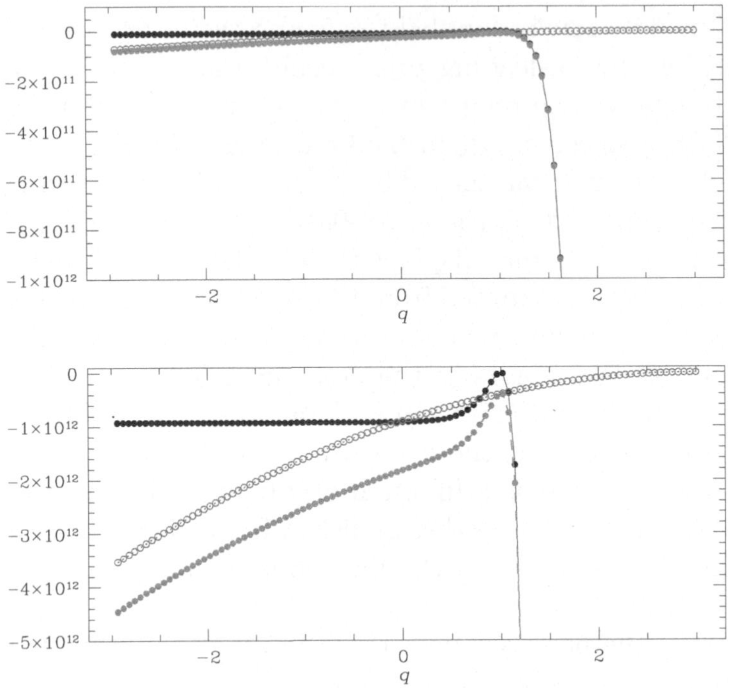

FIGURE 2.1
Plots – against chosen values of Q – of prior probability density (in open circles); likelihood (in black filled circles); joint posterior *pdf* of model parameters Q and R (in light grey filled circles, or in green in the colour version), given data that comprises observations of length, breadth and area of 28 rectangles (top panel), and 56 rectangles (lower panel).

chosen parameter values – is the joint posterior. However, inference on the model parameters comprise learning values of any of these parameters, at chosen values of the marginal posterior *pdf* of the parameter, (given the data at hand). So. we could have generated the marginal of Q (or R) from this extremely simplistic joint posterior *pdf* mentioned above in Equation 2.5, by integrating this posterior over all values of R, (or Q). Such a proposed integration – i.e. proposed technique for computing the marginal posterior of

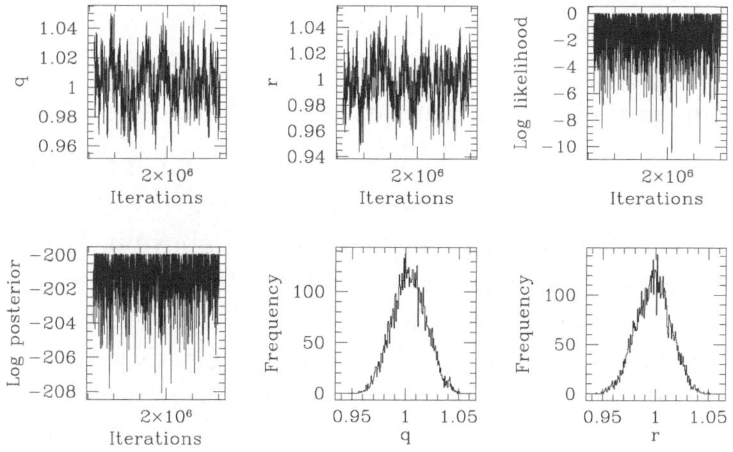

FIGURE 2.2
Top left and middle panels: traces of values of model parameters Q and R learnt using data comprising length, breadth and area observations of 28 rectangles tabulated in Table 2.1. A seed value of 0 was used for both parameters. The Chain employed prior *pdfs* on Q and R that were each Normal, with mean of 3 and variance of 0.04. Trace of logarithm of the likelihood is presented in the top right panel. The log of the joint posterior *pdf* is shown in the lower left panel. Histograms of Q and R given the data, as presented in the lower middle and right panels, are approximations to the respective marginal posterior *pdf* of the parameter, given the data. Using theses, the 95% HPDs are learnt; these suggest $q \in [0.9740, 1.0380]$ and $r \in [0.9652, 1.030]$.

Q and R, respectively given the data – is not closed-form, and numerical integration techniques will need to be undertaken.

This is the situation in this very simple problem with only two unknowns that we hope to learn. Now consider a real-world problem with 20-30 parameters that we wish to infer upon, given the data. Then, the sought multiple numerical integrations will be too cumbersome/difficult to undertake and/or the numerical approximations can pose a problem. So, we still need to make that inference. As we have seen before, we will undertake this using MCMC – and we opt for Random Walk Metropolis Hastings (RWMH) based inference.

2.5.1 Learning model parameters using RWMH

We aim to learn values of the model parameters Q and R that are the powers of the length and the breadth variables respectively, in the model of the posterior *pdf* that is proposed above in Equation 2.5. To undertake this learning, we employ data on 28 values of the observable $\boldsymbol{Y}$, (tabulated in Table 2.1), and choose the prior of each parameter to be $\mathcal{N}(3, 0.04)$. We start the Chain with values of 0 for both Q and R. Thus, the prior on each parameter is quite peaky - peakier by a factor of 400 compared to the prior used in the 2nd run discussed above, in which the aim was to generate the posterior *pdf* at chosen model parameter values. In spite of this prior that (incorrectly) imposes a high probability mass at values around 3, the model parameters are correctly inferred upon given the data; the 95% HPDs for the parameters are learnt as $q \in [0.9740, 1.0380]$ and $r \in [0.9652, 1.030]$. We employ a value of 1 for the variance parameter σ^2 in Equation 2.5. We undertake inference on model parameters Q and R using the Random Walk Metropolis Hastings (RWMH) algorithm. Thus, we propose Q from a Normal proposal density, and R is similarly proposed from a Normal; this Normal has a mean that is the current value of the parameter – as per the design of the RWMH – while its variance is input as 10^{-6}. Results of the learning of the parameters Q and R are displayed in Figure 2.2.

We have undertaken this entire discussion of a suggested Chain, without suggesting source of the value $\boldsymbol{\theta}_i$ – that is the *current* value of $\boldsymbol{\Theta}$, at which the proposal of $\boldsymbol{\theta}_{i+1}$ is tabled for consideration. Since the chain is positive recurrent, we are at liberty to choose any value as the seed value of this variable. Then this Markov Chain is generated from this seed or the 0-th element, and proceeds thereafter sequentially. A seed of 0 was chosen for each parameter that we learnt in this application.

2.6 Empirical illustration – sequential updating

The example provided towards the first empirical illustration discussed in the previous section, indicates the relevance of Bayesian inference on the model parameters, by sampling from the posterior. In particular, that application of MCMC-based inference was included to bring forth the advantage offered by Bayes rule in learning the model given data, which is a harder - though more pertinent – exercise to undertake within Science, than the exercise of churning out values of the observable (*aka* data points), given a model. Such advantage of Bayes rule in inverse learning notwithstanding, sequential updating of the values of the model parameters is facilitated by Bayes rule, in light of updates in the available information. In this section we undertake an illustration of such Bayesian sequential updating of the learning of a model parameter, using MCMC-based inference.

Consider a 6-faced die that is monitored for an outcome of a "6" showing up on its top surface, once rolled. The showing up of a "6" is identified as the event termed "success". We know that the probability p of this "success" is $1/6$ in a fair die. Using observations on the number of rolls $N \in \mathbb{Z}_{\geq 0}$ that fetch a "6", out of the total $N_{max} \in \mathbb{N}$ rolls of the die that are undertaken, we want to learn the value of p. So, we undertake this experiment with sequentially increasing number N_{max} of rolls of this die; our observations on the number N of successes within any considered number of trials are noted below in Table 2.2.

Learning p in 6 stages

In our learning we use sequentially increasing number of observations; thus, we learn p in 6 stages, when M rolls of the die are considered, with M taking values of: 20, 60, 110, 200, 300 and 420 rspectively, in the 6 stages. At each stage, p is learnt using the corresponding value of the number of "success"es to be as enumerated at a given value of N_{max} in the data tabulated in Table 2.2.

TABLE 2.2
Observations on number N of throws that result in a "6", out of N_{max} number of throws of a die.

N_{max}	4	8	10	14	20	22	30	40
N	0	1	1	2	2	3	4	5
N_{max}	50	60	70	80	90	100	110	120
N	7	8	8	9	10	11	11	12
N_{max}	140	160	180	200	220	240	260	280
N	13	15	18	19	20	22	23	26
N_{max}	300	320	340	370	400	420		
N	29	30	32	36	40	41		

Prior

For each of these runs, we employ the prior on p to be $\pi_0(p) = \mathcal{N}(1/6, 0.02^2)$. While the mean of this prior density is motivated by a fair die, the variance is arbitrarily chosen.

Likelihood

The observable here is the number n of "success"es, at any value of N_{max}. Then probability of this observable N, given a value of N_{max} and p is given by the Binomial probability mass function:

$$\Pr(N = n|n_{max}, p) = \binom{n_{max}}{n} p^n (1-p)^{n_{max}-n},$$

s.t. the likelihood is the product of this *pmf* across all data points for which $n_{max} = m$, where m is the value of M relevant for the stage of learning under consideration.

Posterior

Then using this likelihood and the Normal prior mentioned above, the posterior *pdf* of p is computed, given the considered observations relevant to the stage at which the learning is being carried out.

Inference

As p is the probability of success, the constraint that it needs to lie in [0,1] is validated by proposing p in any iteration of an MCMC chain, from the truncated Normal proposal density - truncated on the left at 0, and on the right at 1. Thus, vanilla MCMC is used to undertake the inference on p.

Results

Results of this learning are displayed in Figure 2.3. The panels of this figure depict the trace of p learnt at one of the 6 undertaken stages, given the data that is relevant to the stage at hand. We notice that in the first stage, (i.e. for a total of $M = 20$ throws of the die), the learnt p is closest to 1/6, which is the value of p for a fair die. But it is clear from the data tabulated in Table 2.2 that the number of "6"s is less than a sixth of the total number of throws – and this gets clearer in the data with increasing n_{max}. Thus, as we progress through the 6 stages of our learning of p, the value of p declines from the 1/6 threshold relevant to a fair die. At the last of these stages, we use the number of "success"es for all the throws that are noted in the data set in Table 2.2, to learn the value of p – the histogram of which is shown in the right panel of Figure 2.4. From this histogram – which is equivalently, the approximation of the marginal posterior density of p given the data – we learn the 95% HPDs to p to be [0.0960, 0.1135].

2.7 Summary

This chapter discusses some of the founding concepts that are relevant to the learning of random functions, by modelling them with a relevant stochastic process. Such learning is at the core of this book. In this book, such learning is presented within a Bayesian framework, with inference that is MCMC based, though the implementation of other inference techniques can be amiably accommodated within the scope of the methodologies presented in the late chapters. In this chapter, we appreciated the capacity for se-

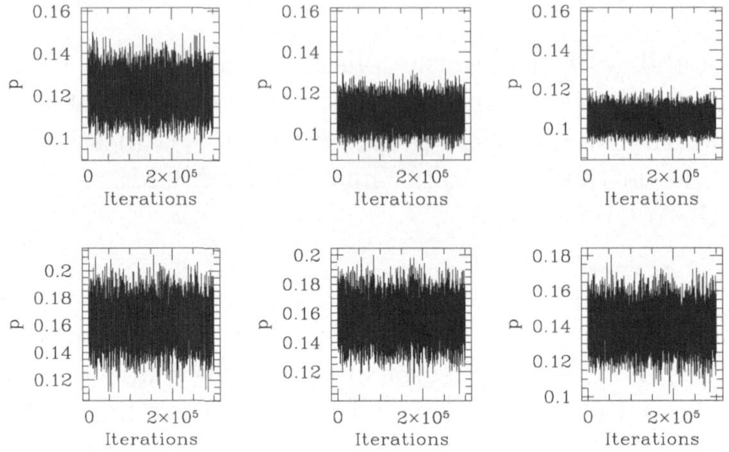

FIGURE 2.3
Here we plot the traces of the learnt values of probability p of a "6" showing up, in N_{max} throws of an unfair die, as N_{max} increases in value. From left to right in the lower panels, results are depicted for runs undertaken to learn p using the maximum number of throws that varies from 20, 60, to 110, while in the upper panels, value of N_{max} increases fom 200 to 420, via 300. All runs employ the same Normal prior with large variance, centred at the p value that would correspond to a fair die, namely a p of 1/6.

quential updating that is available via Bayesiamism, in addition to the learning of the model given the available data. We reviewed sampling-based Bayesian inference and undertook two empirical illustrations to bring forth these concepts.

In this book, we will typically pursue MCMC-based inference using Random Walk Metropolis Hastings (RWMH) or vanilla MCMC – either in itself, or by using this within a Metropolis-within-Gibbs scheme. In other words, inference will often involve proposing a value of a parameter in the $i+1$-th iteration, from a proposal density that is centred at the value of the parameter that is the update in the i-th iteration. Proposing θ_{i+1} from a density that takes θ_i as its mean, is a neat way of ensuring that the proposed value is within probabilistically identifiable distance from

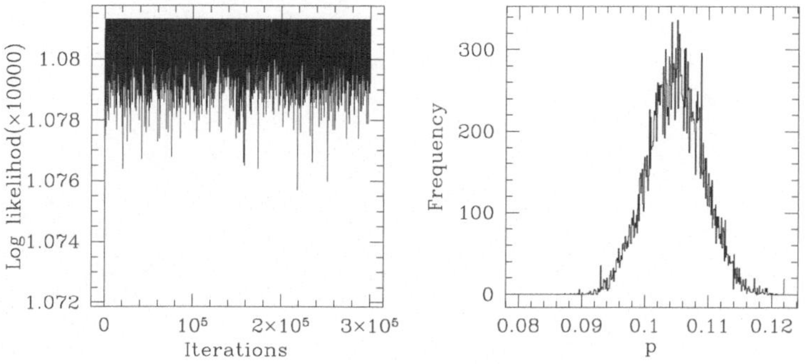

FIGURE 2.4
Right panel: trace of the log likelihood obtained in the MCMC chain, run using data on all values of the number of throws of this die, and the number of "6"s that show up in each such throw, as delineated in Table 2.2. *Left panel*: histogram of the learnt values of p from this chain.

the value of $\boldsymbol{\Theta}$ that was relevant or *current*, until the next sampled value of $\boldsymbol{\Theta}$ became available. In this way, we identify a routine that allows for a value of the to-be-learnt vector of parameters $\boldsymbol{\Theta}$ to be sampled from a proposal density that is centred at the currently accepted value of $\boldsymbol{\Theta}$ – whether this newly proposed value will be rejected or accepted as the updated current value of $\boldsymbol{\Theta}$, will depend on the acceptance ratio computed at the current and the proposed values of $\boldsymbol{\Theta}$. The sequence of the current values of $\boldsymbol{\Theta}$, generated according to this routine, is then a Markov Chain, since the probability for $\boldsymbol{\Theta}$ to be equal to an element of this sequence, is dependent only on the previous element, and no other element of this sequence by construct.

3

Supervised learning & prediction, using Gaussian Processes

3.1 Introduction

In the first chapter, we came across the suggestion that it is extremely difficult to build a high-dimensional function, by sequentially designing individual component functions, while ensuring adherence of each such component, to multiple (probabilistic) interconnected constraints. An alternative to such an approach, was proposed in collecting realisations of the sought function inside a "bag" of functions, where it is the responsibility of said bag, to transfer the correlation structure of the data, onto

— the inter-component correlation (i.e. correlation amongst component functions that constitute the sought high-dimensional function), and

— the inter-realisation correlation (i.e. correlation amongst possible realisations of this function).

This bag is in fact a Gaussian Process ($\mathcal{GP}$) [105, 55, 79, 115], and it feeds information about the empirical correlation structure, to sample functions, s.t. the joint probability density of any finite set of samples drawn from this bag, given the data at hand, is the high-dimensional equivalent of the multivariate Normal density, where the dimension of the data that comprises observations of the output determines this "dimensional-equivalence". Here the output of the sought function is denoted Y, while the input to this function is denoted $\boldsymbol{X}$. Thus, when measurements of the scalar-valued output constitute the data vector, the joint probability for the function variable computed at the design values of the input

DOI: 10.1201/9781003296331-3

– at which such output values are recorded – is multivariate Normal in the data variable. When the output is vector-valued, the joint *pdf* of a finite number of realisations of the sought function is the matrix Normal. Following on, joint of n realisations of a k-th ordered tensor-valued function is the $k+1$-th ordered tensor Normal density in the data variable $\mathcal{D}$. Here this data variable $\mathcal{D}$ comprises the n k-th ordered tensor-valued output variables $\boldsymbol{Y}_1, \ldots, \boldsymbol{Y}_n$, each of which is the output of the function variable, at the respective design value of the input $\boldsymbol{X}$.

Such considerations emanate from the model of the relationship between $\boldsymbol{Y}$ and $\boldsymbol{X}$ as $\boldsymbol{Y} = \boldsymbol{f}(\boldsymbol{x})$, that we have come across in Chapter 1. As stated above, here $\boldsymbol{f}(\cdot)$ is a random function, i.e. a function-valued random variable. We will learn this using the training data $\mathbf{D} := \{(\boldsymbol{x}_i, \boldsymbol{y}_i)\}_{i=1}^n$. Upon acknowledgement of measurement uncertainties in the output $\boldsymbol{Y}$, as well as of the uncertainties that are identified in the learning of this function given the training data, closed-form prediction is performed, of the mean and variance of the output variable at a test input $\boldsymbol{x}^{(test)}$. Additionally, we will discuss the learning of missing values of the input variable $\boldsymbol{X}$, at which a given (noisy) measurement of $\boldsymbol{Y}$ is realised.

3.2 What is the input? What is the output?

There is one remaining question that we will address, before moving on to discussing $\mathcal{GP}$-based learning in detail. Given two r.v.s $\boldsymbol{X}$ and $\boldsymbol{Y}$, how do we decide which variable is the output, and which of these two r.v.s is to be treated as the input? In previous discussions we have treated $\boldsymbol{Y}$ as the output and $\boldsymbol{X}$ the input to the function $\boldsymbol{f}(\cdot)$, where r.v. $\boldsymbol{Y}$ is a k-th order tensor of dimensionality $m_1 \times m_2 \times \ldots \times m_k$, $k, m_i \in \mathbb{Z}^+$; $i = 1, \ldots, k$, and $\boldsymbol{X}$ is a vector-valued r.v. of dimension d, i.e. $\boldsymbol{X} \in \mathbb{R}^d$, $d \in \mathbb{Z}^+$. Now we query the justification behind this choice.

The main issue to be considered here is about the comparative difficulty of the learning exercise when $\boldsymbol{Y}$ is the output, as distinguished from when $\boldsymbol{Y}$ is considered the input. We recall that when performing the supervised learning of the function $\boldsymbol{f}(\cdot)$, by modelling it as a sample function of a $\mathcal{GP}$, closed-form prediction of mean and variance of the output of the function is possible. Value of the input to this function – at which a measured value of the output is realised – can be learnt in principle, within the inferential process; we refer to this learning as *inverse prediction* in this book. Such inverse prediction is not closed-form, but the sought value of the input will be *learnt*, by sampling it from its marginal posterior *pdf*, given the noisy (i.e. uncertainty-included) measurement of the output that is realised at this input, along with the model of the input-output functional relationship, (that has to precede this attempted input learning, given the typically-noisy training data). Then given that we can learn/predict the input/output, which variable – out of $\boldsymbol{X}$ and $\boldsymbol{Y}$ – would be easier to treat as the output of the sought functional relation between these variables?

The answer does depend on what is implied by "easier". If $\boldsymbol{Y}$ is the output of the sought function, i.e. if the vector-valued $\boldsymbol{X}$ is the input, then the covariance matrix that informs on the covariance between a pair of outputs that are realised at the respective input values, can be parametrised using a covariance kernel. Thus, any element of said covariance matrix is then a chosen declining function, or kernel – characterised with unknown parameters – of the difference between the corresponding input values. We will soon discuss such parametrisation of a covariance matrix with a kernel [105, 10, 82, among others]. For now, we appreciate the following.

— If the input is a vector ($\boldsymbol{X}$), the kernel comprises a norm of the difference between a pair of input vectors, and the number of parameters of the kernel function is driven by the dimension of the input vector $\boldsymbol{X}$. We will study details of this assertion soon.

— However, if the input to the sought function is the tensor $\boldsymbol{Y}$, then the kernel is comparatively more complicated. It is then a declining function of a norm of the difference between two values of the tensor-valued $\boldsymbol{Y}$, with the parameters of this kernel function modelled as certain elements of a tensor of order driven by the order and dimensions of the tensor $\boldsymbol{Y}$. The number of parameters (of the kernel) that we then need to learn, is typically higher than for a lower dimensional input variable. We will soon discuss the maths of $\mathcal{GP}$-based learning to appreciate this, and will illustrate this comparatively higher difficulty of kernel parametrisation when the input is high-dimensional.

Thus, from the point of view of ease of kernel parametrisation and learning of kernel parameters, it appears that the lower dimensional of the two r.v.s $\boldsymbol{X}$ and $\boldsymbol{Y}$, would be a better choice for the input to the sought function. In this example case, the vector-valued $\boldsymbol{X}$ is lower dimensional than the tensor-valued $\boldsymbol{Y}$, and it will be easier to treat $\boldsymbol{X}$ as the input variable, leaving $\boldsymbol{Y}$ to be the output of the sought function. Then, a model equation $\boldsymbol{Y} = \boldsymbol{f}(\boldsymbol{x})$ appears appropriate, where we seek $\boldsymbol{f}(\cdot)$. In this chapter, we will consult empirical illustrations of learning exercises in which $\boldsymbol{Y}$ is instead a scalar and $\boldsymbol{X}$ a vector; we will study the comparatively higher difficulty in inference, when treating the vector-valued $\boldsymbol{X}$ as the input, over the scalar-valued $\boldsymbol{Y}$.

However, the difficulty of the learning exercise is also to be considered from the point of view of the complexity of the model.

— When the output is vector-valued, within the paradigm of $\mathcal{GP}$-based learning, likelihood of the model parameters, (which we will soon identify as parameters of the kernel function that parametrises a covariance function), conditional on the data, is rendered matrix Normal. We will soon formalise the model.

— For a tensor-valued sought function, i.e. for a tensor-valued output, (of order k), likelihood is the $k+1$-th ordered, tensor Normal density.

Thus, the model gets more difficult to conceptualise – and therefore, to code – as the dimensionality of the output variable increases. From this point of view, we would prefer $\boldsymbol{X}$ as the output over $\boldsymbol{Y}$.

So it is a trade-off, and as has been promised above, we will appreciate details of this assertion via examination of the model structure, as well as from empirical illustrations performed with the two distinct outputs. However, even at this premature stage of model exposition, we assert that once high-dimensional versions of the multivariate Normal density are internalised, (and coded), the difficulty in learning is driven by difficulty with the inference – and therefore, we will be led to simplify the input to the sought function as much as is possible. Thus, in the example of a vector-valued $\boldsymbol{X}$ and a tensor-valued $\boldsymbol{Y}$ that is potentially-associated with this $\boldsymbol{X}$, we will be better off using $\boldsymbol{X}$ as the input, (and therefore $\boldsymbol{Y}$ as the output) to the sought function. In other words, we will persist with the model equation $\boldsymbol{Y} = \boldsymbol{f}(\boldsymbol{x})$. In this model, the function that we seek to learn, is tensor-valued; it is $\boldsymbol{f} : \mathbb{R}^d \longrightarrow \mathbb{R}^{m_1 \times m_2 \times \ldots \times m_k}$ as we recall from Chapter 1. The training data that we will learn this function using, is $\mathbf{D} = \{(\boldsymbol{x}_i, \boldsymbol{y}_i)\}_{i=1}^n$.

The discussion above pertains to the choice of the output of the function that expresses the relationship between $\boldsymbol{X}$ and $\boldsymbol{Y}$ – from the point of view of ease of learning this function. But the ulterior aim to such learning is prediction. How difficult does our choice of the output of the function, render prediction at a test datum? This important question will be addressed below.

3.3 Gaussian Process-based supervised learning

In our model equation $\boldsymbol{Y} = \boldsymbol{f}(\boldsymbol{x})$, we attempt learning the random function $\boldsymbol{f}(\cdot)$, using the data $\mathbf{D}_{\boldsymbol{Y}} := \{\boldsymbol{y}_1, \ldots, \boldsymbol{y}_n\}$, with $\boldsymbol{y}_i$ realised at the i-th known design point $\boldsymbol{X} = \boldsymbol{x}_i$, $(i = 1, \ldots, n)$. Thus, the information on the n design points and the data $\mathbf{D}_{\boldsymbol{Y}}$ are what comprise the training data $\mathbf{D}$. To abbreviate, we will sometimes state that we seek to learn $\boldsymbol{f}(\cdot)$ using the training data $\mathbf{D}$.

Let us define the r.v. $\boldsymbol{Y}_i$ as the output of the random function $\boldsymbol{f}(\cdot)$, computed at $\boldsymbol{X} = \boldsymbol{x}_i$, $\forall i = 1, 2, \ldots, n$. Then the joint probability distribution of $\boldsymbol{Y}_1, \boldsymbol{Y}_2, \ldots, \boldsymbol{Y}_n$ can be defined. We realise that the joint density of $\boldsymbol{Y}_1, \ldots, \boldsymbol{Y}_n$ is in fact, the density of the $k+1$-th ordered tensor-valued variable $\mathcal{D}$ that is obtained by stacking up n number of the k-th ordered tensor-valued r.v.s $\boldsymbol{Y}_1, \ldots, \boldsymbol{Y}_n$.

We shall denote the joint probability density of $\boldsymbol{Y}_1, \boldsymbol{Y}_2, \ldots, \boldsymbol{Y}_n$, computed at $\boldsymbol{y}_1, \boldsymbol{y}_2, \ldots, \boldsymbol{y}_n$, using the abbreviated notation "$[\boldsymbol{y}_1, \ldots, \boldsymbol{y}_n]$". Here $\boldsymbol{Y}_i = \boldsymbol{y}_i$ at the design input $\boldsymbol{x}_i$; $i = 1, \ldots, n$.

3.3.1 Seeking scalar-valued function

To begin our model development, let us consider the output $\boldsymbol{Y}$ to be a scalar. (We will thereafter consider the case of $\boldsymbol{Y}$ being tensor-valued). So, the function $\boldsymbol{f}(\cdot)$ is rendered scalar-valued, given that the considered scalar-valued output is given by $\boldsymbol{f}(\boldsymbol{x})$. Here, the input to the function $\boldsymbol{f}(\cdot)$ remains a vector in $\mathbb{R}^d$. To emphasise the scalar-valued nature of the output variable and the function, we will in fact refer to these using the notation Y and $f(\cdot)$ respectively; after all, notation in bold typically suggests vector-valued, or higher dimensional still. So in this subsection, the model equation is $Y = f(\boldsymbol{x})$.

As we motivated above, we will consider $f(\cdot)$ to be a realisation from a $\mathcal{GP}$, where parameters of this $\mathcal{GP}$ are the mean function $\mu(\cdot)$ and the covariance function $K(\cdot, \cdot)$, i.e.

$$f(\cdot) \sim \mathcal{GP}(\mu(\cdot), K(\cdot, \cdot)),$$

where the input(s) to these functions are realisations of $\boldsymbol{X}$. Now, a Stochastic Process sampled at a given value of the index parameter is a random variable, (see Section 1.5), and the index parameter for this $\mathcal{GP}$ is $\boldsymbol{X}$. So for $\mathcal{GP}$ $\{f(\boldsymbol{x})\}_{\boldsymbol{x} \in \mathbb{R}^d}$, $f(\boldsymbol{x}_i)$ is a random variable. A (scalar-variate) $\mathcal{GP}$, by definition, provides the joint probability density of a finite set of the random variables to be a multivariate Normal density, [5, 56, 132]. Then by definition of a $\mathcal{GP}$, the joint probability distribution of n realisations of the function $f(\cdot)$ – at each of the n design points $\boldsymbol{x}_1, \ldots, \boldsymbol{x}_n$ – is multivariate Normal, (denoted $\mathcal{MN}$). In other words, the joint *pdf*

$$[f(\boldsymbol{x}_1), f(\boldsymbol{x}_2), \ldots, f(\boldsymbol{x}_n)] = \mathcal{MN}(\boldsymbol{\mu}, \boldsymbol{\Sigma}),$$

where the parameters of this multivariate Normal density are the n-dimensional mean vector $\boldsymbol{\mu}$ and the $n \times n$-dimensional covariance matrix $\boldsymbol{\Sigma}$.

Then the joint probability density mentioned in the LHS of the last equation can be alternatively expressed as

$$[Y_1, Y_2, \ldots, Y_n] = \mathcal{MN}(\boldsymbol{\mu}, \boldsymbol{\Sigma}).$$

Now, following on from the above discussion, stacking the n random variables $Y_1, \ldots, Y_n$, we achieve an n-dimensional vector-valued random variable $\mathcal{D}$. A realisation of this data variable $\mathcal{D}$ is the vector $\mathbf{D_Y} = (y_1, y_2, \ldots, y_n)^T$, i.e. $\mathcal{D}$ is the data variable.

Now, by the last equation, the joint probability density of $Y_1, \ldots, Y_n$, computed at $y_1, y_2, \ldots, y_n$ is

$$\mathcal{MN}(\boldsymbol{\mu}, \boldsymbol{\Sigma}).$$

But the joint density of $Y_1, \ldots, Y_n$, computed at $y_1, y_2, \ldots, y_n$ is equivalently the *pdf* of $\mathcal{D}$, computed at $\mathbf{D_Y}$, i.e.

$$f_{\mathcal{D}}(\mathbf{D_Y}) = \mathcal{MN}(\boldsymbol{\mu}, \boldsymbol{\Sigma}).$$

However, this equation is in fact not altogether correct since the *pdf* on the LHS clearly changes if the model parameters $\boldsymbol{\mu}$ and $\boldsymbol{\Sigma}$ vary, but the notation on the LHS does not reflect this dependence. Thus, in the corrected notation, this equation is:

$$f_{\mathcal{D}}(\mathbf{D_Y} | \boldsymbol{\mu}, \boldsymbol{\Sigma}, \boldsymbol{x}_1, \ldots, \boldsymbol{x}_n) = \mathcal{MN}(\mathbf{D_Y}; \boldsymbol{\mu}, \boldsymbol{\Sigma}), \qquad (3.1)$$

where we have included the dependence of the probability of the data variable $\mathcal{D}$ on the choice of the design inputs. Indeed, the input values at which the n observations of the data vector are realised, need to be included in the LHS, since the variance-covariance matrix $\boldsymbol{\Sigma}$ that lies on the RHS, depends on the input values, via kernel parametrisation of this matrix, as we will see more clearly below.

The last equation has a ready interpretation. When we examine the LHS of this last equation, it tells us that the LHS is in fact, the likelihood – after all, the *pdf* of the data given model parameters is the likelihood function of the model parameters in the data.

Equation 3.1 provides the likelihood of $\boldsymbol{\mu}, \boldsymbol{\Sigma}$ given the data on the output $\boldsymbol{Y}_i$ at the design point $\boldsymbol{x}_i$, $\forall i = 1, \ldots, n$. Such data, along with the design points, is basically the information contained in the training data $\mathbf{D}$. So in the depiction of the likelihood, we may equivalently condition on the training data $\mathbf{D}$.

3.3.2 Kernel parametrisation – preliminary considerations

The covariance matrix $\boldsymbol{\Sigma}$ in Equation 3.1 can be parametrised with a kernel function, to render its learning easier.

Here $\boldsymbol{\Sigma}^{(n \times n)}$ is the covariance matrix of the multivariate Normal likelihood that emerges when the sought function $f(\cdot)$ is modelled using a $\mathcal{GP}$. Then matrix $\boldsymbol{\Sigma} = [\sigma_{ij}] = [Cov(Y_i, Y_j)]$, s.t. if we opt to parametrise the covariance between a pair of outputs using a covariance kernel, we state

$$\boldsymbol{\Sigma} = [\sigma_{ij}] = [Cov(Y_i, Y_j)] = [K(\boldsymbol{x}_i, \boldsymbol{x}_j)],$$

where $K(\cdot, \cdot)$ is a declining function of the difference between $\boldsymbol{x}_i$ and $\boldsymbol{x}_j$.

The idea behind kernel parametrisation of a covariance matrix is that the covariance between a pair of output r.v.s that are realised at identified input values, (i.e. design points), is a declining function of the difference between these input values. The higher this difference – i.e. further $\boldsymbol{x}_i$ is from $\boldsymbol{x}_j$ – lower is the covariance between the outputs Y_i and Y_j at these design inputs. This covariance kernel $K(\cdot, \cdot)$ is typically assigned a parametric form, which is parametrised using multiple parameters.

We realise that upon standardising the data on the output, with the sample mean $(\bar{y})$ and standard deviation (s) of the observed sample $\{y_1, \ldots, y_n\}$ of output values, the mean of Y_i is 0 $\forall i = 1, \ldots, n$, s.t. mean vector $\boldsymbol{\mu}$ of the multivariate Normal likelihood is rendered an n-dimensional null vector. Following such standardisation, the variance-covariance matrix $\boldsymbol{\Sigma}$ of this vector-valued data variable, reduces to being the correlation matrix, the ij-th element of which is the correlation between Y_i and Y_j, which are realised respectively at $\boldsymbol{x}_i$ and $\boldsymbol{x}_j$, $(i, j = 1, \ldots, n)$. So subsequent to the standardisation of the observations of the output, $\boldsymbol{\mu}$ is set to $\underline{0}$ and the matrix $\boldsymbol{\Sigma}$ is considered to be the correlation matrix.

3.3.3 Why kernel parametrise?

The advantage of kernel parametrisation of the correlation matrix $\boldsymbol{\Sigma}$ of the multivariate Normal likelihood, is that kernel parametrisation reduces the burden of learning: instead of learning the $(n^2 - n)/2$ off-diagonal elements of the upper (or lower) triangle of the symmetric correlation matrix, we only need to learn the few parameters of the kernel $K(\cdot, \cdot)$. Indeed, the number of parameters of the kernel increases with the dimensionality of the input, but typically, the number of such kernel parameters is much lower than the number of (generally) distinct elements of the (symmetric) covariance matrix. As the observed output values are standardised, the diagonal elements of $\boldsymbol{\Sigma}$, are 1 each, and it is indeed the off-diagonal elements of this correlation matrix $\boldsymbol{\Sigma}$ that we are required to learn.

The pursuit of the few kernel parameters, is easier than learning the $(n^2 - n)/2$ number of distinct off-diagonal elements of the upper (or lower) triangle of $\boldsymbol{\Sigma}$. For example, for a training data set with only $n = 20$ pairs of a design point and the corresponding realisation of the output variable, the number of such off-diagonal elements of the correlation matrix $\boldsymbol{\Sigma}$ that we need to learn, is 190, while resorting to kernel parametrisation of correlation matrix $\boldsymbol{\Sigma}$ using the simplest and popular Square Exponential kernel, will require the learning of d kernel parameters, where d is the dimensionality of the input vector $\boldsymbol{X}$ that is likely to be ≤ 190. So kernel parametrisation is often likely to lead to a less difficult learning task, and also, the difficulty of the learning exercise does not increase just because newer data points are added to the training set.

By the way, the kernel parameters are better referred to as

"hyperparameters" as these are parameters of the parameter that is the kernel itself; hereon, we refer to parameters of the kernel function as hyperparameters. Thus, we will sometimes denote a kernel with the notation

$$K(\cdot, \cdot; \boldsymbol{\theta}),$$

where $\boldsymbol{\theta}$ is the vector of hyperparameters of this kernel function. Additionally, all reference to "kernel parameters" above, should in fact be reread as "kernel hyperparameters".

3.3.4 Following on from multivariate Normal likelihood

In Equation 3.1 we noted the likelihood of the model parameters $\boldsymbol{\mu}$ and $\boldsymbol{\Sigma}$, in the data $\mathbf{D}_{\boldsymbol{Y}}$ on the output variable, realised at the identified set of design points that reside in the training data $\mathbf{D}$. In recognition of the kernel hyperparameters $\boldsymbol{\theta}$ that parametrise the covariance kernel – that in turn parametrises the multivariate Normal likelihood – we update the likelihood $\mathcal{L}(\cdot|\cdot)$ to be declared a function of $\boldsymbol{\theta}$, in the data $\mathbf{D}_{\boldsymbol{Y}}$ on the output and the considered set of design points:

$$f_{\mathcal{D}}(\mathbf{D}_{\boldsymbol{Y}}|\boldsymbol{\mu}, \boldsymbol{\Sigma}, \boldsymbol{x}_1, \ldots, \boldsymbol{x}_n) \equiv \mathcal{L}(\boldsymbol{\theta}; \mathbf{D}_{\boldsymbol{Y}}, \boldsymbol{x}_1, \ldots, \boldsymbol{x}_n), \tag{3.2}$$

with this density $f_{\mathcal{D}}(\cdot|\cdot)$ identified as $\mathcal{MN}(\mathbf{D}_{\boldsymbol{Y}}; \boldsymbol{\mu}, \boldsymbol{\Sigma})$. So we notice that the learning of the function $f(\cdot)$ – the burden of which has been shifted to the learning of the parameters of the stochastic process, (to be precise, a $\mathcal{GP}$) that this unknown function is a sample function of – has reduced to the learning of the kernel hyperparameters $\boldsymbol{\theta}$, given the information in the training data $\mathbf{D} = \{(\boldsymbol{x}_i, y_i)\}_{i=1}^{n}$. Once $\boldsymbol{\theta}$ is learnt given the information in this training data, this $\mathcal{GP}$ that $f(\cdot)$ is a sample function of, is specified. Then we sample $f(\cdot)$ from it, and in fact, perform closed-form prediction of the mean and variance of the sample functions computed at $\boldsymbol{X} = \boldsymbol{x}, \forall \boldsymbol{x} \in \mathbb{R}^d$. So all that remains is for us to know how to learn the hyperparameters $\boldsymbol{\theta}$ of the kernel function.

To appreciate this, we rewrite the likelihood of $\boldsymbol{\theta}$ in training data $\mathbf{D}$ – this time in gory details, by invoking the form of the multivariate Normal density.

Definition 3.1 *By modelling the sought function $f(\cdot)$ as a sample function of a $\mathcal{GP}$ with mean function $\mu(\cdot)$ and covariance function that is parametrised with the covariance kernel $K(\cdot, \cdot; \boldsymbol{\theta})$, as in:*

$$f(\cdot) \sim \mathcal{GP}(\mu(\cdot), K(\cdot, \cdot; \boldsymbol{\theta}),$$

the joint pdf of the n number of realisations of $f(\cdot)$ at the n design points $\boldsymbol{x}_1, \ldots, \boldsymbol{x}_n$ that reside in the training data $\mathbf{D}$ is

$$[f(\boldsymbol{x}_1), \ldots, f(\boldsymbol{x}_n)] = \mathcal{MN}(\boldsymbol{\mu}, \boldsymbol{\Sigma}),$$

where:

— *the n-dimensional vector $\boldsymbol{\mu} = \mathbb{E}(Y_1, Y_2, \ldots, Y_n)^T$ with Y_i representing the r.v. $f(\boldsymbol{x}_i)$, $\forall i = 1, \ldots, n$, and*

— *the $n \times n$-dimensional matrix*

$$\boldsymbol{\Sigma} = [\sigma_{ij}] = [K(\boldsymbol{x}_i, \boldsymbol{x}_j; \boldsymbol{\theta})],$$

is the covariance matrix of the random vector $(Y_1, \ldots, Y_n)^T$.

Then likelihood of the kernel hyperparameters – collated into vector $\boldsymbol{\theta}$ – in the training data $\mathbf{D} = \{(\boldsymbol{x}_i, y_i)\}_{i=1}^n$, is:

$$\mathcal{L}(\boldsymbol{\theta}; \mathbf{D}) = \frac{1}{\sqrt{|2\pi\boldsymbol{\Sigma}|}} \exp\left(-\frac{(\boldsymbol{x} - \boldsymbol{\mu})^T \boldsymbol{\Sigma}^{-1} (\boldsymbol{x} - \boldsymbol{\mu})}{2}\right). \qquad (3.3)$$

Here $|\boldsymbol{\Sigma}|$ denotes the determinant of $\boldsymbol{\Sigma}$.

Then invoking priors $\pi_0(\boldsymbol{\theta})$ on $\boldsymbol{\theta}$, and porting $\pi_0(\boldsymbol{\theta})$ and likelihood $\mathcal{L}(\boldsymbol{\theta}; \mathbf{D})$ into Bayes rule, we formulate the posterior *pdf* of the parameter $\boldsymbol{\theta}$ given training data $\mathbf{D}$, i.e formulate

$$\pi(\boldsymbol{\theta}|\mathbf{D}) \propto \mathcal{L}(\boldsymbol{\theta}; \mathbf{D}) \pi_0(\boldsymbol{\theta}),$$

where the constant of proportionality can remain unknown, without affecting inference on $\boldsymbol{\theta}$, since inference is undertaken using MCMC.

3.3.4.1 MCMC-based inference on kernel hyperparameters

In such MCMC-based inference,

1. we will need to start with a "seed" value of each component of $\boldsymbol{\theta}$.
2. Then in any iteration, we propose a value of each component of $\boldsymbol{\theta}$, by sampling from a chosen proposal density.
3. The likelihood and priors are computed at the proposed values of each component of $\boldsymbol{\theta}$, and the ensuring posterior is compared – within the defined acceptance ratio – against the posterior computed at the current values of the components of $\boldsymbol{\theta}$.
4. If the criterion for acceptance of the proposed values of $\boldsymbol{\theta}$ is passed, the proposed candidates are accepted as the current ones; else the existing current values persist.

Following this protocol, the MCMC-based learning of the kernel hyperparameters can be undertaken. We will soon see an illustration of this on a real-world data set. Once the kernel hyperparameters are learnt, i.e. the generative $\mathcal{GP}$ is specified, we will then make predictions of the value of Y at a test value $\boldsymbol{x}^{(test)}$ of the input variable.

3.3.5 Prediction

First we discuss prediction of the output Y at $\boldsymbol{x}^{(test)}$. Later in the subsection, we will explore inverse prediction. In reality, we would be interested in predicting the values of Y at – not just a single, but – a series of test inputs: $\boldsymbol{x}_1^{(test)}, \boldsymbol{x}_2^{(test)}, \ldots, \boldsymbol{x}_m^{(test)}$. for $m \in \mathbb{N}$.

3.3.5.1 Predicting output at test inputs

To perform prediction of Y at $\boldsymbol{x}^{(test)}$, we will need to construct the "augmented" correlation matrix $\boldsymbol{\Sigma}^{/}$ that includes correlations between the r.v.s representing the outputs at the n design inputs, in addition to correlations between r.v.s $Y_1^{(test)}, Y_2^{(test)}, \ldots, Y_m^{(test)}$ that represent the outputs at the test inputs $\boldsymbol{x}_1^{(test)}, \boldsymbol{x}_2^{(test)}, \ldots, \boldsymbol{x}_m^{(test)}$ respectively, in addition to correlations between outputs at test and design inputs. We indicate this "augmented" correlation matrix $\mathbf{\Sigma}^{/}$ below:

$$\mathbf{\Sigma}^{/} =$$

$$\begin{bmatrix} c(Y_1, Y_1) & \ldots & c(Y_1, Y_n) & c(Y_1, Y_1^{(test)}) & c(Y_1, Y_m^{(test)}) \\ c(Y_2, Y_1) & \ldots & c(Y_2, Y_n) & c(Y_2, Y_1^{(test)}) & c(Y_2, Y_m^{(test)}) \\ \vdots & & & & \\ \vdots & & & & \\ c(Y_n, Y_1) & \ldots & c(Y_n, Y_n) & c(Y_n, Y_1^{(test)}) & c(Y_n, Y_m^{(test)}) \\ c(Y_1^{(test)}, Y_1) & \ldots & c(Y_1^{(test)}, Y_n) & c(Y_1^{(test)}, Y_1^{(test)}) & c(Y_1^{(test)}, Y_m^{(test)}) \\ \vdots & & & & \\ c(Y_m^{(test)}, Y_1) & \ldots & c(Y_m^{(test)}, Y_n) & c(Y_m^{(test)}, Y_1^{(test)}) & c(Y_m^{(test)}, Y_m^{(test)}) \end{bmatrix}$$

Here "c" stands for "correlation".

As we can see, the augmented correlation matrix $\mathbf{\Sigma}^{/}$ is built with the older correlation matrix $\mathbf{\Sigma}$, and the $m \times m$-dimensional correlation matrix $\mathbf{\Sigma}^{(test)}$ that informs on the correlation between pairs of r.v.s in the set $\{Y_1^{(test)}, Y_2^{(test)}, \ldots, Y_m^{(test)}\}$; the remaining two partitions of the block matrix $\mathbf{\Sigma}^{/}$ are rectangular matrix $\mathbf{\Sigma}_{nm}$ of dimensions $n \times m$ and rectangular matrix $\mathbf{\Sigma}_{mn} = (\mathbf{\Sigma}_{nm})^T$ of dimensions $m \times n$. These $n \times m$-sized partitions inform on correlations between one of the n r.v.s that represent the output at a design input, and one of the m r.v.s that represent the output at a test input. Thus, both of these partitions – that are rectangular matrices – bear the same information; one is the transpose of the other.

Definition 3.2 *The joint pdf of the $n+m$ realisations of the random function $f(\cdot)$, at the design and test inputs, is multivariate Normal, with a zero mean vector and correlation matrix $\mathbf{\Sigma}^{/}$, i.e.*

$$[f(\boldsymbol{x}_1), f(\boldsymbol{x}_2), \ldots, f(\boldsymbol{x}_n), f(\boldsymbol{x}_1^{(test)}), \ldots, f(\boldsymbol{x}_m^{(test)})] = \mathcal{MN}(\underset{\sim}{0}, \mathbf{\Sigma}^{/}).$$

Then the joint density of $f(\boldsymbol{x}_1^{(test)}), \ldots, f(\boldsymbol{x}_m^{(test)})$, given the information in the training data $\mathbf{D}$ (i.e. information on each of the n realisations of $f(\cdot)$ at the corresponding design input), and the test inputs $\boldsymbol{x}_1^{(test)}, \ldots, \boldsymbol{x}_m^{(test)}$, is:

$$[f(\boldsymbol{x}_1^{(test)}), \ldots, f(\boldsymbol{x}_m^{(test)}) | \mathbf{D}_{\boldsymbol{Y}}, \boldsymbol{x}_1, \ldots, \boldsymbol{x}_n, \boldsymbol{x}_1^{(test)}, \ldots, \boldsymbol{x}_m^{(test)}] =$$

$$\mathcal{MN}(\mathbf{\Sigma}_{mn}\mathbf{\Sigma}^{-1}\mathbf{D}_{\boldsymbol{Y}}, \mathbf{\Sigma}^{(test)} - \mathbf{\Sigma}_{mn}\mathbf{\Sigma}^{-1}\mathbf{\Sigma}_{nm}).$$

Thus, the expectation of the vector of output values predicted at the m test points is $\mathbf{\Sigma}_{mn}\mathbf{\Sigma}^{-1}\mathbf{D}_{\boldsymbol{Y}}$, and the variance-covariance matrix of this vector is $\mathbf{\Sigma}^{(test)} - \mathbf{\Sigma}_{mn}\mathbf{\Sigma}^{-1}\mathbf{\Sigma}_{nm}$, [105].

3.3.6 Acknowledging noise in measurements

Till now, we have ignored the noise in the value of Y observed at a given design point. However it is more likely than not, that there is noise in such an observation.

Definition 3.3 *Let the variable ϵ_i that represents the error in output Y_i, (attained at the i-th design point $\boldsymbol{X} = \boldsymbol{x}_i$) be s.t.*

$$\epsilon_i \sim g_{\epsilon_i}(\boldsymbol{\lambda}_i), \ \ for \ i = 1, \ldots, n$$

i.e. the error density of the noise on y_i is $g_{\epsilon_i}(\boldsymbol{\lambda}_i)$ where $\boldsymbol{\lambda}_i$ is the vector of parameters of this error density.

Then we model the value of Y_i at $\boldsymbol{X} = \boldsymbol{x}_i$ to lie in the interval of $[y_i - \alpha_1\sigma_i, y_i + \alpha_2\sigma_i]$, where σ_i^2 is the variance of r.v. ϵ_i, computed using the error density $g_{\epsilon_i}(u; \boldsymbol{\lambda}_i)$. Here α_1 and α_2 are s.t. $\int_{\alpha_1\sigma_i}^{\alpha_2\sigma_i} g_{\epsilon_i}(u; \boldsymbol{\lambda}_i)du = 0.95$. Of course, the value of 0.95 under the *pdf*, bounded by $\alpha_1\sigma_i$ and $\alpha_2\sigma_i$, is a choice, and if tweaked, definition of the bounds will change accordingly. Here $i = 1, \ldots, n$. Typically the parametric form of this error density is known, and α_1, α_2 are also known.

A situation that we often encounter in this book is when the error density on the measurements of Y_i is Normal with zero mean and variance σ_i^2, i.e. $\boldsymbol{\lambda}_i = (0, \sigma_i^2)^T$. Then $\alpha_1 = -\alpha_2 = -1.96$ satisfies the condition that $\int_{\alpha_1\sigma_i}^{\alpha_2\sigma_i} g_{\epsilon_i}(u; \boldsymbol{\lambda}_i)du = 0.95$.

In general, the error density on Y_i is convolved with the *pdf* of Y_i conditional on the model parameters, and the likelihood is formulated using the result of such convolutions.

Then it follows that when the error density is Normal, and the error variable ϵ_i ($\sim \mathcal{N}(0, \sigma_i^2)$) of the output attained at the i-th design input, is independent of ϵ_j, (for $i \neq j; i, j \in \{1, \ldots, n\}$), convolution of the density of error in $(Y_1, \ldots, Y_n)^T$ with the multivariate Normal *pdf* of $\mathcal{D}_Y$, given the model parameters $\boldsymbol{\theta}$ – i.e. the likelihood – *is multivariate Normal again*, with

- the same mean vector $\boldsymbol{\mu}$ as in the zero-error case, (with $\boldsymbol{\mu}$ set to $\underset{\sim}{0}$ when output is standardised), and
- the covariance matrix $\boldsymbol{\Sigma} + \boldsymbol{E}$, where $\boldsymbol{E}$ is an $n \times n$-dimensional diagonal matrix, with diagonal elements $\sigma_1^2, \ldots, \sigma_n^2$.

If the error on the output attained at all design points is identically distributed, then $\sigma_1^2 = \ldots = \sigma_n^2 = \sigma^2$. Then the covariance matrix of the noise-included likelihood of $\boldsymbol{\theta}$ given training data $\mathbf{D}$ is $\mathcal{MN}(\boldsymbol{\mu}, \boldsymbol{\Sigma} + \sigma^2 \boldsymbol{I}_n)$, where $\boldsymbol{I}_n$ is the n-dimensional identity matrix.

Definition 3.4 *For the case of Normal $\mathcal{N}(0, \sigma^2)$, independent and identically distributed error variables, the joint of $f(\boldsymbol{x}_1^{(test)}), \ldots, f(\boldsymbol{x}_m^{(test)})$ - given training data $\mathbf{D}$ and test inputs $\boldsymbol{x}_1^{(test)}, \ldots, \boldsymbol{x}_m^{(test)}$ – changes from what was suggested for the zero-error case (in Definition 3.2) to the following:*

$$[f(\boldsymbol{x}_1^{(test)}), \ldots, f(\boldsymbol{x}_m^{(test)}) | \mathbf{D}, \boldsymbol{x}_1^{(test)}, \ldots, \boldsymbol{x}_m^{(test)}] =$$
$$\mathcal{MN}(\boldsymbol{\Sigma}_{mn}(\boldsymbol{\Sigma} + \sigma^2 \boldsymbol{I}_n)^{-1} \mathbf{D}_Y, \boldsymbol{\Sigma}^{(test)} - \boldsymbol{\Sigma}_{mn}(\boldsymbol{\Sigma} + \sigma^2 \boldsymbol{I}_n)^{-1} \boldsymbol{\Sigma}_{nm}).$$

Thus, the expectation of the vector of output values predicted at the m test points is $\boldsymbol{\Sigma}_{mn}(\boldsymbol{\Sigma} + \sigma^2 \boldsymbol{I}_n)^{-1} \mathbf{D}_Y$, and the variance-covariance matrix of this vector is $\boldsymbol{\Sigma}^{(test)} - \boldsymbol{\Sigma}_{mn}(\boldsymbol{\Sigma} + \sigma^2 \boldsymbol{I}_n)^{-1} \boldsymbol{\Sigma}_{nm}$.

3.3.7 Inverse prediction – aka learning of input at which test output is realised

Inverse prediction becomes relevant when for the purpose of easing kernel-parametrised based learning, the lower-dimensional of the two variables is chosen as the input, where the functional link between the two said variables is being pursued. So if variable $\boldsymbol{X}$ is the input and variable $\boldsymbol{Y}$ is the output to the function $\boldsymbol{f}(\cdot)$ – that we model as a sample function from a $\mathcal{GP}$, which induces a kernel-parametrised covariance matrix $\boldsymbol{\Sigma}_n$ on the likelihood – then we seek the joint *pdf* of the kernel hyperparameter vector $\boldsymbol{\Theta}$ and the variable $\boldsymbol{X}^{(test)}$. given the training data $\mathbf{D}$ and the recorded output value of $\boldsymbol{y}^{(test)}$ that is realised at the sought value of the variable $\boldsymbol{X}^{(test)}$.

Then to signify that the covariance matrix $\boldsymbol{\Sigma}_n$ is kernel-parametrised, where the kernel is distinguished by the hyperparameters that are collated into vector $\boldsymbol{\Theta}$, we will in fact use the notation $\boldsymbol{\Sigma}_n(\boldsymbol{\theta})$ for this matrix that is computed at the value $\boldsymbol{\theta}$ of the vector of hyperparameters $\boldsymbol{\Theta}$.

We aim at the learning of $\boldsymbol{\Theta}$ and the input $\boldsymbol{X}^{(test)}$ at which a test datum $\boldsymbol{y}^{(test)}$ is realised, given the training data $\mathbf{D}$ that contains all the information of $\mathbf{D}_{\boldsymbol{Y}}$, as well as the respective design inputs $\boldsymbol{x}_1, \ldots, \boldsymbol{x}_n$. Thus, we seek the joint posterior density

$$\pi(\boldsymbol{x}^{(test)}, \boldsymbol{\theta} | \boldsymbol{y}^{(test)}, \mathbf{D}_{\boldsymbol{Y}}, \boldsymbol{x}_1, \ldots, \boldsymbol{x}_n),$$

given the training data $\mathbf{D}$ and the recorded output $\boldsymbol{y}^{(test)}$.

Now, at the value $\boldsymbol{x}^{(test)}$ of the input variable $\boldsymbol{X}^{(test)}$, we augment the data $\mathbf{D}_{\boldsymbol{Y}}$ with the data point $\boldsymbol{y}^{(test)}$ that is realised at this value of the input variable $\boldsymbol{X}^{(test)}$. Then the posterior density of $\boldsymbol{\Theta}$ given the augmented data on the outputs, and the respective inputs, is

$$\pi(\boldsymbol{\theta} | \mathbf{D}_{\boldsymbol{Y}}, \boldsymbol{y}^{(test)}, \boldsymbol{x}_1, \ldots, \boldsymbol{x}_n, \boldsymbol{x}^{(test)}).$$

Since $\{(\boldsymbol{x}_1, \boldsymbol{y}_1), \ldots, (\boldsymbol{x}_n, \boldsymbol{y}_n), (\boldsymbol{x}^{(test)}, \boldsymbol{y}^{(test)})\}$ is the augmented training data at $\boldsymbol{X}^{(test)} = \boldsymbol{x}^{(test)}$, likelihood of the kernel hyperparameters in this data is multivariate Normal - or its high-dimensional counterpart – depending on the dimensionality of the output variable in the model.

Let us first invoke the simple low-dimensional paradigm, within which the sought function is a univariate, scalar-valued function, s.t. likelihood of the kernel hyperparameters given the augmented training data at $X^{(test)} = x^{(test)}$ is multivariate Normal, with a covariance matrix $\mathbf{\Sigma}(\boldsymbol{\theta})$ that is kernel-parametrised with parameters collated into the vector $\mathbf{\Theta}$.

But the joint posterior of $X^{(test)}$ and $\mathbf{\Theta}$ given $\mathbf{D}_Y$, $y^{(test)}$ and $x_1, \ldots, x_n$, is related to the density of $\mathbf{\Theta}$ conditional on all the knowns and $X^{(test)} = x^{(test)}$, as in:

$$\pi(\boldsymbol{\theta}, x^{(test)}|\mathbf{D}_Y, y^{(test)}, x_1, \ldots, x_n) = \\ \pi(\boldsymbol{\theta}|\mathbf{D}_Y, y^{(test)}, x_1, \ldots, x_n, x^{(test)}) \times f(\mathbf{D}_Y, y^{(test)}, x_1, \ldots, x_n, x^{(test)}). \tag{3.4}$$

Then recalling that $Y^{(test)}$ is the name of the variable that attains the value $y^{(test)}$ at the sought input variable $X^{(test)}$,

$$f(\mathbf{D}_Y, y^{(test)}, x_1, \ldots, x_n, x^{(test)}) = \\ f_{Y^{(test)}|X^{(test)}}(y^{(test)}|\mathbf{D}_Y, x_1, \ldots, x_n, x^{(test)}) \times \\ f(\mathbf{D}_Y, x_1, \ldots, x_n, x^{(test)}),$$

where

$$f(\mathbf{D}_Y, x_1, \ldots, x_n, x^{(test)}) = \text{constant} \ ,$$

since the prior density of $X^{(test)}$ is Uniform in our consideration, and value of the input $X^{(test)}$ at which the test datum on $Y^{(test)}$ is realised, is not affected by choices of the n design inputs, (or the output values realised at these design inputs). So, $f(\mathbf{D}_Y, x_1, \ldots, x_n, x^{(test)}) = f(\mathbf{D}_Y, x_1, \ldots, x_n) f_{X^{(test)}}(x^{(test)}) \propto f(\mathbf{D}_Y, x_1, \ldots, x_n)$.

From Definition 3.2, we can recall that

$$f_{Y^{(test)}|X^{(test)}}(y^{(test)}|\mathbf{D}_Y, x_1, \ldots, x_n, \boldsymbol{x}^{(test)}, \boldsymbol{\theta}) = \\ \mathcal{MN}(\mathbf{\Sigma}_{1,n}(\boldsymbol{\theta})(\mathbf{\Sigma}_n(\boldsymbol{\theta}))^{-1}\mathbf{D}_Y, \mathbf{\Sigma}^{(test)}(\boldsymbol{\theta}) - \mathbf{\Sigma}_{1,n}(\boldsymbol{\theta})(\mathbf{\Sigma}_n(\boldsymbol{\theta}))^{-1}\mathbf{\Sigma}_{n,1}(\boldsymbol{\theta})),$$

up to a constant, where $\mathbf{\Sigma}_{\cdot,\cdot}(\boldsymbol{\theta})$ is the covariance matrix induced by the underlining $\mathcal{GP}$, and is kernel-parametrised using kernel hyper-parameters that are components of vector $\mathbf{\Theta}$. Also, the augmented covariance matrix $\mathbf{\Sigma}^{/}(\boldsymbol{\theta})$ that is $(n+1) \times (n+1)$-dimensional, can be partitioned into the $n \times n$-dimensional matrix $\mathbf{\Sigma}_n(\boldsymbol{\theta})$; the $1\times$

1-dimensional matrix $\boldsymbol{\Sigma}^{(test)}(\boldsymbol{\theta}) = K(x^{(test)}, x^{(test)}; \boldsymbol{\theta})$; the $1 \times n$-dimensional matrix $\boldsymbol{\Sigma}_{1,n}(\boldsymbol{\theta}) = K(x^{(test)}, x_i; \boldsymbol{\theta})$, for $i = 1, \ldots, n$; and the $n \times 1$-dimensional matrix $\boldsymbol{\Sigma}_{n,1}(\boldsymbol{\theta}) = (\boldsymbol{\Sigma}_{1,n}(\boldsymbol{\theta}))^T$.

Therefore, the density of the test datum, conditional on the value of the input $\boldsymbol{X}^{(test)}$, and the available data, is given by marginalising over all values of $\boldsymbol{\Theta}$, i.e.

$$f_{Y^{(test)}|X^{(test)}}(y^{(test)}|\mathbf{D}_Y, x_1, \ldots, x_n, x^{(test)}) =$$

$$\int_{\boldsymbol{\theta}} \mathcal{MN}(\boldsymbol{\Sigma}_{1,n}(\boldsymbol{\theta})(\boldsymbol{\Sigma}_n(\boldsymbol{\theta}))^{-1}\mathbf{D}_{\boldsymbol{Y}}, \boldsymbol{\Sigma}^{(test)}(\boldsymbol{\theta}) - \boldsymbol{\Sigma}_{1,n}(\boldsymbol{\theta})(\boldsymbol{\Sigma}(\boldsymbol{\theta}))^{-1}\boldsymbol{\Sigma}_{n,1}(\boldsymbol{\theta}))d\boldsymbol{\theta},$$

up to a constant factor.

Then collating all these results together, and recalling that

$$\pi(\boldsymbol{\theta}|\mathbf{D}_Y, y^{(test)}, x_1, \ldots, x_n, x^{(test)}) = \mathcal{MN}(\underline{0}, \boldsymbol{\Sigma}^{/}(\boldsymbol{\theta})) \times \pi_0(\boldsymbol{\theta}),$$

in Equation 3.4, for standardised values of the outputs,

$$\begin{aligned} \pi(\boldsymbol{\theta}, x^{(test)}|\mathbf{D}_Y, y^{(test)}, x_1, \ldots, x_n) = \\ \mathcal{MN}(\underline{0}, \boldsymbol{\Sigma}^{/}(\boldsymbol{\theta}))\pi_0(\boldsymbol{\theta})\times \\ \int_{\boldsymbol{\theta}} \mathcal{MN}(\boldsymbol{\Sigma}_{1,n}(\boldsymbol{\theta})(\boldsymbol{\Sigma}_n(\boldsymbol{\theta}))^{-1}\mathbf{D}_{\boldsymbol{Y}}, \boldsymbol{\Sigma}^{(test)}(\boldsymbol{\theta}) - \boldsymbol{\Sigma}_{1,n}(\boldsymbol{\theta})(\boldsymbol{\Sigma}(\boldsymbol{\theta}))^{-1}\boldsymbol{\Sigma}_{n,1}(\boldsymbol{\theta}))d\boldsymbol{\theta}. \end{aligned} \tag{3.5}$$

The integral on the RHS of the last equation over all possible values of the hyperparameters is then the sum of the values of multivariate Normal densities parametrised by all possible means, and variances – computed at a given observation, namely, $y^{(test)}$. The integral then yields a constant at a given $y^{(test)}$.

While the above is developed for a single test datum, the concept can be generalised, so that we learn components of the $\boldsymbol{\Theta}$ vector, and values of the inputs at which observed test data points are respectively realised, given the available data. Once the joint posterior of the sought variables is formulated as in the last equation, we sample from this posterior using MCMC, permitting the learning of the marginal posterior probability density on each such variable, given the available data. The marginals are then employed to learn the 95% HPD credible region on each variable, in this scheme in which we simultaneously learn the inputs (at which test output values are realised), and the kernel hyperparameters.

This way, we could undertake inverse prediction – which really should be called learning and not prediction – of the input values at given test and training data. This follows the formulation of the model s.t. the lower dimensional of $\boldsymbol{X}$ and $\boldsymbol{Y}$ is selected as the input in the model, to facilitate learning. Below we discuss the implementation of such inverse prediction in a real-world problem in which the input is a scalar, while the output to the sought function is a vector.

3.4 Covariance kernels

As we have seen above, the covariance function of the $\mathcal{GP}$ that generates the sought function $f(\cdot)$ is kernel parametrised by the kernel function $K(\cdot, \cdot; \boldsymbol{\theta})$, s.t. the ij-th element of the covariance matrix $\boldsymbol{\Sigma}$ of the multivariate Normal likelihood is $K(\boldsymbol{x}_i, \boldsymbol{x}_j; \boldsymbol{\theta})$. Here $\boldsymbol{\theta}$ is defined as the vector of hyperparameters of the kernel.

Various parametric kernels exist in the literature, and in the following chapter, we are going to explore the possibility of working with a non-parametric kernel. We will discuss the pitfalls of kernel parametrisation in the following subsection, but for now, let us look at some examples of covariance kernels.

A simple kernel is the SQuare Exponential (SQE) that states:

$$Cov(Y_i, Y_j) = \exp\left(-((\boldsymbol{x}_i - \boldsymbol{x}_j)^T \boldsymbol{Q}(\boldsymbol{x}_i - \boldsymbol{x}_j))\right),$$

where the matrix $\boldsymbol{Q}$ is a $d \times d$-dimensional diagonal matrix, the c-th diagonal element of which is the reciprocal of the squared length scale hyperparameter along the c-th direction in input space (i.e. space of input $\boldsymbol{X} \in \mathbb{R}^d$). Thus,

$$\boldsymbol{Q} = [\delta_{cc'}/\ell_c^2], \quad c, c' \in \{1, \ldots, d\}.$$

An interpretation of ℓ_c is that it is the length scale along the c-th direction in input space, over which correlations change, [105, 10].

The kernel $K(\cdot,\cdot;\boldsymbol{\theta})$ is equivalently denoted $K(\cdot,\cdot;\ell_1,\ldots,\ell_d)$ in this example of a d-dimensional input variable, and an SQE kernel. Thus, the learning of the covariance matrix $\boldsymbol{\Sigma}$ is reduced to the learning of the length scale hyperparameters $\ell_1,\ldots,\ell_d$. Once these hyperparameters are learnt given the data – using the posterior that follows from chosen priors and the multivariate Normal likelihood declared in Definition 3.4 – the kernel is specified, and thereby, we learn the $\mathcal{GP}$ that generates the sought function $f(\cdot)$. We will then be in a position to predict the output at a test input.

We note that the SQE kernel raised to any positive power, amounts to a modulation of the length scale hyperparameters. In this sense it is referred to as infinitely divisible. Also, as [105] shows in Chapter 4 of their book, for the scalar-valued input X, expanding the correlation function $Corr(Y_i,Y_j) = K(x_i,x_j;\ell)$ in an infinite series over an index k, of the product of the two Gaussians in x_i and x_j – both centred at k, with common variance of ℓ^2 – gives rise to the SQE form for $K(x_i,x_j;\ell)$, in the limit that k is in $\mathbb{R}$. In other words, the expansion of the kernel function using Gaussian basis functions, give rise to the SQE kernel.

3.4.1 $\mathcal{GP}$-generated spatio-temporal function

Often in real-world applications, need arises for the learning of a variable that depends on spatial coordinates, as well as on time. Then the sought functional relationship between such a variable, and the spatio-temporal input variable, can be modelled using a $\mathcal{GP}$. The invoked $\mathcal{GP}$ is then a space-time process specified by a covariance function that is defined for a pair of realisations of the relevant spatio-temporal vector-valued input [64]. A generic model of the sought function emerges from acknowledging possible asymmetry in this covariance structure, (in addition to non-stationarity, which we comment on in the following paragraph); it is also desired that non-separability on space and time dependence of the sought functional relationship, be accommodated [102]. By "symmetry" in the covariance function, is implied its invariance to a toggling of the sign of a change in the spatial coordinates and/or the time lag [125, 120], where the latter work also proposes acknowledging non-separability of the time and space dependence. Often, the physical processes relevant in atmospheric physics, or geostatistics – where spatio-temporal $\mathcal{GP}$s are invoked – are s.t. asymmetry becomes relevant.

By non-stationarity of a process is implied that for random variables $f(\boldsymbol{x}_i)$ and $f(\boldsymbol{x}_j)$ – each of which is one of the sequence of r.v.s that comprise this process - the correlation between these variables is a function of $\boldsymbol{x}_i$ and $\boldsymbol{x}_j$, not just via a distance between $\boldsymbol{x}_i$ and $\boldsymbol{x}_j$. Here, the mean of each such constituent r.v. is the same. The covariance between pairs of constituent r.v.s of such a process is then a non-stationary covariance. A stationary covariance between $f(\boldsymbol{x}_i)$ and $f(\boldsymbol{x}_j)$ depends on $\boldsymbol{x}_i$ and $\boldsymbol{x}_j$, via a function of $\boldsymbol{x}_i - \boldsymbol{x}_j$.

Indeed, for a (time-indexed) stationary process – as we have motivated in Chapter 2 for a Markov process – we would be unable to distinguish if the process is evolving forwards in time, or backwards, i.e. a stationary process is invariant to a shift in time.

We now consider the process $\{f(\boldsymbol{x})\}_{\boldsymbol{X}\in\mathbb{R}^d}$ indexed by a purely spatial variable $\boldsymbol{X} \in \mathbb{R}^d$, with $f(\boldsymbol{x})$ a sample function s.t.

$$(\kappa^2 - \boldsymbol{\Delta})^{\alpha/2} f(\boldsymbol{x}) = W(\boldsymbol{x}).$$

- Here $\alpha = \nu + d/2$ is the smoothness parameter with $\nu > 0$, where qualifications of "smoothness" (and "scale") as used here, is motivated below;

- $\kappa > 0$ is a length scale that represents the Euclidean distance that one needs to move by in input space, for the correlation between two realisations of $f(\cdot)$ to fade by a such a factor that these two realisations are "almost independent" [77];

- $W(\cdot)$ is Gaussian white noise, given by a spatial $\mathcal{GP}$ with zero mean and unit variance.

- Lastly, $\boldsymbol{\Delta}$ is the Laplacian operator, i.e.

$$\boldsymbol{\Delta} = \sum_{i=1}^{d} \frac{\partial^2}{\partial x_i^2}.$$

Then [129] showed that such an $f(\cdot)$ is generated by a spatial $\mathcal{GP}$, the covariance function of which is referred to as the Matern covariance function:

$$Cov(\boldsymbol{x}_i, \boldsymbol{x}_j) = \frac{\sigma^2_{Marginal}}{2^{\nu-1}\Gamma(\nu)}(\kappa s)^\nu K_\nu(\kappa s), \tag{3.6}$$

where

$$s \equiv \| \boldsymbol{x}_i - \boldsymbol{x}_j \|,$$

and $K_\nu(u)$ is the "modified Bessel function of the 2nd kind" in u, at ν. Such a special function is given as

$$K_\nu(u) = \frac{(\pi/2)(I_{-\nu}(u) - I_\nu(u))}{\sin(\alpha\pi)},$$

where $I_\nu(u)$ is a modified Bessel function of the 1st kind, that is given as $I_\nu(u) = i^{-\nu} J_\nu(iu)$, where $i^2 = -1$ and $J_\nu(\nu)$ is the Bessel function of the first kind that is the solution to Bessel's equation:

$$u^2 \frac{d^2 y(u)}{du^2} + u\frac{dy(u)}{du} + (u^2 - \nu^2)y(u) = 0.$$

In the declaration of the Matern covariance function above, $\Gamma(\cdot)$ is a Gamma function and $\sigma^2_{Marginal}$ is the marginal variance, defined as

$$\sigma^2_{Marginal} = \frac{\Gamma(\nu)}{\Gamma(\nu + d/2)(4\pi)^{d/2}\kappa^{2\nu}}.$$

Now, the reason for calling ν the smoothness parameter above, is that ν drives the smoothness of any sample function of a $\mathcal{GP}$ with a Matern covariance function; such a sample function is differentiable $\lfloor \nu - 1 \rfloor$ times.

The function $f(\cdot)$ in the Stochastic Partial Differential Equation (SPDE) $(\kappa^2 - \boldsymbol{\Delta})^{\alpha/2} f(\boldsymbol{x}) = W(\boldsymbol{x})$, is sampled from a $\mathcal{GP}$ with a Matern covariance defined in Equation 3.6, except when $\kappa \longrightarrow 0$

or $\nu \longrightarrow 0$; (then well-defined solutions still exist for this SPDE, as discussed in [77]). When ν is not an integer, (unlike in the work by [77]), the above SPDE is referred to as "fractional". With the smoothness parameter $\nu \longrightarrow \infty$, the Matern covariance function approaches the SQE covariance in shape. On the other hand, when $\nu = 1/2$, the Matern covariance reduces to the exponential covariance function, which we discuss below (in Section 3.4.2).

Let us attempt developing the motivation for the covariance function that is of the form of the Matern covariance discussed above, by solving the SPDE $(\kappa^2 - \boldsymbol{\Delta})^{\alpha/2} f(\boldsymbol{x}) = W(\boldsymbol{x})$ for the case of $\boldsymbol{X} \in \mathbb{R}^2$, (i.e. $d = 2$) and $\nu = 1$, s.t. $\alpha = 2$. Then, in the resulting equation, we take the spatial Fourier transform on $\mathbb{R}^2$ on both sides; solve for $F(\xi_1, \xi_2)$; obtain the spectral density; and follow by taking the Fourier transform, to lead to the covariance function between $\boldsymbol{x} = (x_1, x_2)^T$ and $\boldsymbol{x}' = (x_1', x_2')^T$, i.e. $Cov(x_1, x_1'; x_2, x_2')$. Here. the spatial frequency vector is $(\xi_1, \xi_2)^T$.

By taking a Fourier transform of the LHS of this SPDE, for $\alpha = 2$, we get the spectral density to be $1/(\kappa^2 + \xi_1^2 + \xi_2^2)^2$. Now, the Fourier transform of a cirularly symmetric or radial function of two variables can be reduced to its Hankel transform [17]. Thus, the Fourier transform of a function of x_1 and x_2 with a radially symmetric input $r := \sqrt{x_1^2 + x_2^2}$ – such as the function $\psi(r)$ – is its Hankel transform. Then the Fourier transform of the function $\psi(\sqrt{\xi_1^2 + \xi_2^2}) := 1/(\kappa^2 + \xi_1^2 + \xi_2^2)^2$, is the corresponding Hankel transform, which is proportional to

$$\sqrt{x_1^2 + x_2^2} K_1(\kappa\sqrt{x_1^2 + x_2^2})/2\kappa \propto$$

$$\sigma^2_{Marginal}(\kappa\sqrt{x_1^2 + x_2^2}) K_1(\kappa\sqrt{x_1^2 + x_2^2}),$$

where we have defined $\sigma_{marginal}$ above. We recall that $K_\nu(\cdot)$ is the modified Bessel function of the 2nd kind, at ν. Using this result, the sought covariance between $\boldsymbol{x}$ and $\boldsymbol{x}'$ is computed to be proportional to

$$\sigma^2_{Marginal}(\kappa\sqrt{(x_1 - x_1')^2 + (x_2 - x_2')^2}) K_1(\kappa\sqrt{(x_1 - x_1')^2 + (x_2 - x_2')^2}).$$

This holds for all $(x_1, x_2)^T, (x_1', x_2')^T \in \mathbb{R}^2$. We recognise this covariance function to have the same shape as the Matern covariance in this case of $\nu = 1; d = 2$.

In general, the original SPDE could be recast as $\mathcal{L}f(\boldsymbol{x}) = W(\boldsymbol{x})$, where the linear operator $(\kappa^2 - \boldsymbol{\Delta})^{\alpha/2}$ is called $\mathcal{L}$. Then the solution to this SPDE is a $\mathcal{GP}$, with a covariance function $(\mathcal{L}^*\mathcal{L})^{-1}$, where $\mathcal{L}^*$ is the conjugate of $\mathcal{L}$. Other numerical methods have been suggested for the approximate solution of $\mathcal{L}$ at generic α values.

For completion's sake, we mention the modelling of the spatio-temporally varying function $f(\boldsymbol{X}, T)$, where the time variable is T. The white noise term is then dependent on both $\boldsymbol{X}$ and T. [125] have advanced solutions to a generic class of such (linear) spatio-temporal SPDEs, using Fourier transforms.

3.4.2 Orstein-Uhlenbeck or exponential kernel

Consider a particle that is undergoing Brownian motion in $\mathcal{X} \subset \mathbb{R}$, s.t. the position coordinates of this particle at time $T = t$ is $X(t)$.

Definition 3.5 *The Stochastic Process $\{X(t)\}_{t\geq 0}$ is called a Brownian motion if at $t = 0$, $X(0) \in \mathbb{R}$; increments $X(t_2) - X(t_1), X(t_3) - X(t_2), \ldots, X(t_n) - X(t_{n-1})$ are independent random variables, with the r.v. $X(t+\delta_t) - X(t) \sim \mathcal{N}(0, \delta_t)$, $\forall t > 0$, $\forall \delta_t \geq 0$; and almost surely, $\mathcal{T} \longrightarrow \mathcal{X}$ is continuous, for $T \in \mathcal{T} \subseteq \mathbb{R}_{\geq 0}$.*

Then for any given time $T = t$, almost surely, Brownian motion is not differentiable [87]. So it would appear that for a particle undergoing Brownian motion, its velocity cannot be computed. Indeed, the velocity can then be modelled directly as comprising the Stochastic Process $\{V(t)\}_{t\geq 0}$ that is a solution of the following Stochastic Differential Equation (SDE):

$$dV_t = \kappa V_t dt + \sigma dB_t,$$

where we have denoted $V(t) \equiv V_t$; κ and σ are constants, (matrices for a vector-valued $V(t)$); and $\{B_t\}_{t\geq 0}$ is a Brownian motion that starts from 0, i.e. $B_0 = 0$. This SDE is referred to as the Orstein-Uhlenbeck equation [37]. This function

$$V(\cdot) \sim \mathcal{GP}(0, \sigma^2 \exp(-\kappa|\cdot - \cdot|)).$$

with this covariance kernel referred to as the Orstein-Uhlenbeck kernel, [31].

It is a stationary kernel, that is a special case of the Matern kernel, as motivated in the previous subsection.

3.5 Output is vector-valued

Earlier in this chapter, (in Section 3.2 to be precise), we motivated the advantage of opting to use the higher-dimensional of the two variables – $\boldsymbol{X} \in \mathcal{X} \subseteq \mathbb{R}^d$ and $Y \in \mathcal{Y} \subseteq \mathbb{R}$ – as the output of the sought functional relationship between $\boldsymbol{X}$ and Y. We recall that this owes to the greater difficulty that will accompany the learning of the higher-dimensional kernel, if the higher-dimensional of the two variables is cast as the input, (rather than the output). Indeed such difficulty was apprehended to be more potent than the difficulty of conceptualisation and coding of the more complex likelihood that will be relevant when the higher-dimensional of $\boldsymbol{X}$ and Y is treated as the output.

For example, if we persist with the model equation $Y = f(\boldsymbol{x})$ from the previous section, (in which the output and inputs are evidently Y and $\boldsymbol{X}$ respectively), when we parametrise the covariance matrix of the multivariate Normal likelihood with a kernel – say the simple SQE kernel – we will need to learn the d length scale hyperparameters $\ell_1, \ldots, \ell_d$ that define the diagonal elements of a matrix $\boldsymbol{Q}$, [23].

Alternatively, if we opt to use Y as the input and $\boldsymbol{X}$ as the output, s.t. we seek to learn the function $\boldsymbol{g}(\cdot)$ by modelling it with a $\mathcal{GP}$, where $\boldsymbol{X} = \boldsymbol{g}(y)$, then by virtue of the vector-valued output, the sought function $\boldsymbol{g}(\cdot)$ is vector-valued and its generative $\mathcal{GP}$ is vector-variate s.t. likelihood is matrix Normal on that occasion, (as we will soon see below). This likelihood is then parametrised by

— an "inter-$\boldsymbol{x}$" covariance matrix, that informs on covariance between pairs of output vectors attained at the corresponding pair of Y values;

— and by the inter-component covariance matrix, (or "intra-$\boldsymbol{x}$" covariance matrix), an element of which is the covariance between a pair of components of the vector-valued output $\boldsymbol{X}$.

Then the inter-$\boldsymbol{x}$ covariance matrix would be kernel parametrised, s.t. its ij-th element is expressed as the kernel function of the difference between the i-th and j-th values of the input Y, at which the i-th and j-th outputs are respectively attained. This kernel – if again modelled as an SQE kernel – will have only a single length scale parameter (for standardised data on the output), that we will need to learn, to specify the $\mathcal{GP}$ that the sought function $\boldsymbol{g}(\cdot)$ is modelled to be a sample path of.

Indeed, in this model, the matrix Normal likelihood is potentially more complicated with 2 covariance matrices and a matrix-variate mean, than in the previous case when the likelihood was multivariate Normal. But once we are no longer intimidated by the complexity of the likelihood – i.e. have the computational paraphernalia prepared to undertake learning – it is the learning of the kernel hyperparameters that then drives the appreciation of model difficulty. Thus, the learning of d kernel length scale hyperparameters under the model equation $Y = f(\boldsymbol{x})$ is a harder learning task to undertake, than the learning of the single length scale hyperparameter relevant when we state the relationship between $\boldsymbol{X}$ and Y as $\boldsymbol{X} = \boldsymbol{g}(y)$. There is an unaddressed concern about prediction in this model; we shall consider this in Section 3.3.7 below.

However, we have not discussed the learning of vector-valued functions, but have simply put forward the notion that when such a random function is modelled as sample function of a $\mathcal{GP}$, likelihood is matrix Normal. We proceed to discuss such learning next.

3.6 Sought function is vector-valued

In this section we attempt learning $\boldsymbol{g}(\cdot)$, where $\boldsymbol{X} = \boldsymbol{g}(y)$, i.e. $\boldsymbol{g} : \mathcal{Y} \subseteq \mathbb{R} \longrightarrow \mathcal{X} \subseteq \mathbb{R}^d$. We model the sought function $\boldsymbol{g}(\cdot)$ as a sample path of a vector-variate $\mathcal{GP}$ with mean function $\boldsymbol{\mu}(\cdot)$, and covariance function $\boldsymbol{K}(\cdot, \cdot)$. So to clarify, the output variable in this learning is $\boldsymbol{X}$, and the input is Y. We attempt this learning using the training data $\mathbf{D}^{/} := \{(y_i, \boldsymbol{x}_i)\}_{i=1}^{n}$, that consists of values of the output $\boldsymbol{X}$, "generated at" the designed input value $Y = y_i$. So this training data $\mathbf{D}^{/}$ could bear the same information as the training data $\mathbf{D}$ that we have seen above, except that this time, it is the input variable (Y), that is assigned non-zero measurement noise. Y was indeed the variable that we have noted above, to be noisy.

In that alternative model - invoked within the learning exercise that was undertaken above, (with $\boldsymbol{X}$ the input and Y the output) – had the chosen or design values of $\boldsymbol{X}$ been noisy, we would have needed to have acknowledged such errors in the input variable. That learning exercise appeared to not have worried about the prospect of working with noise in the identified values of the input, since we chose to ignore such a prospect. However, in reality, one might come across a situation in which input values, as well as the output values, are known only within respective intervals that can be interpreted as noise-driven.

If we can figure out how to acknowledge noise in both input and output variables in general, in our non-parametric $\mathcal{GP}$-based supervised learning approach, we can proceed with the learning endeavour that considers Y as the input and $\boldsymbol{X}$ as the output variable.

3.6.1 Noise in the input values as well

Given the model equation $A = h(B)$, where A and B are the output and input random variables, and $h(\cdot)$ is the unknown functional relation between A and B, we recall that noise in the output

A is acknowledged in the learning of $h(\cdot)$ that is modelled as a realisation from a $\mathcal{GP}$, given the training data $\mathbf{C} := \{(b_j, a_j)\}_{j=1}^{N}$, by convolving the joint *pdf* of $A_1, A_2, \ldots, A_N$ - which is multivariate Normal in our current considerations – with the density of $(\epsilon_1, \epsilon_2, \ldots, \epsilon_N)^T$, where ϵ_j is the measurement error on A_j, with ϵ_j independent of $\epsilon_{j'}$; $j \neq j'; j, j' = 1, \ldots, N$. If $\epsilon_j \sim \mathcal{N}(0, \sigma_j^2)$, then the convolution results in the addition of the $N \times N$-dimensional diagonal matrix: $\Sigma_{err}=$

$$\begin{bmatrix} \sigma_1^2 & 0 & \ldots & \ldots & 0 \\ 0 & \sigma_2^2 & 0 & \ldots & 0 \\ \vdots & & & & \\ \vdots & & & & \\ 0 & 0 & \ldots & 0 & \sigma_N^2. \end{bmatrix}$$

to the covariance matrix $\mathbf{\Sigma}$ of the multivariate Normal likelihood, i.e. in presence of noise in the measured values of the output, the likelihood is

$$\mathcal{MN}(\underset{\sim}{0}, \mathbf{\Sigma} + \mathbf{\Sigma}_{err}).$$

Here $\mathbf{\Sigma}$ is the variance of the multivariate Normal joint of $A_1, A_2, \ldots, A_N$, (that we consider here, as stated before).

If errors on all output variables obey the same error density, then $\sigma_j = \sigma_0, \forall j = 1, \ldots, N$, and then $\mathbf{\Sigma}_{err} = \sigma_0^2 \boldsymbol{I}_N$, where $\boldsymbol{I}_N$ is the N-dimensional identity matrix. Then convolution of the *pdf* of $A_1, A_2, \ldots, A_N$, with the error density, results in the addition of $\mathbf{\Sigma}_{err} = \sigma_0^2 \boldsymbol{I}_N$ to the covariance matrix of the multivariate Normal likelihood, i.e. the likelihood is updated to

$$\mathcal{MN}(\underset{\sim}{0}, \mathbf{\Sigma} + \sigma_0^2 \boldsymbol{I}_N).$$

Such acknowledgement of noise in the measured values of the output is clear. But how do we account for noise that may affect values of the input? Then, the (designed) values of the input lie in an interval that is driven by the density of the noise in the measurement of an input value; thus, the j-th input is a random variable, $j = 1, \ldots, N$. Let the N inputs that reside inside the training data, be the random variables $B_1, B_2, \ldots, B_N$. Randomness in the r.v. B_j emerges from the fact that B_j takes values in an interval around the fixed (or designed) mean of the j-th input - denoted b_{j0} – where the non-zero width of said interval owes to the noise in the measurement of B_j. So in summary, B_j is a random variable, but its mean is known as b_{j0}.

Similar to the modelling of the measurement error of the output variable A_j – according to a known error density, e.g. Normal – allowed for superposition of the measurement error on the observation of A_j, so too, in this case, we will invoke a parametric form of the error density on measurements of B_j. Thus if the error density on B_j is Gaussian with mean 0 and variance s_j^2, then $\varepsilon_j \sim \mathcal{N}(0, s_j^2)$. Then B_j is modelled to attain values in the interval that is symmetric about b_{j0}, where width of this interval is set by the standard deviation s_j of this error density. (Here $j = 1, \ldots, N$).

In the absence of errors in the measured value of B_j, $b_j = b_{j0}$, but in the presence of Normally-distributed errors, b_j is likely to be in the interval $[b_{j0} - 3s_j, b_{j0} + 3s_j]$, since about 99% of the area under the Normal density is bounded by the ± 3 standard deviations. The superposition of error on the designed mean of B_j is depicted by the convolution of a Delta function $\delta(b - b_{j0})$ and the Gaussian $\mathcal{N}(b; 0, s_j^2)$. From properties of the Delta function, this convolution results in the Gaussian $\mathcal{N}(b; b_{j0}, s_j^2)$.

Thus, if the error density is identical for each input variable, i.e. $s_j = s \ \forall j = 1, \ldots, N$, then density of B_j is $f_{B_j}(b) = \mathcal{N}(b; b_{j0}, s^2)$, $\forall j = 1, \ldots, N$, s.t.

$$f_{B_1,\ldots,B_N}(b_1, \ldots, b_N) = (2\pi s^2)^{-N/2} \exp\left(-\frac{\sum\limits_{j=1}^{N} (b - b_{j0})^2}{2s^2} \right), \quad (3.7)$$

where we have recalled that choices of the inputs are independent of each other; so $f_{B_1,\ldots,B_N}(b_1, \ldots, b_N) = f_{B_1}(b_1) f_{B_2}(b_2) \ldots f_{B_N}(b_N)$.

In order to formulate the likelihood, we seek the joint *pdf* of the input and output observable random variables: $A_1, A_2, \ldots, A_N$ and $B_1, \ldots, B_N$, conditional on the model parameter $\mathbf{\Sigma}$ that is induced by the covariance kernel of the $\mathcal{GP}$ that generates the functional relation between this input and output. But

$$f_{A_1,\ldots,A_N,B_1,\ldots,B_N|\boldsymbol{\Sigma}}(a_1,\ldots,a_n,b_1,\ldots,b_N|\boldsymbol{\Sigma})$$
$$= f_{A_1,\ldots,A_N|B_1,\ldots,B_N,\boldsymbol{\Sigma}}(a_1,\ldots,a_n|b_1,\ldots,b_N,\boldsymbol{\Sigma})$$
$$f_{B_1,\ldots,B_N|\boldsymbol{\Sigma}}(b_1,\ldots,b_N|\boldsymbol{\Sigma}). \quad (3.8)$$

However, the inputs $B_1\ldots,B_N$ are independent of the covariance matrix $\boldsymbol{\Sigma}$, as we see in Equation 3.7, in which the joint of the inputs is given by the (designed) means of the inputs and the variance of the *pdf* of the variable that is the error of an input. Therefore, the second term on the RHS of the last equation is $f_{B_1,\ldots,B_N}(b_1,\ldots,b_N)$, which is given in Equation 3.7,

Also, here $f_{A_1,\ldots,A_N|B_1,\ldots,B_N,\boldsymbol{\Sigma}}(a_1,\ldots,a_n|b_1,\ldots,b_N\boldsymbol{\Sigma})$ is multivariate Normal with zero mean and covariance matrix $\boldsymbol{\Sigma}$, given that the function that represents the relation between the output and input is generated from a Gaussian Process. Then recalling $f_{B_1,\ldots,B_N}(b_1,\ldots,b_N)$ from above, acknowledgement of (Gaussian) noise in the inputs as well as the outputs results in the likelihood:

$$f_{A_1,\ldots,A_N,B_1,\ldots,B_N}(a_1,\ldots,a_n,b_1,\ldots,b_N|\boldsymbol{\Sigma})$$
$$= \mathcal{MN}(\underset{\sim}{0},\boldsymbol{\Sigma}+\sigma^2\boldsymbol{I}_N)(2\pi s^2)^{-N/2}\exp\left(-\frac{\sum\limits_{j=1}^{N}(b-b_{j0})^2}{2s^2}\right)\times K. \quad (3.9)$$

Indeed, as the measurement error on an input variable approaches 0, i.e. $s \longrightarrow 0$, the exponential term on the RHS of this equation factorises to forms approaching Delta functions s.t. the multivariate Normal likelihood is computed at the fixed design values $b_{10}, b_{20},\ldots,b_{N0}$ then. We can drop the constant K (> 0) from this likelihood hereafter, since our method of performing inference – namely, sampling from the posterior using MCMC – is oblivious to the presence of a global constant in the posterior.

3.7 Tensor-variate output

Equipped with the know-how involving acknowledgement of measurement errors in the input Y of the model $\boldsymbol{X} = \boldsymbol{g}(Y)$, in which the output $\boldsymbol{X}$ is in general an $m-1$-th order tensor, we now embark upon the learning of the tensor-valued (in general) function $\boldsymbol{g}(\cdot)$, as a sample function of a tensor-variate GP, s.t. the joint density of N_{data} realisations of this function – at each of the N_{data} design values of input Y – is the m-th order tensor Normal density.

Data

Let the output realised at the i-th design input $Y = y_i$ be $\boldsymbol{x}_i$, $i = 1, \ldots, N_{data}$, implying that $\{\boldsymbol{x}_1, \vdots, \ldots, \vdots, \boldsymbol{x}_{N_{data}}\}$ is an m-th ordered tensor of dimensionality $k_1 \times k_2 \times \ldots \times k_m$; $k_i \in \mathbb{N}$. Thus, here $k_m \equiv N_{data}$. We denote the data variable $\mathbf{D}_{output}$; then this variable is composed of the outputs at each of the design inputs $y_1, \ldots, y_{N_{data}}$. Then the realisation of variable $\mathbf{D}_{output}$, given by the training set is denoted $\mathbf{d}_{output} := \{\boldsymbol{x}_1, \vdots, \ldots, \vdots, \boldsymbol{x}_{N_{data}}\}$. Then from the above, *pdf* of $\mathbf{D}_{output}$ is m-th order tensor Normal – with the mean tensor ($\boldsymbol{\mu}$) and the m number of covariance matrices ($\boldsymbol{\Sigma}_1, \ldots, \boldsymbol{\Sigma}_m$). If realisations of such parameters of the tensor Normal density change, then the *pdf* of $\mathbf{D}_{output}$ will change, i.e. this *pdf* of $\mathbf{D}_{output}$ should in fact be treated as conditional on the parameters of the tensor Normal density, [80].

Likelihood

Thus, we write the likelihood of the model parameters as:

$$\mathcal{L}(\boldsymbol{\mu}, \boldsymbol{\Sigma}_1, \ldots, \boldsymbol{\Sigma}_m; \mathbf{d}_{output}, y_1, \ldots, y_{N_{data}}) = (2\pi)^{-k/2}(\prod_{i=1}^{m} |\boldsymbol{\Sigma}_i|^{-k/2k_i})$$
$$\times \exp(-\|(\mathbf{d}_{output} - \boldsymbol{\mu}) \times_1 \boldsymbol{A}_1{}^{-1} \times_2 \boldsymbol{A}_2{}^{-1} \times \ldots \times_m \boldsymbol{A}_m^{-1}\|^2/2), \quad (3.10)$$

where $k = \prod_{i=1}^{m} k_i$ and $\boldsymbol{\Sigma}_i = \boldsymbol{A}_i \boldsymbol{A}_i^T$, $i = 1, \ldots, m$ Thus, $\boldsymbol{A}_i$ is the unique positive definite square root of the positive definite covariance matrix $\boldsymbol{\Sigma}_i$, [63].

Covariance matrices of the likelihood

We note that the $k_i \times k_i$-dimensional matrix $\boldsymbol{\Sigma}_i$ here is such, that its p_i, q_i-th element is the covariance between the p_i-th and q_i-th, $k_1 \times k_2 \times \ldots \times k_{i-1} \times k_{i+1} \ldots \times k_m$-dimensional slices of the tensor $\mathbf{D}_{output}$; $i = 1, \ldots, m$, $p_i, q_i = 1, \ldots, k_i$. Then $\boldsymbol{\Sigma}_m$ is the covariance matrix, the p_m, q_m-th element of which is the covariance between the variables ($\boldsymbol{X}_{p_m}$ & $\boldsymbol{X}_{q_m}$) that represent the realisations of the output at the input $Y = y_{p_m}$ and $Y = y_{q_m}$, respectively. Here $p_m, q_m = 1, \ldots, k_m \equiv N_{data}$.

Standardisation of data on output

The likelihood of the mean tensor and covariance matrices, (given the data $\tilde{\mathbf{d}}_{output}$ on the standardised realisations of the output $\boldsymbol{X}$ and choice of the design inputs), will be rewritten from Equation 3.10, to the likelihood of the parameters that the specification of the tensor-variate $\mathcal{GP}$ will reduce to, in light of the assorted techniques that will be invoked, to learn/estimate such parameters. For example, upon standardising the data on the output – to $\tilde{\mathbf{d}}_{output}$ – the underlying $\mathcal{GP}$ will be assigned a null mean function, s.t. $\boldsymbol{\mu} = \underset{\sim}{0}$ then.

Kernel parametrisation

Also, we will attempt kernel parametrising as many of the covariance matrices as we can. It will be possible to kernel parametrise $\boldsymbol{\Sigma}_m$, where this kernel function $K(\cdot, \cdot)$ is s.t.

$$Cov(\boldsymbol{X}_{p_m}, \boldsymbol{X}_{q_m}) = K(y_{p_m}, y_{q_m}; \boldsymbol{\theta}),$$

where $\boldsymbol{\theta} \in \mathbb{R}^H$ is the vector of the H-number of hyperparameters of the kernel $K(\cdot, \cdot)$. For example,

— if using the SQE kernel for standardised data on $X_1, \ldots, X_d$, for $Y \in \mathcal{Y} \subseteq \mathbb{R}$, $\boldsymbol{\theta}$ is in fact the scalar valued length-scale hyperparameter,

— $H = 2$ for the Matern kernel, with the two components of this $\boldsymbol{\theta}$ vector then given by the length scale hyperparameter and the roughness or regularity parameter (denoted ν above).

Learning elements of covariance matrices directly

In addition, there may exist other covariance matrices the distinct upper (or lower) triangle elements of which are learnt directly from MCMC. This can only be undertaken for covariance matrices that are low in dimensionality, (i.e. if the matrix size does not exceed $\lesssim 10 \times 10$). Let such directly learnt elements of the various covariance matrices be $\psi_1, \ldots, \psi_D$.

Estimating some covariance matrices

Let any covariance matrix that we have not kernel parametrised or directly learnt using MCMC, be learnt by estimating the sample correlation between the relevant slices, after collapsing each such high-dimensional slice of the m-th ordered tensor $\tilde{\mathbf{d}}_{output}$, into a vector.

Such empirical estimates of covariance between a pair of slices are noted to be least satisfactory as this method of computing the inter-slice covariance leads to loss of information that results from the collapsing (or projecting) of the slice to a vector. Also, the empirically-estimated covariance holds only in the available sample, and not in the population, in general.

Parameters to be learnt

Thus,

- the only parameters of the m-th ordered tensor Normal density that are to be learnt are: the hyperparameters $\theta_1, \ldots, \theta_H$ of the Kernel-parametrised,

- and the parameters $\psi_1, \ldots, \psi_D$ that are the elements of the covariance matrices that are directly learnt using MCMC,

- with the remaining covariance matrices estimated empirically.

Likelihood – gory details

Given such reduction in the learning of the parameters of the tensor Normal density – that is induced by the tensor-variate $\mathcal{GP}$ that models the sought function $g(\cdot)$ – we recast the likelihood relevant to this learning exercise as

$$\mathcal{L}(\theta_1, \ldots, \theta_H, \psi_1, \ldots, \psi_D; \tilde{\mathbf{d}}_{output}, y_1, \ldots, y_{N_{data}}) \propto \\ \exp(-\|(\tilde{\mathbf{d}}_{output}) \times_1 \boldsymbol{A}_1{}^{-1} \times_2 \boldsymbol{A}_2{}^{-1} \times \ldots \times_m \boldsymbol{A}_m^{-1}\|^2/2), \tag{3.11}$$

Once the likelihood is computed, we can then proceed towards the formulation of the posterior probability density of the unknowns $\theta_1, \ldots, \theta_H, \psi_1, \ldots, \psi_D$ given the data at hand, subsequent to judicious selection of priors. But one issue has remained unaddressed so far is that we have not discussed the notation "$\times_i$", for $i = 1, \ldots, m$. We remedy this now.

3.7.1 Mode-n multiplication of a tensor and a matrix

In the tensor Normal likelihood, the core tensor $\tilde{\mathbf{D}}_{output}$ is mode-1 multiplied by the square root $\boldsymbol{A}_1$ of the covariance matrix $\boldsymbol{\Sigma}_1$ – the result of which is in turn mode-2 multiplied by $\boldsymbol{A}_2$, and so on. Here, the tensor $\tilde{\mathbf{D}}_{output}$ is $k_1 \times k_2 \times \ldots \times k_m$-dimensional.

To appreciate the result of mode-1 multiplication, we first consider the vector along the first of the dimensions in the m-th ordered tensor $\tilde{\mathbf{D}}_{output}$. In other words, *consider the k_1-dimensional vector that is such that its components are generically identified with the index that is (an m-tuple):*

$$(\star, p_2, p_3, \ldots, p_m), \text{ where } \star \in \{1, \ldots, k_1\}.$$

$\forall p_i \in \{1, \ldots, k_i\}$, for $i = 2, 3, \ldots, m$. We refer to this k_1-dimensional vector as $\boldsymbol{D}_1$, and for brevity's sake, denote $\boldsymbol{D}_1 = (d_1, \ldots, d_{k_1})^T$.

Now,

- the $\times_1$ or mode-1 multiplication of the tensor $\tilde{\mathbf{D}}_{output}$ is possible with a $\star_1 \times k_1$-dimensional matrix that we call $\mathbf{V}_1$,

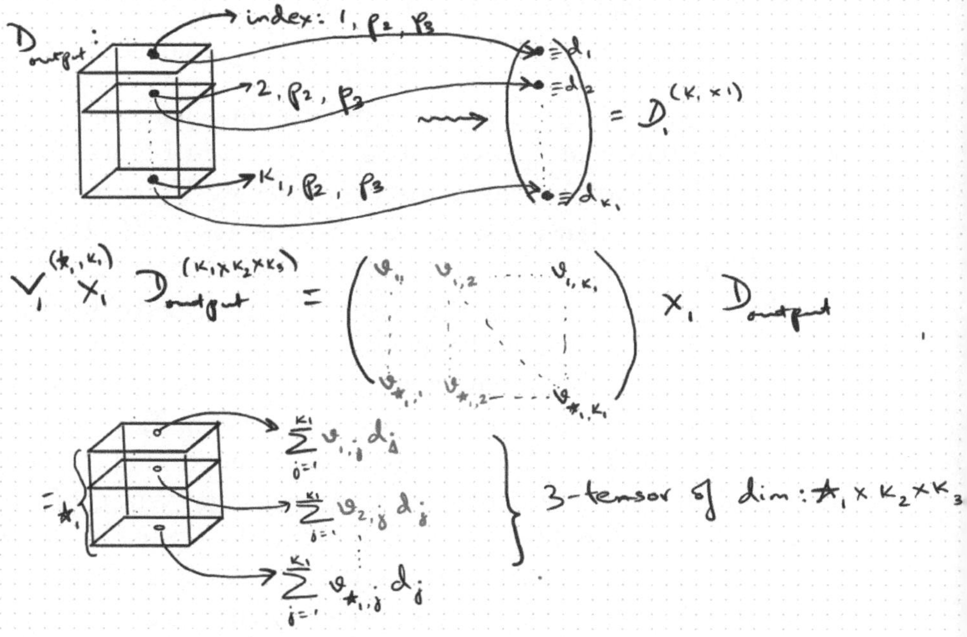

FIGURE 3.1
Cartoon showing the schematic implementation of "$\times_1$" multiplication of a matrix $\mathbf{V}_1^{(\star_1 \times k_1)}$ into the example 3-tensor $\mathbf{D}_{output}^{(k_1 \times k_2 \times k_3)}$.

- s.t. the result of this mode-1 multiplication is an m-th ordered tensor of size $\star_1 \times k_2 \times \ldots \times k_m$, where $\star_1 \in \mathbb{N}$.
- Here $\mathbf{V}_1 = [v_{u,j}]$ where $u \in \{1, \ldots, \star_1\}$ and $j \in \{1, \ldots, k_1\}$.
- In fact, the result of this multiplication is that the vector $\boldsymbol{D}_1$ is replaced within the tensor $\tilde{\mathbf{D}}_{output}$ by the vector $(\sum_{j=1}^{k_1} v_{1,j} d_j, \sum_{j=1}^{k_1} v_{2,j} d_j, \ldots, \sum_{j=1}^{k_1} v_{\star_1,j} d_j)^T$.

The Figure 3.1 depicts a schematic representation of $\times_1$ or mode-1 multiplication of a matrix with a toy 3rd order tensor $\mathbf{D}_{output}$.

To continue with the example of this 3rd-order toy, we understand that mode-2 multiplying a 3-tensor $\mathbf{D} \in \mathbb{R}^{p \times q \times r}$ with a matrix $\mathbf{V} \in \mathbb{R}^{s \times q}$. yields the tensor $\mathcal{C} \in \mathbb{R}^{p \times s \times r}$ as: $\mathcal{C} = \mathbf{D} \times_2 \mathbf{V}$, where the i, j, k-th element of $\mathcal{C}$ is $c_{i,j,k} = \sum_t d_{i,t,k} v_{j,t}$, where $i = 1, \ldots, p$, $j = 1, \ldots, s$, $k = 1, \ldots, r$. This mode multiplications for a toy 3rd-ordered tensor are presented as a cartoon in Figure 3.2.

In our application that is discussed below, $\mathbf{V}_1$ is the $k_1 \times k_1$-dimensional covariance matrix $\mathbf{\Sigma}_1$. So the result of the mode-1 multiplication is still a tensor of size $k_1 \times k_2 \times \ldots \times k_m$.

The $\star_1 \times k_2 \times \ldots \times k_m$-dimensional tensor, (say $\mathbf{D}_1$) that results from the mode-1 multiplication of $\tilde{\mathbf{D}}_{output}$ and $\mathbf{V}_1$, is then mode-2 multiplied by a matrix $\mathbf{V}_2$. (Indeed, in our work, this matrix is the $k_2 \times k_2$-dimensional covariance matrix $\mathbf{\Sigma}_2$). Again, this mode-2 multiplication implies matrix multiplication of the ($\star_2 \times k_2$-dimensional) matrix $\mathbf{V}_2$ with the k_2-dimensional vector ($\boldsymbol{D}_2$) that is generated within $\mathbf{D}_1$, at given indices along all dimensionalities except the 2nd. Here $\star_2$ is any natural number. (In our work, $\star_2 = k_2$). So the result of this mode-2 multiplication is the replacement of the $\boldsymbol{D}_2$ vector inside tensor $\mathbf{D}_1$, with the vector that results from multiplying matrix $\mathbf{V}_2$ with $\boldsymbol{D}_2$. The mode-i multiplications are continued $\forall i = 1, \ldots, m$, in order to define the tensor Normal density, [52].

3.7.2 Inference on unknowns, given the data

Priors

Once the likelihood of the parameters $\theta_1, \ldots, \theta_H, \psi_1, \ldots, \psi_D$ is formulated as per Equation 3.11, given the data $\tilde{\mathbf{D}}_{output}$ and the design inputs $y_1, \ldots, y_{N_{data}}$, we consider prior probability densities on these parameters that we seek to learn. For a given application, it might be that strong priors are available on all, or some of the parameter. Information that is available on values that a parameter could probably attain, could translate to strong priors on the parameter in general, but if such information is not available in an application, weak priors could be used.

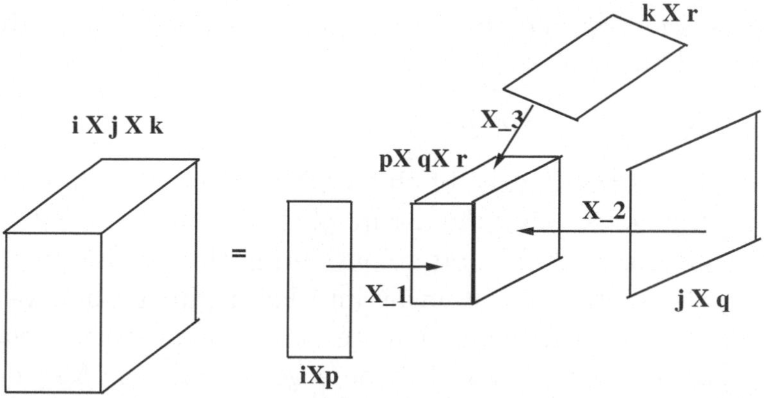

FIGURE 3.2
Cartoon depicting the scheme of $\times_1$, $\times_2$ and $\times_3$ multiplication of a toy 3rd-order tensor with three matrices of compatible dimensions.

Uninformative Gaussian priors

In the illustrative applications that are discussed in the following sections, weak priors are typically used since these considered applications are not blessed by information on the sought parameters. These densities are typically Gaussian, with an arbitrarily-chosen, or loosely-motivated central tendency value, commitment towards which is loose, as manifest in the high variance of such a Normal prior. In real-world applications, the attempt is towards the thorough usage of all available information regarding values that a parameter could attain.

Posterior computation and Bayesian inference

Following the choice of the prior probability density on each unknown, such priors are input into Bayes rule, along with the likelihood of these model parameters that we wish to learn, in order to formulate the joint posterior probability density $\pi(\theta_1, \ldots, \theta_H, \psi_1, \ldots, \psi_D|\tilde{\mathbf{D}}_{output}, y_1, \ldots y_{N_{data}})$, of all the unknowns, given the data at hand. As stated before, we will undertake Bayesian inference on each parameter given the data, by generating posterior samples from this joint posterior, using MCMC techniques [107].

Bayesian inference using MCMC

We will choose the flavour of the MCMC algorithm that we will employ, after considering the nature of the parameters that we are aiming to learn, as well as after considering the possibility of partitioning the vector of unknowns into lower-dimensional vectors, the components of which might share a commonality in correlation. If such partitioning is judged to be a good idea, the Metropolis-within-Gibbs algorithm might be considered. Such inference might still be considered relevant to the work, with the aim of speeding up the attainment of convergence of the undertaken MCMC chain. Using MCMC, the marginal posterior probability density of each parameter will be computed, given the data at hand, and subsequetly, the 95% Highest Probability Density credible region (95% HPDs) is learnt for each parameter, given this available data.

3.8 Learning the relationship between two variables – an illustration

In this section, we discuss an application of the technique of learning the function that represents the relationship between two random variables $\boldsymbol{X}$ and $\boldsymbol{Y}$, using training dataset that comprises the set of pairs of values of $\boldsymbol{X}$, and $\boldsymbol{Y}$. In our illustrations discussed within this section, $\boldsymbol{X}$ is the meteorologically-relevant vector of parameters recorded over a given period of time in a city, and $\boldsymbol{Y}$ is the total energy that is consumed over this temporal interval, over the three zones that locations of this city's energy-consumers are partitioned into. Thus, the variable $\boldsymbol{Y}$ is in fact a scalar in our illustrations, and we will resort to the notation Y for this variable hereon. The data on such an $\boldsymbol{X}$ and Y are available for Tetouan city, which is in Morocco. This dataset is publicly-available, at `https://archive.ics.uci.edu/ml/datasets/Power+consumption+of+Tetouan+city`. The available data, [113], consists of observations of the following metereological parameters: temperature, (denoted X_1); humidity, (denoted X_2); wind speed, (denoted X_3); general diffuse flows, (denoted X_4); and diffuse flows, (denoted X_5). These parameters were recorded every 10 minutes, for 365 days in 2017, starting on the 1st of Jauary of 2017. We call $\boldsymbol{X} = (X_1, X_2, X_3, X_4, X_5)^T$. Each value of $\boldsymbol{X}$ is noted in this dataset along with the date and time stamp of when this value was realised. The data also includes the power consumption at each of these time points, in each of the three distribution zones that the power distribution in this city is partitioned into.

Some details follow.

- We use this data to sum the power consumption in the three zones of the city, into one variable Y, which we refer to as the (integrated) power consumption in the city.
- We also randomly sample N rows from this dataset, while ensuring that there is no duplication of the date and the time of the realisations of $\boldsymbol{X}$ and Y, within any of the N selected rows. Here N=253.
- The sampled data points $\{(\boldsymbol{x}_i, y_i)\}_{i=1}^{N}$ comprise the dataset $\mathbf{D}$, which is then split into a training set and a test dataset, where the size of the training data is N_{data} =203 and the remaining test dataset is then a sample with N_{test} =50 data points.

The scatter plots of Y against X_i, for $i = 1, \ldots, 5$ are shown in the four different panels of Figure 3.3.

Broadly, we undertake 3 exercises:

1. In the first exercise (or the "1st-exercise"), we learn the relationship between the integrated energy consumption Y, and temperature X_1, using the training dataset $\mathbf{D}_1 := \{(x_{1,i}, y_i)\}_{i=1}^{N_{data}}$, which is a subset of $\mathbf{D}$. We express the relationship between X_1 and Y as $Y = f(X_1)$, where we model $f(\cdot)$ as a realisation from a $\mathcal{GP}$, that induces a multivariate Normal likelihood on the parameters of the model, which we learn using the methodology delineated above, given data $\mathbf{D}_1$. These model parameters

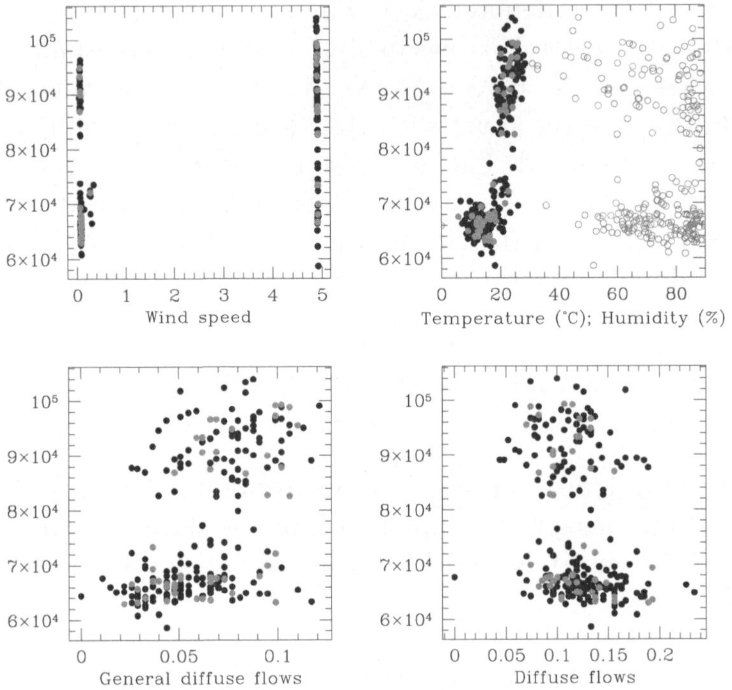

FIGURE 3.3

Top right: scatterplot of output Y (integrated power consumption in Tetouan city), against X_1, or atmospheric temperature (in black filled circles), and against X_2, which represents atmospheric humidity (in open grey circles that will appear red in the coloured version) The points that are part of the test data are depicted as filled circles in light grey, (or green in the coloured version). *Bottom right:* plot of Y against X_3 which represents wind speed; again, the points that are in the test data set, are shown in grey (or green) filled circles. *Bottom left & top left:* plots of Y against X_4 (diffused flows) and X_5 (general flows). The training-test split of the data is shown in black and grey (or green).

will include – as motivated during the discussion of the learning strategy in Section 3.7 – the hyperparameters of the kernel that we invoke, to parametrise the covariance matrix, any element of which informs on the covariance between the output variable that is realised at one design point, and at another design point, out of the N_{data} design points that lie within $\mathbf{D}_1$. We will thereafter predict the mean and variance values of Y at the N_{test} number of test inputs $x_{1,1}^{(test)}, x_{1,2}^{(test)}, \ldots, x_{1,N_{test}}^{(test)}$, and compare this prediction against the values of the output variable Y that are known to be realised, within the dataset $\mathbf{D}$. Multiple kernels – including parametric non-stationary kernels – will be consulted in this exercise. In the next chapter, we will consider a new non-parametric non-stationary covariance kernel, to undertake the learning and prediction of the 1st-exercise.

2. In the "2nd-exercise", we will learn the nature of the association between the inegrated power consumption (Y) and all the metereological parameters: $X_1, \ldots, X_5$. Thus, we suggest the model $Y = g(\boldsymbol{X})$, where this multivariate function $g(\cdot)$ is modelled as a sample function of a $\mathcal{GP}$, leading to a multivariate Normal likelihood of five length scale hyperparameters that are relevant to a (stationary) kernel, with which we will parametrise the covariance matrix of this multivariate Normal density. Again, prediction of values of Y realised at a given test input $\boldsymbol{x}_i^{(test)}$ will be undertaken, ($i = 1, \ldots, N_{test}$) and these predictions compared against the value that Y is known to attain at this input, within the dataset $\mathbf{D}$. The SQE kernel will be used for this exercise.

3. In the "3rd-exercise", we will model the relationship between $\boldsymbol{X}$ and Y as $\boldsymbol{X} = \boldsymbol{f}(Y)$, where $\boldsymbol{f}(\cdot)$ is a vector-valued function of the same dimensionality as $\boldsymbol{X}$. This function will be modelled as a realisation from a vector-variate $\mathcal{GP}$, leading to a matrix Normal likelihood. There are N_{data} realisations of the output $\boldsymbol{X}$ in the training data $\mathbf{D}_2 := \{(y_i, \boldsymbol{x}_i)\}_{i=1}^{N_{data}}$. Of this matrix Normal den-

sity, one of the covariance matrices – an element of which informs of the covariance between pairs of realisations of $\boldsymbol{X}$ (at two different values of input Y) – is kernel-parametrised. There is only one length scale hyperparameter that is to be learnt of the kernel that parametrises this inter-realistion covariance matrix. The other covariance matrix – the one, the ij-th element of which informs on the covariance between pairs of components X_i, X_j $(j, i = 1, \ldots, 5)$ of $\boldsymbol{X}$ – will need to be empirically-estimated. The empirical estimate will involve computing the correlation between X_i and X_j, $(i, j = 1, \ldots, 5)$, each of which has N_{data}=203 number of observations included within the training set $\mathbf{D}_2$. The values of Y at which a given $\boldsymbol{X} = \boldsymbol{x}^{(test)}$ is realised, is learnt within 95% HPDs.

In all exercises, data on the output is standardised s.t. the covariance matrices reduce to the respective correlation matrices, and a zero-mean $\mathcal{GP}$ is used.

3.8.1 1st-Exercise

As indicated above, in the 1st-Exercise, the aim is to learn the relation between the (integrated) energy consumption Y and the temperature X_1, using training data $\mathbf{D} = \{(x_{1,i}, y_i)\}_{i=1}^{N_{data}}$. We express this relation as $Y = f(X_1)$, and model the sought function $f(\cdot)$ as a sample function of a $\mathcal{GP}$, i.e. model the sought $f(\cdot)$ as a random function $f(\cdot) \sim \mathcal{GP}(\mu(\cdot), Cov(\cdot, \cdot))$, s.t. the likelihood of the mean vector $\boldsymbol{\mu}$ and covariance matrix $\boldsymbol{\Sigma}$ that are induced by this $\mathcal{GP}$ is

$$\mathcal{L}(\boldsymbol{\mu}, \boldsymbol{\Sigma} | \mathbf{d}_{output}, x_{1,1}, \ldots, x_{1,N_{data}}) = \\ (2\pi)^{-n/2} |\boldsymbol{\Sigma}|^{-1/2} \exp\left(-\left((\boldsymbol{x}_1 - \boldsymbol{\mu})^T \boldsymbol{\Sigma}^{-1} (\boldsymbol{x}_1 - \boldsymbol{\mu})\right)\right),$$

where

- the standardised observations on Y comprise the set $\mathbf{d}_{output}$;
- $\boldsymbol{\mu} = \underset{\sim}{0}$ as we model the function-valued random variable $f(\cdot)$ as a realisation from a zero-mean $\mathcal{GP}$;

– the covariance matrix is kernel parametrised, i.e. $\boldsymbol{\Sigma} = [Cov(f(x_{1,i}), f(x_{1,j})] = [K(x_{1,i}, x_{1,j})]$, with the kernel $K(\cdot,\cdot)$ given by an SQE form in this application, s.t. $K(x_{1,i}, x_{1,j}) = \exp(-(x_{1,i} - x_{1,j})^2/\ell^2)$.

Here $i, j \in \{1, \ldots, N_{data}\}$. Then, the likelihood of the model parameters reduces to being the likelihood of the kernel length scale hyperparameter $\ell \in \mathbb{R}$, i.e.

$$\mathcal{L}(\ell|\mathbf{d}_{output}, x_{1,1}, \ldots, x_{1,N_{data}}) =$$

$$(2\pi)^{-n/2}|\boldsymbol{\Sigma}|^{-1/2} \exp\left(((\boldsymbol{x}_1 - \boldsymbol{\mu})^T \boldsymbol{\Sigma}^{-1}(\boldsymbol{x}_1 - \boldsymbol{\mu}))\right).$$

Then using this likelihood and Gaussian prior on ℓ, we formulate the posterior *pdf* of ℓ given data $\mathbf{d}_{output}$ on Y and the design inputs $x_{1,1}, \ldots, x_{1,N_{data}}$, at each of which, observations of Y are recorded.

The mean and variance of the Normal prior on ℓ are chosen by hand. As we do not have information on the likely values of the length scale hyperparameter of the covariance kernel, we set the prior mean to 1, under the impression that correlation fades between energy consumption at 2 time points, when temperatures are separated by 1 unit; however, such a prior mean is not well informed upon, and is an *ad hoc* choice. This implies that the prior variance needs to be chosen as high; in this application, we choose the prior variance as ≥ 10.

In the undertaking of inference on ℓ, we sample from the posterior *pdf* of ℓ given the data, using MCMC. Thus, in any iteration, a value of ℓ is proposed (from a Normal with an experimentally-chosen variance, and mean given by the current value of ℓ), and the posterior *pdf* of ℓ is computed at this proposed value. The posterior values computed at the current and proposed values of ℓ are compared in the computation of the acceptance ratio in this Random Walk Metropolis Hastings implementation of MCMC.

At the proposed value of ℓ, the covariance matrix $\boldsymbol{\Sigma}$ is computed, and its inverse evaluated, in order to compute the likelihood – which in turn allows for the posterior *pdf* of ℓ given the data. Thus, the computation of the likelihood demands inversion of a correlation matrix. If the coding for the application is undertaken in a language/software that automates matrix inversion, then this

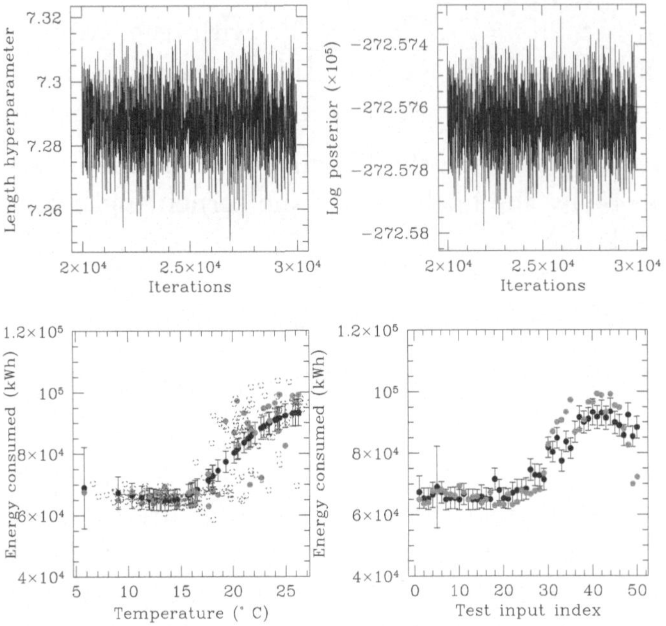

FIGURE 3.4
Top left: trace of the length scale hyperparameter of the kernel that parametrises the covariance function of the multivariate Normal likelihood. *Top right:* trace of log posterior *pdf* of ℓ given the training data. *Bottom right:* predicted mean of the output Y (i.e. integrated energy consumption in Tetouan city) is plotted in filled black circles, as plotted against the index of the test input at which this prediction is made. ± 1-standard deviation – predicted by the model – is superimposed as error bar. The true value of the output at the given input index is shown in grey circles (or in green, in the coloured version). *Bottom left:* mean and ± 1 standard deviation of Y as predicted at a given test input, (i.e. a test temperature), is plotted against the test temperature, with the true value of Y superimposed in grey (or green). The training data points are also plotted in black open circles.

aspect of the likelihood computation is not another computational challenge. If the need arises for the coding of matrix inversion, that too is not difficult; a code for inverting an example covariance matrix is provided at the ed of this chapter. *Indeed, all codes used in this book invoke matrix inversions – amongst other matrix operations – and are available on request.*

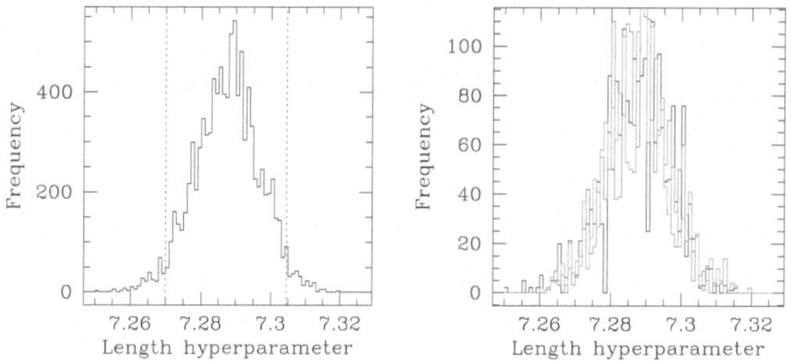

FIGURE 3.5
Left: histogram of the length scale hyperparameter (ℓ) constructed using the post-burnin samples of ℓ obtained using MCMC. The edges of the 95% Highest Probability Density credible region are marked on this histogram. *Right:* Histograms of ℓ values sampled from its posterior, during last, second last, and third last non-overlapping blocks of 5000 iterations, in the undertaken MCMC chain, plotted respectively in red, green and black, (or in two shades of grey and black, in the monochromatic version).

In an iteration, once ℓ and the covariance matrix $\boldsymbol{\Sigma}$ are updated, we invoke Definition 3.2 to compute the closed-form prediction of the mean and variance of the output at each of the test inputs: $x_{1,1}^{(test)}, x_{1,2}^{(test)}, \ldots, x_{1,N_{test}}^{(test)}$. At the same time, it is possible to learn ℓ in the MCMC chain, and at a summary of the marginal posterior of ℓ given the data – as learnt using MCMC – we compute the mean and variance of the output value predicted (in a closed-form way, as indicated in Definition 3.2), at each of the $N_{test} = 50$ test inputs. We will construct plots of trace of ℓ - to check on convergence of the MCMC chain. Another way to display parameter values that have been inferred upon, is by converting the trace to a histogram. Such a histogram is an approximation of the marginal posterior *pdf* of this parameter, and it can be employed to learn the 95% Highest Probability Density credible region (HPD) on the parameter ℓ. For this chain, this 95% HPD o ℓ is learnt as [7.2700, 7.3045]. While parameter histograms will be presented are reported in Figure 3.5, traces of the ℓ parameter and of logarithm of the posterior *pdf*, are displayed in the top panels of Figure 3.4, with the predictions plotted in the lower panels of the figure. The predicted variance is used to compute the standard deviation, and the ± 1 standard deviation is depicted as the error bar about the predicted mean value of the output at a given test input. The true values of the output variable at each of these test inputs are also known, and these are superimposed. Figure 3.5 presents histograms of ℓ computed using samples generated during 3 independent (non-overlapping) parts of the MCMC chain, each comprising the same number of iterations. Checking for concurrence of such histograms allows for a means of testing convergence of the chain; indeed, had the chain not attained convergence, the centroids of these histograms would not have concurred.

3.8.1.1 Algorithm & code

This is the first example discussed in the book, to illustrate the method of learning a function – that depicts the relation between an output and an input - by modelling it as a random realisation from a $\mathcal{GP}$, in order to make predictions on the output variable, at test inputs, with inference undertaken using MCMC. So we will take this opportunity to present the sketch of the algorithm used in this learning + prediction exercise, and thereafter, discuss an annotated version of the code.

Here is the enumeration of the steps that comprise the algorithm that is coded to undertake the learning of ℓ given the training data $\mathbf{D}_1$, and thereafter predict values of the output at the N_{test} number of test inputs.

1. Standardise every element of the dataset $\mathbf{d}_{output}$ with the sample mean and sample standard deviation. Let the values of the output Y – that were in the original training set – upon standardisation, constitute the dataset $\tilde{\mathbf{d}}_{output} = \{\tilde{y}_1, \ldots, \tilde{y}_{N_{data}}\}$, where the (standardised) realisation $\tilde{y}_i$ of Y is attained at the i-th design point $x_{1,i}$.

2. Choose a seed value of ℓ and populate the covariance matrix $\mathbf{\Sigma} = [K(x_{1,i}, x_{1,j}; \ell)]$ at this seed value of ℓ. In this 1st exercise, $K(x_{1,i}, x_{1,j}; \ell) = \exp(-(x_{1,i} - x_{1,j})^2/\ell^2)$. Set the likelihood and prior to a large negative number; in particular, logarithm of the likelihood and of the prior probability density on ℓ is set as -2 $\times$ 10^8. Now we are ready to start the Metropolis Hastings chain.

3. In the k-th iteration of the chain, in which the currently accepted value of ℓ, (as accepted during the $k-1$-th iteration), is $\ell^{(k-1)}$, we propose a value of ℓ as:

$$\ell^{(k,\star)} \sim \mathcal{N}(\ell^{(k-1)}; \sigma^2),$$

   where the variance σ^2 of this Normal proposal density is experimentally chosen. Thus, this implementation of Metropolis Hastings is in fact, Random Walk.

4. Populate the covariance matrix given the data $\mathbf{d}_{output}$ on Y – or rather, the correlation matrix $\mathbf{\Sigma}^{(k,\star)}$ given the standardised data $\tilde{\mathbf{d}}_{output}$, using the proposed ℓ, i.e. using $\ell^{(k,\star)}$. Compute the inverse of $\mathbf{\Sigma}^{(k,\star)}$, in anticipation of its usage in the computation of the likelihood at the proposed ℓ.

5. At this proposed ℓ, likelihood is computed:

$$\mathcal{L}(\ell^{(k,\star)}|\mathbf{d}_{output}, x_{1,1}, \ldots, x_{1,N_{data}}) =$$

$$\frac{1}{\sqrt{|2\pi\mathbf{\Sigma}^{(k,\star)}|}} \exp\left(-\frac{(\boldsymbol{x}-\boldsymbol{\mu})^T(\mathbf{\Sigma}^{(k,\star)})^{-1}(\boldsymbol{x}-\boldsymbol{\mu})}{2}\right).$$

6. A Normal prior is used on ℓ. Thus, we suggest that the prior on $\ell^{(k,\star)}$ is:

$$\pi_0(\ell^{(k,\star)}) = \mathcal{N}(1, 10).$$

7. Then from Bayes rule, the posterior *pdf* of $\ell = \ell^{(k,\star)}$, given the training set comprising standardised output values, (at each design values), is:

$$\pi(\ell^{(k,\star)}|\mathbf{d}_{output}, x_{1,1}, \ldots, x_{1,N_{data}}) = \mathcal{MN}(\underline{0}, \mathbf{\Sigma}^{(k,\star)}) \times \mathcal{N}(1, 10).$$

8. Ratio of the posterior *pdf*s computed at $\ell = \ell^{(k,\star)}$ and $\ell = \ell^{(k-1)}$, is used in the acceptance ratio. Indeed, if this ratio is $\geq u$, where $U \sim Uniform[0, 1]$, we accept this proposed ℓ, i.e. we set $\ell^{(k)} = \ell^{(k,\star)}$; else we reject the proposed ℓ and set the current value of the parameter to its value that has been the current value till this point in the algorithm, i.e. the value $\ell^{(k-1)}$. The likelihood and prior values are also updated to the same computed at the proposed ℓ, if the acceptance ratio $\geq u$; else the current likelihood and prior (computed at $\ell^{(k-1)}$), are retained.
9. We maintain a record of all current samples of ℓ; of the current likelihood; and of the prior. We employ this set to draw the trace of ℓ as the chain progresses – which we monitor for trendlessness – along with trace of the log posterior. The accepted values of ℓ are also employed in constructing the histogram of ℓ, using which, the 95% HPD on the inferred values is learnt.

The code that is used to implement this step-wise course of actions, is included over the following 3 pages, in Figure 3.6. The comments made within the code show up in this figure, highlighted in yellow, or light grey in the monochromatic version.

3.8.2 2nd-exercise – learning the relation between multiple metereological variables & energy consumption

In the 1st-exercise, we learnt the function that connects the input temperature to the output Y that is the energy consumed across Tetouan city, by modelling said function with a $\mathcal{GP}$. The aim behind this learning is the (closed-form) prediction of the (mean) output at a test input. Then the learning was reduced to the learning of the single kernel hyperparameter that was relevant.

Following this learning, we now embark upon the learning of the relation between the same output Y, and all the metereological parameters $(X_1, \ldots, X_5)$, observations of which are included in the Tetouan city data, (from which our training set is sampled). Thus, the input appears to be a 5-dimensional vector-valued variable, except, X_3 appears to be a linear transformation of X_4, (as corroborated by the lower 2 panels of Figure 3.3 that display the scatterplots of Y against X_3, and Y against X_4). Given this, we learn the relation between the output Y and the vector-valued input $\boldsymbol{X} = (X_1, X_2, X_4, X_5)^T$, $\boldsymbol{X} \in \mathcal{X} \subseteq \mathbb{R}^4$, where the state of relationship between the input and output is expressed as $Y = f(\boldsymbol{X})$, in which $f(\cdot)$ is a random function, which we will treat as a sample function of a $\mathcal{GP}$. The output Y is standardised using the sample mean and standard deviation of the observations on Y, s.t. the training data that we will employ in the learning of $f(\cdot)$ is $\tilde{\mathbf{D}} = \{(\boldsymbol{x}_i, \tilde{y}_i\}_{i=1}^{N_{data}}$, where $\tilde{y}_i$ is the standardised observation of Y, attained at the i-th design point $\boldsymbol{x}_i$. The test inputs are $\boldsymbol{x}_1^{(test)}, \ldots, \boldsymbol{x}_{N_{test}}^{(test)}$. At each test input, the value of the output will be predicted and the prediction compared to the true value of Y that is known to have been realised at that value of the input. For this exercise, N_{data}=203 and N_{test}=50.

```
---------------DECLARATIONS MADE-------------
--------FUNCTIONS THAT PERFORM CHOLESKY DECOMPOSITION INVOKED------------

main()
{
 FILE CONTAINING TRAINING DATA, OPENED TO BE READ;
 FILE CONTAINING TEST INPUTS AND TRUE OUTPUT VALUES AT SUCH TEST
   INPUTS, OPENED TO BE READ;
 FILES THAT VARIOUS OUTPUTS FROM THE CODE ARE TO BE WRITTEN INTO (FILES
   REFERRED TO AS fout and fout2) OPENED TO BE WRITTEN INTO.

 sum=0;sum_sq=0;
 for(n=1; n<=N_data; n++)
  {
   fscanf(fin, "%*d %lf %*f %*f %*f %*f %lf\n",        TRAINING DATA ON INPUT (temp) AND OUTPUT OF
          &temp,&consum);                              ENERGY CONSUMPTION (consum), READ IN.
   arr_r[n] = temp;
   arr_med[n] = consum;
   sum += arr_med[n];
   sum_sq += arr_med[n]*arr_med[n];
  }
 fclose(fin);
 mean = sum/(1.*N_data);
 sd = pow(sum_sq/(1.*N_data -1) - mean*mean, 0.5);

                                                SAMPLE MEAN & SD COMPUTED
 for(n=1; n<=N_data; n++)                       FOR DATA ON OUTPUT; OUTPUT
  arr_med[n] =  (arr_med[n] -mean)/sd;          STANDRDAISED.

 for(n=1; n<=N_test; n++)
  {
   fscanf(fin2, "%*d %lf %*f %*f %*f %*f %lf\n", &r, &true_consum);
   arr_r_test[n] = r;
   arr_true_consum[n] = true_consum;             TEST INPUTS READ IN; TRUE VALUES OF OUTPUT READ
  }                                              IN, AT EACH TEST INPUT.
 fclose(fin2);

 /*Starting MCMC*/
 ell=ell_0;
 likeli=-2e18;                    INITIALISE THE UNKNOWN ell, AND LIKELIHOOD AND PRIOR.
 prior = -2e18;
 for(q=1; q <= N_iter; q++)
  {
   prop_ell = ell + sigma_ell*gasdev(&idum);
   sigma_func_first(prop_ell);
   prop_likeli=0;                 PROPOSE A VALUE prop_ell OF ell. COMPUTE
   for (i = 1; i <= N_data; i++)  FUNCTION sigma_func_first AT ell=prop_ell, ie.
       {                          POPULATE CORRELATION MATRIX USING KERNEL
        likeli_i= 0;              COMPUTED AT ell=prop_ell; CHOLESKY
                                  DECOMPOSE THIS MATRIX.
        prop_likeli_i=0;
        for (j = 1; j <= N_data; j++)    LIKELIHOOD COMPUTED AT ell=prop_ell.
         {
          prop_likeli_i += -0.5*arr_med[j]*arr_sigma_inv_first[j][i];
         }
```

```
            prop_likeli += (prop_likeli_i) * arr_med[i];
          }
     prop_likeli += -0.5*lg_det;
```

LIKELIHOOD COMPUTATION AT PROPOSED ell IS DONE; lg_det IS LOGARITHM OF DETERMINANT OF CORRELATION MATRIX AT THIS ell..

```
     prop_prior = -pow(prop_ell - prior_mean, 2)/(fac*sigma_prior*sigma_prior);
```

PRIOR COMPUTED AT PROPOSED ell; prior_mean & sigma_prior are chosen, and declared above; fac is 2.

```
     tired = log(ran1(&idum));
     lg_acc_ratio = prop_likeli - likeli + (prop_prior - prior);
     if(lg_acc_ratio > tired)
          {
          printf("here\n");
          prior = prop_prior;
          likeli = prop_likeli;
          ell = prop_ell;
          printf("ditio %d %10.4e %10.4e %f %f\n",
                    q, prior, likeli, ell, tired);
          }
```

ACCEPTANCE RATIO IS COMPUTED – IN THIS RWMH IMPLEMENTATION, IT IS THE RATIO OF POSTERIOR AT PROPOSED ell TO CURRENT ELL. IF PROPOSED ell IS ACCEPTED, THE CURRENT VALUES OF ell, LIKELIHOOD AND PRIOR ARE UPDATED TO prop_ell or VALUES

```
     if(q > burnin && q%1==0)
          {
          sum_l += ell;
          sum_flag += 1;
          fprintf(fout2, "%d %10.4e %0.4e %10.4e\n", q, likeli, ell, prior);
          printf("%d %10.4e %10.4e %10.4e %10.4e %10.4e %10.4e\n",
                    q,likeli,ell,prior,curr_med[2],curr_med[3],curr_med[4]);
          }
   }
```

ONCE PAST BURNIN, RECORD CURRENT VALUES OF LIKELIHOOD, PRIOR & ell, (AMONGST OTHER PARAMETERS).

MCMC CONDUCTED TO LEARN ell, IS DONE. PREDICTION STARTS.

```
mean_ell = sum_l/(1.*sum_flag);
printf("mean_ell = %f   sum_flag=%d\n",mean_ell, sum_flag);
sigma_func_first(mean_ell);
for (i = 1; i <= N_test; i++)
  {
   for (j = 1; j <= N_data; j++)
        {
         K[i][j] = exp(-pow(arr_r[j] - arr_r_test[i],2)/
                           (mean_ell*mean_ell));
        }
  }
```

MEAN (mean_ell) OF ell VALUES FROM THE CONVERGED PART OF THE CHAINIS USED, i.e.. POST-BURNIN SAMPLES ARE USED FOR PREDICTION. CHOLESKY DECOMPOSITION IS UNDERTAKEN OF CORRELATION MATRIX COMPUTED AT mean_ell.

COMPUTATION OF CLOSED-FORM MEAN OF OUTPUT - AT ONE OF N_test TEST INPUTS – IS FIRST COMPUTED. THIS CALLS FOR THE INVERSE OF THE CORRELATION MATRIX COMPUTED AT mean_ell.

```
for (i = 1; i <= N_test; i++)
  {
   for (j = 1; j <= N_data; j++)
        {
         arr_a[i][j] =0;
         for (k = 1; k <= N_data; k++)
           {
            arr_a[i][j] += K[i][k] *
                  arr_sigma_inv_second[k][j];
           }
        }
```

```
  }
  for (i = 1; i <= N_test; i++)
  {
    arr_mean[i] =0;
    for (j = 1; j <= N_data; j++)
          arr_mean[i] += arr_a[i][j] * arr_med[j];
  }
```

PREDICTED MEAN OF OUTPUT IS IN THE ARRAY arr_mean. COMPUTATION OF CLOSED-FORM VARIANCE OF OUTPUT - AT ONE OF N_test TEST INPUTS - IS NOW COMPUTED. THE PREDICTED VARIANCE-COVARIANCE MATRIX IS IN THE ARRAY arr_cov[][].

```
  for (i = 1; i <= N_test; i++)
  {
    for (j = 1; j <= N_test; j++)
          {
           arr_b[i][j] =0;
           for (k = 1; k <= N_data; k++)
             {
              arr_b[i][j] += arr_a[i][k] *
                    exp(-pow(arr_r[k] - arr_r_test[j],2)/
                       (mean_ell*mean_ell));
             }
          }
  }
  for (i = 1; i <= N_test; i++)
  {
    for (j = 1; j <= N_test; j++)
          {
           if(i==j)
             {
              if(arr_b[j][j] >= 1.)
                    arr_cov[i][j] = 1.;
              if(arr_b[j][j] < 1.)
                    arr_cov[i][j] = 1 + 0*arr_sigma_test[i] - arr_b[i][j];
             }
           else
             arr_cov[i][j] = exp(-pow(arr_r_test[i] - arr_r_test[j],2)/
                                   (mean_ell*mean_ell)) - arr_b[i][j];
          }
  }
```

PREDICTED VARIANCE AND MEAN OF OUTPUT AT TEST INPUT arr_r_test[] ARE RECORDED, AFTER UNSTANDARDISING BY THE SAMPLE MEAN AND STANDARD DEVIATION USED INITIALLY.

```
  for (j = 1; j <= N_test; j++)
  {
    pred_med[j] = arr_mean[j] ;
    fprintf(fout, "%d %10.4e %10.4e %10.4e %10.4e %10.4e %10.4e %10.4e\n",
              j,mean_ell,
              arr_r_test[j], pred_med[j]*sd+mean, sd*sqrt(arr_cov[j][j]),
              pred_med[j]*sd+mean - sd*sqrt(arr_cov[j][j]),
              pred_med[j]*sd+mean + sd*sqrt(arr_cov[j][j]),
              arr_true_consum[j]
              );
  }
  fclose(fout2);
  fclose(fout);
  exit(0);
}
```

FIGURE 3.6
Annotated code used to perform learning of ℓ and prediction of Y at different test values of X_1.

Modelling the unknown function $f(\cdot)$ with a zero-mean $\mathcal{GP}$, results in a multivariate Normal likelihood with a covariance matrix $\boldsymbol{\Sigma}$ that is kernel parametrised. Standardisation of the output results in $\boldsymbol{\Sigma}$ being rendered the correlation matrix, an element of which informs on the correlation between the pair of outputs at two given values of the input.

Parametrisation of $\boldsymbol{\Sigma}$ by a covariance kernel is undertaken in this exercise; in other words, any element of $\boldsymbol{\Sigma}$ is modelled to express the correlation between the outputs realised at two given input values, via a decreasing function of a distance between the input values. As discussed above, typically, a parametric form is assigned to this kernel, though in the next chapter, we will encounter the parametrisation of a covariance matrix with a non-parametrically learnt kernel.

For our current purpose, we assign a parametric form to the kernel, and in this exercise, we use the SQE form. As we have discussed above, parameters of a kernel are referred to as its hyperparameters, and the SQE kernel bears only one type of a hyperparameter – namely, length scale hyperparameters – with which the kernel is specified so that a correlation matrix is parametrised. Each of the relevant length scale hyperparameters informs on the interval along the respective direction in input space, that needs to be traversed, in order for correlation between the outputs – realised at the edges of this interval – to fall by an identified factor.

In the application that we address here, the input is a 4-dimensional vector. This implies that there are four length scale hyperparameters: $\ell_1, \ell_2, \ell_4, \ell_5$, that are required, to specify the kernel: $K(\boldsymbol{x}_i, \boldsymbol{x}_j) = \exp(-(\boldsymbol{x}_i - \boldsymbol{x}_j)^T \boldsymbol{Q}(\boldsymbol{\xi} - \boldsymbol{x}_j))$, where $\boldsymbol{Q}$ is a 4×4-dimensional diagonal matrix that is the inverse of the (diagonal) length scale matrix, the i-th diagonal element of which is ℓ_i, $i = 1, 2, 4, 5$. This $\boldsymbol{Q}$ is the diagonal matrix, the diagonals of which are, $1/\ell_1, 1/\ell_2, 1/\ell_4, 1/\ell_5$. We may refer to $1/\ell_i$ as q_i, which we can then interpret as "smoothness" parameters.

So the 2nd exercise boils down to the learning of the four length scale hyperparameters $\ell_1, \ell_2, \ell_4, \ell_5$, supplemented by the prediction of the mean and variance values of the output variable that is realised at each of 50 test inputs.

In Figure 3.7, the results of this learning and prediction are displayed. Histograms of the four length scale hyperparameters are depicted, as are the traces of the prior and the likelihood. The length scale hyperparameter over which correlation fades along the direction of X_5 in the space $\mathcal{X}$, is found to be the smallest, compared to the other four length scales. It was in fact, relatively harder to learn this length scale hyperparameter than the other 3; the data distribution across X_5 is strongly bimodal, (as noticed in Figure 3.3). Indeed each mode of this distribution is peaky, with very low spread across values of X_5 within a mode. Given this, it was anticipated that the learning of ℓ_5 would be difficult, and that the value of the sought ℓ_5 is likely to be low. The distribution of the observations on Y against X_4, also tells us that the length scale ℓ_4 is likely to be low ($\lesssim 0.1$).

The predictions performed on the output, at the 50 values of the 1st component of the test input – namely the component X_1 that represents temperature – are depicted in black, in Figure 3.8, superimposed on the true values of the output, shown in green. While the left panel displays the plot of the output values against the index of the test input – at which the output is predicted/observed – the plot of the output against the test value of X_1 is shown in the right panel. The quality of prediction is not parametrised by a single scalar in this exercise, as this quality varies from one test input to another. However, from the plot of the predictions, we can conclude that the method attempts to correctly predict, by converging at length scales (at least along the X_1 direction in input space), that allow for acknowledgement of the following two peculiarities of the data:

1. observed multiple values of the output at the same, or nearly same values of the input;
2. heterogeneities in the correlation between the outputs generated at pairs of given values of X_1.

However, it is not easy to understandat why the predictions indicate comparatively higher errors in the prediction –

i.e. higher variance in the output – at some values of the input, as distinguished from other inputs. We conject that the roughness that is ascribed to a sample function (of the underlying $\mathcal{GP}$) – by the inferential procedure that learns the length scales – renders the sample function "less incorrect" for the data over one subset of the test inputs, than another subset, leading to larger error bars on predictions made at the inputs that are relevant to the latter of the two subsets mentioned above. Other length scales should then also be available, at which the sample function fits data over a different subset of input values preferentially "less incorrect"ly, than the data that the currently inferred length scales do. Then allowing for length scales that vary with input location should be the key to improving redictions. In other words, it is anticipated that prediction will improve for this data, if we invoke non-stationary kernels, which by definition have location-dependent hyperparameters, s.t. correlation between outputs realised at a pair of inputs, depends on that input pair. We will discuss the relevance of such kernels in real-world datasets (such as the Tetouan city data), in the next chapter.

3.9 3rd-exercise; learning the input at a test output

In this exercise we recall the suggestion that there are advantages in treating the lower dimensional – of the two variables, the relationship between which is being sought – as the input variable, if we are attempting to learn this relationship by modelling it as generated by a GP, the covariance structure of which is kernel parametrised. Using a lower-dimensional input implies that we need to learn fewer hyperparameters of the kernel. Thus, in the current application, we propose the input to be the scalar-valued variable Y that represents the energy consumed across all three zones, that power distribution in Tetouan city is par-

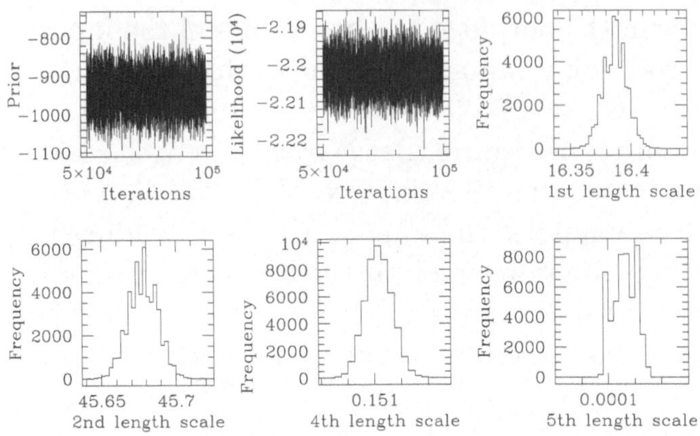

FIGURE 3.7
Top left and middle: traces of the prior and likelihood. *Top right:* histogram of values of ℓ_1, as learnt in the post-burnin iterations of the MCMC chain. *Bottom panels, left to right:* histograms of post-burnin samples of ℓ_2, ℓ_4, ℓ_5.

titioned into. Correspondingly, the output is the vector-valued $\boldsymbol{X} = (X_1, X_2, X_4, X_5)^T$, where $\boldsymbol{X}$ and X_j are discussed above; $j = 1, 2, 4, 5$. So we attempt the learning of the function $\boldsymbol{g}(\cdot)$, where $\boldsymbol{X} = \boldsymbol{g}(Y)$, using the training data set $\mathbf{D}_{inv}$ that consists of pairs of values of Y and the (standardised) value of $\boldsymbol{X}$ that is realised at the chosen value of Y.

By opting to learn $\boldsymbol{g}(\cdot)$ – instead of $f(\cdot)$ - we find that it suffices to learn the single length scale hyperparameter ℓ of the kernel that parametrises the inter-instance correlation matrix of the data on the vector-valued output $\boldsymbol{X}$. (The inter-component correlation matrix cannot be kernel parametrised, since we do not have an input variable at which the distinct components of the $\boldsymbol{X}$ vector are realised). On the other hand, the learning of $f(\cdot)$ had motivated the need to learn four length scale hyperparameters, $\ell_1, \ell_2, \ell_4, \ell_5$. It is always challenging to learn hyperparameters of covariance kernels.

One logistical downside to learning $\boldsymbol{g}(\cdot)$ instead of $f(\cdot)$ – i.e. to treating $\boldsymbol{X}$ as the output instead of the input – is that model complexity is higher in the 3rd-exercise than in the 2nd-exercise.

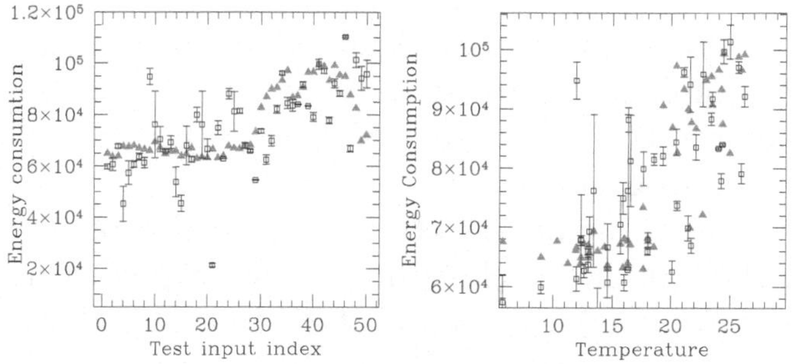

FIGURE 3.8
Figure to present predictions of the output (Y, or energy consumption in kWh, in Tetouan city) at test values of X_1, i.e. temperature, in the exercise in which the relation between Y and $\boldsymbol{X} = (X_1, X_2, X_4, X_5)^T$ is sought. The ± 1 standard deviation that is predicted – along with the mean – of the output at a given test input, is depicted as the error bar, with the predicted mean value shown with an open square. The true value of the output is depicted in green.

To be precise, the likelihood is matrix Normal when the output is a vector, while it is multivariate Normal when the output is a scalar. Thus, the coding (of the likelihood), when learning $\boldsymbol{g}(\cdot)$ is more complex, than when learning $f(\cdot)$, where we learn a function by modelling it as a random realisation from a $\mathcal{GP}$ of respective dimensionality. However, this increased complexity is easily surmounted, and once the coding of a matrix Normal likelihood is completed, this computational problem ceases to be of concern.

What will persist to trouble us however, is the problem of learning the two covariance matrices of the matrix Normal distribution. Indeed, as indicated above, in this exercise, we cannot learn the correlation matrix that informs on the correlation between pair of components of the output $\boldsymbol{X}$, as there is no other variable that is such that the different components of $\boldsymbol{X}$ are realised at different values of this secondary variable. Then we could either learn the distinct (upper/lower triangle) elements of this correlation matrix directly from MCMC, or populate this matrix using the unbiased sample-estimate. The larger is the number of instances of observing $\boldsymbol{X}$, the better we might consider such an empirical or plugin estimate of the inter-component correlation to be. In this exercise, we choose to not undertake the learning of this matrix directly from MCMC, but use its sample-based estimate.

Arguably, the bigger worry in this attempt at inverse prediction, appears to be the lack of any closed-form way for predicting the input Y that is realised at a given value of $\boldsymbol{X}$. Does this shortcoming then annul our ambition of predicting Y at a given value of the metereological parameter vector $\boldsymbol{X}$? That is not the case. Since we are performing MCMC-based inference, we can learn the value $y^{(test)}$ of Y – instead of predicting it – at a given value of the output $\boldsymbol{X} = \boldsymbol{x}^{(test)}$.

We formulate the joint posterior of ℓ and $\boldsymbol{x}^{(test)}$, given the values of Y and $\boldsymbol{X}$ in $\mathbf{D}_{inv}$, and $y^{(test)}$, using Equation 3.4. Then we sample from this joint posterior using MCMC, to compute the marginal posterior density of each unknown, given the data. Each marginal is then employed to learn the 95% HPD on each variable, given the data. In fact, instead of considering just a single test value of $\boldsymbol{X}$ – at which values of Y are learnt – we can learn values of Y that are realised at $\boldsymbol{X} = \boldsymbol{x}_1^{(test)}, \ldots, \boldsymbol{x}_N^{(test)}$. We choose to learn the Y values at each of 19 test values of $\boldsymbol{X}$, along with the length scale hyperparameter ℓ of the inter-instance correlation matrix of the matrix Normal likelihood.

It was our choice to learn the energy consumption at $N = 19$ values of $\boldsymbol{X}$, so that we learn 20 parameters from one MCMC chain. At the same time we could always attempt this learning for other values of N as well. Thus, Figure 3.9 depicts the traces of ℓ and the single $y^{(test)}$ that is learnt, in 2 different MCMC chains, using the data on the $N_{data} = 203$ number of Y, and $\boldsymbol{X}$ values that comprise the training set $\mathbf{D}_{inv}$, in addition to a single known $\boldsymbol{X}$ ($=\boldsymbol{x}^{(test)}$) at which Y attains the value of $y^{(test)}$. Thus, in each of these two MCMC chains, the value of Y that we learn, is realised at a distinct value of $\boldsymbol{X}$. The results displayed in the lower panel of Figure 3.9 pertain to the case of $N = 1$, when ℓ and $y^{(test)}$ are learnt, given $\mathbf{D}_{inv}$ and $\boldsymbol{x}^{(test)} = (20.43, 61.87, 0.099, 0.082)^T$. On the other hand the results shown in the upper panels are from a chain that is similar in all respect, except that in this case, $\boldsymbol{x}^{(test)} = (21.61, 59.59, 0.095, 0.074)^T$. In Figure 3.10, we display the results of learning $\ell, y_1^{(test)}, \ldots, y_{19}^{(test)}$, given the information in data $\mathbf{D}_{inv}$, and 19 known values $(\boldsymbol{x}_1^{(test)}, \ldots, \boldsymbol{x}_{19}^{(test)})$ of $\boldsymbol{X}$ – each of which is distinct from the $\boldsymbol{X}$ values discused in the last two sentences. Traces of the learnt value of Y are plotted in all the panels except for the two right-most panels of the top row; these show the traces of the joint posterior and of ℓ. The index of the test value of $\boldsymbol{X}$, the trace of the corresponding Y at which, is plotted in a panel, is mentioned below the panel. The true value of the energy consumption, (i.e. of Y), at the chosen value of $\boldsymbol{X}$, is overplotted in red. At most of the chosen $\boldsymbol{X}$ values, the learning of the corresonding Y, is noticed to be not bad, though the learnt values compare variously with the truth, at the different $\boldsymbol{X}$ values.

We would like to recall the statement made above about the difficulty in learning kernel hyperparameters; we will in fact, revise this statement to say that it is relatively easier to learn $y_j^{(test)}$ than ℓ – and other kernel hyperparameters. Indeed, from values of Y noted as realised at given values of $\boldsymbol{X}$ in the training set, one is likely to have information, at least about the order of magnitude of $y_j^{(test)}$ that is attained $\boldsymbol{X} = \boldsymbol{x}_j^{(test)}$. On the other hand, in generic applications, information on a length scale hyperparameter is hard to come by. It is through a process of experimentation, that learning of such hyperparameters can be undertaken. Thus, we prefer to learn a single kernel hyperparameter along with (one or multiple) values of Y that are attained at corresponding known values of $\boldsymbol{X}$, over learning four kernel hyperparameters.

Indeed the motivation for this preference becomes all the more clear as one or both of the two variables – the input and the output – become high-dimensional. Thus, for a matrix-valued $\boldsymbol{Y}$ and a vector-valued $\boldsymbol{X}$, (the relation between which is modelled with a GP, the covariance structure of which is kernel parametrised), we will definitely opt for $\boldsymbol{X}$ as the input variable, since kernel parametrisation is easier using a vector-valued input, than a matrix-valued one.

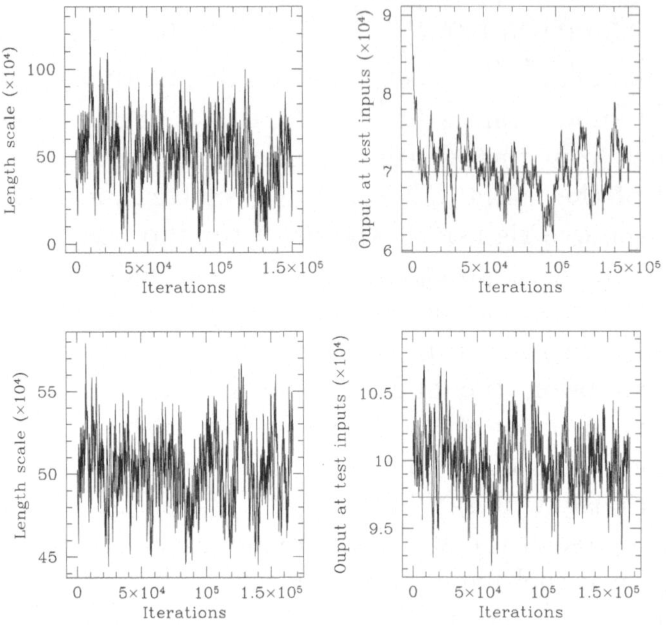

FIGURE 3.9
Top: trace of the length scale hyperparameter ℓ is shown in the left, while the trace of $y^{(test)}$ that is learnt at $\boldsymbol{x}^{(test)} = (20.43, 61.87, 0.099, 0.082)^T$ is in the right. *Bottom:* same as the top panels, except that in this case, results are displayed from the MCMC chain that is employed to perform the learning of ℓ and $y^{(test)}$ at $\boldsymbol{x}^{(test)} = (21.61, 59.59, 0.095, 0.074)^T$. The true value of Y at the respective value of $\boldsymbol{X}$, is overplotted in red, on the trace of the learnt value of Y. The values of Y and ℓ are scaled by a factor of 10^4.

3.10 Using other kernels

In the sections above, we have consistently used the stationary SQE covariance kernel. We change tactic, and now, invoke multiple other kernels to illustrate the learning of the relationship between the energy consumed all over Tetouan city and the atmo-

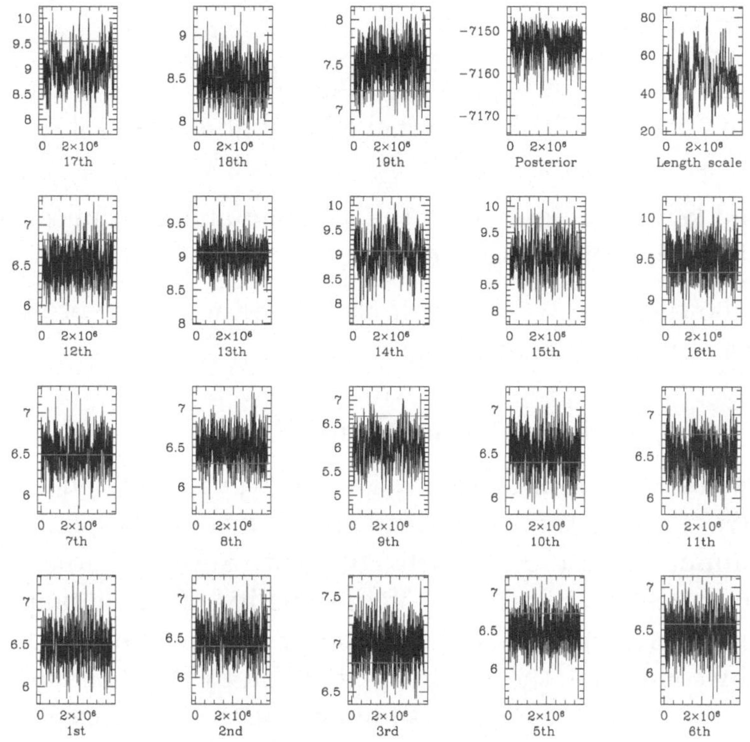

FIGURE 3.10
Figure presenting result of learning ℓ and values of Y at 19 distinct values of $\boldsymbol{X}$. Traces of the learnt Y values are shown in every panel – except the right-most two in the top row – and the index of the test value of $\boldsymbol{X}$, at which the trace of the learnt Y value is displayed, is markd at the bottom of the panel. The true value of Y at the given $\boldsymbol{X}$ value, is depicted as the red broken like that is superimposed on the trace. The right panel in the top row shows the trace of ℓ while the second right-most panel in the top row displays the trace of the joint posterior of ℓ and values of Y learnt at all the chosen test values of $\boldsymbol{X}$, given the training set comprising 203 pairs of values of $\boldsymbol{X}$ and the corresponding Y, along with the 19 chosen test values of $\boldsymbol{X}$.

spheric temperature in the city. Our choice of kernels will include stationary kernels,

— namely, Matern – parametrised variously – inclusive of the Orstein Uhlenbech kernel arrived at a given value 1/2 of the ν parameter of the Matern kernel.

— and kernels produced via the addition and multiplication of multiple stationary kernels.

In the following chapters, we will also have the chance of visiting the effect of non-stationary kernels.

3.10.1 How good are the predictions?

From the above, it is clear that it is always possible to choose the parametric form of the kernel, and the quality of the predictions at test inputs is not paticularly different. Such a statement of course requires an assessment of the quality of prediction. One possibility is of course to compute a scalar parametrisation of such quality by computing the Mean Squared Error (MSE) or the Mean Absolute Error or some other summary of the distance between the prediction and the truth. That is however, not optimal. For example, there could be some test inputs at which the prediction is particularly bad, while the same at all other test points may be good. On the other hand, it is possible that the predictions are consistently off the true values of the output, in such a way that the MSE values are nearly the same in the two cases. Indeed, the question is not merely of parametrisation of the quality of the predictions, but of our judgement of said quality. Here, we would be best placed to compute the MSE, but supplement that with the visualisation of the predicted outputs at each of the test input. (We also identify subsets of the set of test inputs, distinguished by similarity in correlation structure – which we can identify in this illustration – and compute the MSE for the predictions made over each such identified subset).

The question of a scalar parametrisation of the quality of predictions is rendered more involved, when the predictions are accompanied by the mean of the output at a given test point, in addition to the uncertainty on the output, offered as the variance on the value of the variable that is the output at this test input. In our visualisation of the predictions, we include these uncertainties in the form of an error bar that includes ± 1 standard deviation about the predicted mean. If the true value of the output at a given test point lies within this error bar, we suggest that the prediction is "correctly made".

We forward the parameter C for the quality of predictions for a sample of N_{test} test inputs as

$$C = \text{ Number of "correctly made" predictions}/N_{test}.$$

So effectively, our C parameter is the relative frequency of "correctly made" predictions in the given test sample. We consider this an informative parametrisation of the quality of predictions that is not susceptible to misinterpretation, in light of differential quality of predictions across the set of N_{test} number of test inputs.

We will wrap up this chapter with a discussion on Neural Networks which are a popular tool in the learning of inter-variable relationships. The work on the learning and prediction using the Tetouan city data, was undertaken by Kane Warrior.

3.11 Supervised learning using Neural Networks

The basic concept behind Neural Networks is that a complicated, non-linear relation between the input variable vector $\boldsymbol{X}$ and output Y could be mimicked by (non-linear and parametric generalisations of) multiple linear regressions in which each intermediate output variable – that lies in a constructed set or "layer" of such intermediate variables – is linearly regressed against the variables that lie in the previous layer. Indeed, the regression coefficients in this multiple linear regression model are specific to the covariates and the output – or response – variable under consideration, while the intercept terms in this regression model is specific to the output. In fact, it is a chosen function of the output from this multiple linear regression, that then offers the effective value of the variable

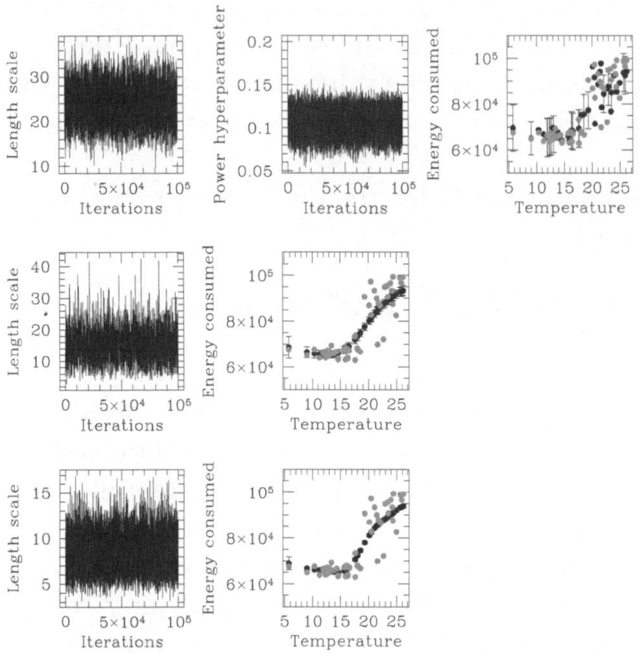

FIGURE 3.11

Top row: figures presenting predictions – at 50 test temperatures – undertaken with a Matern covariance kernel, the length scale hyperparameter ℓ and the power hyperparameter ν of which are learnt, given the training data $\mathbf{D}_1$ that comprises 203 randomly chosen design temperatures from the Tetouan city data, (and the corresponding output that is the energy consumed across all three zones of the city). Traces of the learnt ℓ and ν variables are shown in the left and middle panels while the prediction of the mean output at each test temperature is depicted with filled black circles, with the ± 1 predicted standard deviation on the output overlaid as an error-bar. The true value of the output is superimposed in green. *Middle row:* as in the top row, except this time ν in the Matern kernel is fixed at 3/2. *Bottom row:* same as the panels in the middle row, except this time ν is fixed to 5/2. All of the presented results were plotted using results obtained by Kane Warrior.

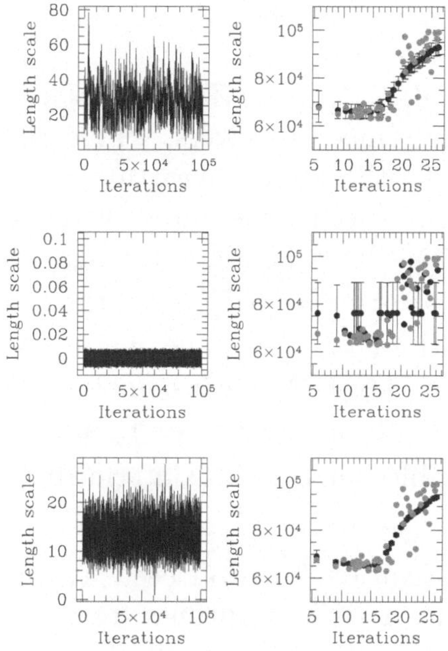

FIGURE 3.12

Top row: as in a row of panels in Figure 3.11 except here results from using a Orsten Uhlenbeck kernel are displayed, with the hyperparameter ℓ learnt. Trace of ℓ is in the left panel, while the predictions at the test temperatures are plotted in the middle panel. *Middle row:* here the panels are similar to those in the top row, except that the periodic kernel is used, in which the hyperparameter ℓ is learnt. *Bottom row:* again, the panels in this row are similar to those in the top row, except that the kernel used here, emerges from the multiplication of the SQE kernel with hyperparameter ℓ, with a Matern kernel that is also assigned a hyperparameter ℓ and ν in which is fixed at 3/2.

that lies in the intermediate layer. This function is referred to as the *activation function*, and its output is the *activated* version of the currently considered variable in the intermediate layer. The activated variables in this intermediate layer, then linearly combine to produce a value that is assigned to a variable in the subsequent layer.

Thus N_1 number of variables, $(Z_1^{(1)}, Z_2^{(1)}, \ldots, Z_{N_1}^{(1)})$ are placed in the first of these constructed intermediate layers, where the "intermediacy" of such a layer is to the input layer that comprises the input variables $X_1, \ldots, X_d$. Here the input variable is $\boldsymbol{X} = (X_1, \ldots, X_d)^T$. Then $Z_i^{(1)}$ is linearly regressed against $X_1, X_2, \ldots, X_d$, $\forall i = 1, \ldots, N_1$. Let $A_i^{(1)} = g(Z_i^{(1)})$ be the activated version of $Z_i^{(1)}$, with $g(\cdot)$ the chosen parametric activation function in the first intermediate layer. A linear combination of $A_1^{(1)}, \ldots, A_{N_1}^{(1)}$ produces the value of the unactivated variable $Z_j^{(2)}$, where $Z_1^{(2)}, \ldots, Z_{N_2}^{(2)}$ are variables in the 2nd intermediate layer. Thus, $j = 1, \ldots, N_2$. Then the activated version of $Z_j^{(2)}$ is $A_j^{(2)} = h(Z_j^{(2)})$, with $h(\cdot)$ the activation function used for the current (2nd) layer. In this way, the activated outputs from one intermediate layer are passed to the next layer via multiple linear regression models, to predict – as the output from the last of the intermediate layers – the value of output variable Y of the model.

To train this neural network, (NN) comprising the set of variables residing in the input and output layers, in addition to the intermediate – or *hidden* - layers, information from the input layer is passed forwards sequentially from one hidden layer to the next, towards the prediction of the value of Y. It is clear that such information passage is undertaken via the linear combination of a set of variables, where each such linear combination is set apart by a specific set of weights and a bias term, with the parametric form of the activation function specified by choice. The predicted value of Y is compared to the known value – corresponding to design inputs – to achieve the training of the NN.

Then the unknowns that require computation are the weights and biases used in this information passage. To estimate these, the error function is minimised within an optimisation technique, (such as undertaken with the gradient descent method) in the "backpropagation" phase of an iteration. Values of weights and biases computed during this phase are fed in, for the forward passage of information to be undertaken again, in the next iteration. Updating of each weight or bias term depends on the *learning rate* which fixes the amplitude of an update. The forward phase is again followed by backpropagation, during which the weights and biases are then updated, over their previously estimated values. Here, the loss function is given by the difference between the predicted and true values of the output variable Y, that is known (for each realisation of $\boldsymbol{X}$), while we train the NN; such a difference could be expressed as proportional to the squared difference between the predicted and true values of Y. The training stops once no further improvement occurs to this loss function, or the set number of iterations is attained. Then using the estimated weights and biases, the value of Y at a test input is predicted. Each forward and backward phase comprise an epoch, (or iteration – as we have been referring to it).

Then the NN architecture is specified by a number of parameters including the number of hidden layers; the number of variables – or neurons – in each hidden layer; the activation functions used in the different layers; the number of epochs; the learning rate. In Figure 3.13, we display results of learning the relation between the energy consumed across all zones of Tetouan city (Y) and the atmospheric temperature (X_1); predictions at the 50 test temperatures are depicted in black, while the true value of Y at the corresponding temperatures are overplotted in green (or grey in the monochromatic version). The figure shows results of the learning undertaken with different learning rates and hidden layers.

Overall, the quality of predictions is worse with NNs, than when the input-output relation is performed by invoking GPs. Indeed, one of the problems with predicting using NNs is that there is a sensitivity on the architecture used in the prediction, and prediction results corresponding to a lower loss function, are not always better. The work on the NN-based learning reported here, was undertaken by Kane Warrior.

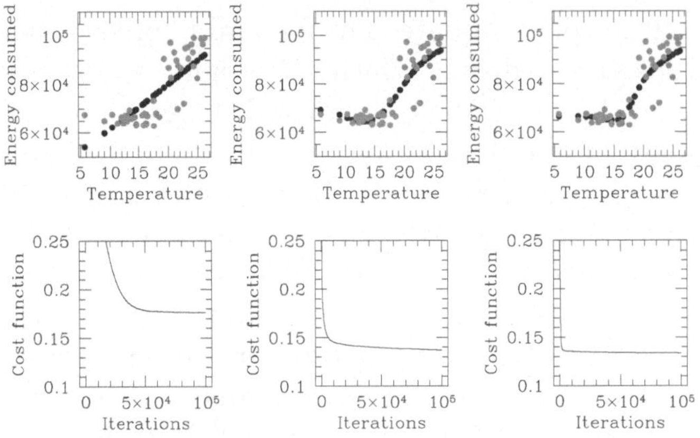

FIGURE 3.13
Top panels predictions on energy consumption values, at the 50 test temperatures from the Tetouan city data, as obtained with fully-connected neural networks. All results have been undertaken using the ReLU activation function, using 50 neurons per hidden layer. *Top left:* in this panel, the results are from using an NN with 2 hidden layers and a learning rate of 0.01. In the *top middle* panel, displayed results on prediction are obtained using an NN with 2 hidden layers again, but a learning rate of 0.3, while the *top right* panel displays results from a NN with 6 hidden layers and a learning rate of 0.1. In all panels, the true value of the energy consumption at a given temperature, is overplotted in grey (or green in the coloured version). *Bottom panels:* variation of the loss function with iteration for the NN, predictions from which are plotted in the corresponding panel in the upper row. Data for these plots was produced and kindly shared by Kane Warrior.

3.12 Looking forward

From the earlier discussion about the different kernels that have been invoked to undertake the learning of the functional relation between the output Y (energy consumption) and X_1, (temperature), it remains unclear as to which kernel we should be using for this example. In fact, we would like to identify the *best* kernel as one that is

- most generic, such that no matter what the training set is, the learning of the input-output relationship is performed well;
- not too hard to learn the hyperparameters of, i.e. fewer the number of hyperparameters that are required to be learnt, the better the kernel;
- as data-driven as possible, as distinguished from being model-driven.

The first of these points suggests that the kernel is *best* if it correctly captures the correlation structure of the data on the output that resides inside the training set. So even when the values of the output Y are distributed inhomogeneously across input values, the kernel should work.

— It is possible that different parts of the data are sourced differently, and the inherent correlation structure of the full training set that we work with, bears inhomogenieties.

— Furthermore, the input-output relationship might not be explainable unless this relation is sampled from a non-stationary Stochastic Process. This would imply that the correlation between the variable Y_i that is the output at the i-th input $X = x_i$, and the output Y_j at x_j, is specific to i and j, and is different from the correlation between the outputs realised at another equally-distant pair of inputs; here $i, j \in \{1, \ldots, N_{test}\}$.

— Additionally, the output variable could be high-dimensional.

Given such data, learning the inter-variable relation is a challenging task, and a kernel that works – without us needing to learn a large number of hyperparameters of such a coveted kernel – is the holy grail of mechanistic learning. To top these demands, we seek a kernel that is not model-driven. Thus, ideally the kernel needs to be non-parametric by nature, the (hyper)parameters of which we will learn.

It does indeed appear that we seek to perform non-parametric learning of the function that represents the relation between the input and the output, by modelling this function with a GP, but then we back this up within a parametric kernel. Such seems a like a compromise on the promise of learning the input-output function non-parametrically. In the following chapter, we will introduce such a non-parametric kernel that maybe motivated by a simple form, but will be shown to dually manifest a non-stationary nature.

Appendix: code for matrix inversion

A simple C code for inverting covariance matrices. The "main" function in the code below calls in the relevant function, to invert an example covariance matix. In the code, an example covariance matrix – called "`matrix_sigma[][]`" - is parametrised with an SQE kernel, inside the function "`sigma_func_first(l)`", that is computed in this example code below, at length scale "`l`"=100. In the function `sigma_func_first(l)`, the ij-th element of `matrix_sigma` is SQE-parametrised as $\exp(-(i-j)^2/l^2)$. `sigma_func_first(l)` inverts the kernel-parametrised `matrix_sigma` matrix, into the matrix "`arr_sigma_inv_first[][]`". Product of this inverted matrix, and the original matrix is sought in `arr_try[][]`. This product should then yield an $N_{data} \times N_{data}$-dimensional diagonal matrix, since the original matrix has this dimensionality. Change the original matrix to any of your choice, and check for accuracy of its inversion – as well as of its determinant.

```
#include <stdio.h>
#include <stdlib.h>
#include <math.h>
#include <bits/stdc++.h>
using namespace std;

#include "ran1.c"
#include "sort.c"
#include "gasdev.c"
#include "cdf_gauss.c"

#define PI 3.14159265

int n,i,j,k;
int const N_data=803;
int const MAX = N_data;
double const tol = 1e-6;
double sum=0.;
double arr_try[N_data+1][N_data+1];
double arr_sigma_inv_first[N_data+1][N_data+1],
    matrix_sigma[N_data+1][N_data+1];
double lg_det;
double lower[MAX+1][MAX+1], lower_inv[MAX+1][MAX+1],
    lower_transp_inv[MAX+1][MAX+1];

void sort(unsigned long n, float arr[]);
double ran1(long *idum);
double gasdev(long *idum);
long idum=-1;

void sigma_func_first(double l)
{
  memset(lower, 0, sizeof(lower));

  for(i=1; i<=N_data; i++)
    {
      for(j=1; j<=N_data; j++)
  {
    if(i==j)
      matrix_sigma[i][j] = 1 + tol;
```

```
      if(i != j && i <=N_data && j<=N_data)
        matrix_sigma[i][j] = exp(-pow(i-j,2)/(l*l));
  }
      }
  //Cholesky_Decomposition(matrix, n);
  for (i = 1; i <= N_data; i++)
    {
      for (j = 1; j <= N_data; j++)
        {
    lower[i][j] = 0;
        }
    }
  for (i = 1; i <= N_data; i++)
    {
      for (j = 1; j <= i; j++)
        {
    sum = 0;
    if (j == i) // summation for diagonals
      {
        for (k = 1; k <= j; k++)
          sum += pow(lower[j][k], 2);
        if(matrix_sigma[j][j] > sum)
          lower[j][j] = sqrt(matrix_sigma[j][j] - sum);
        else
          lower[j][j] = 0;
      }
    else
      {
        // Evaluating L(i, j) using L(j, j)
        for (k = 1; k <= j; k++)
          sum += (lower[i][k] * lower[j][k]);
        if(fabs(lower[j][j]) > 1e-6)
          lower[i][j] = (matrix_sigma[i][j] - sum) /
      lower[j][j];
        else
          lower[i][j] = 0;
      }
        }
    }
  lg_det = 0;
  for (i = 1; i <= N_data; i++)
    lg_det += 2*log(lower[i][i]);

  //Invert the lower triangular matrix.
  for (i = 1; i <= N_data; i++)
```

```
    {
      for (j = 1; j < i; j++)
        {
    lower_inv[i][j] = 0;
        }
    }
  for (i = 1; i <= N_data; i++)
    {
      if(fabs(lower[i][i]) > 1e-6)
        lower_inv[i][i] = 1.0 / lower[i][i];
      else
        lower_inv[i][i] = 0;
      for (j = 1; j < i; j++)
        {
    sum = 0.0;
    for (k = j; k < i; k++)
      {
        sum += lower[i][k] * lower_inv[k][j];
      }
    lower_inv[i][j] = - sum * lower_inv[i][i];
        }
    }
  //Invert the transpose of the lower triangular matrix.
  for (i = 1; i <= N_data; i++)
    {
      for (j = 1; j <= N_data; j++)
        lower_transp_inv[i][j] = lower_inv[j][i];
    }
  //Invert the matrix.
  for (i = 1; i <= N_data; i++)
    {
      for (j = 1; j <= N_data; j++)
        {
    sum = 0.0;
    for (k = 1; k <= N_data; k++)
      {
        sum += lower_transp_inv[i][k] * lower_inv[k][j];
      }
    arr_sigma_inv_first[i][j] = sum;
        }
    }
}

main()
{
```

```
/*A simple code for matrix inversion , and for checking
   the inversion.*/
sum=0;

sigma_func_first(100.);

for (i = 1; i <= N_data; i++)
 {
   for (j = 1; j <= N_data; j++)
     {
 sum = 0;
 for (k = 1; k <= N_data; k++)
   sum += arr_sigma_inv_first[i][k] * matrix_sigma[k][
 j];
 arr_try[i][j] = sum;
 printf("%d %d %10.4e\n",i,j,arr_try[i][j]);
     }
 }
exit(0);
}
```

4

Covariance kernels suitable for real-world data

4.1 Introduction

In the last chapter we acquainted ourselves with the idea of parametrising a covariance matrix with a function $K(\cdot)$, that takes as its input, the difference, (or the distance) between a pair of input values – $\boldsymbol{x}_i$ and $\boldsymbol{x}_j$ – such that (s.t.) the function computed at this difference, is proportional to the correlation between the output variables $\boldsymbol{Y}_i$ and $\boldsymbol{Y}_j$, that are realised at $\boldsymbol{x}_i$ and $\boldsymbol{x}_j$ respectively. Here we recall that $\boldsymbol{X} \in \mathcal{X}$ is the input to the function $\boldsymbol{f} : \mathcal{X} \longrightarrow \mathcal{Y}$ that we seek, where output variable $\boldsymbol{Y} \in \mathcal{Y}$ is realised at a value of $\boldsymbol{X}$. The function $K(\cdot)$ mentioned above, is a particular kind of a covariance kernel, and in general, it bears multiple hyperparameters.

Leaving covariance kernels for now, let us return to the fundamentals. We have seen before that the sought function $\boldsymbol{f}(\cdot)$ is modelled as an unknown – and therefore as a (function-valued) random variable – so that it is ascribed a probability distribution. Probability distribution over the space of functions is given by a Stochastic Process, and we invoke a Gaussian Process (GP) to model $\boldsymbol{f}(\cdot)$, i.e. model $\boldsymbol{f}(\cdot)$ as a sample function of this GP. The GP induces a likelihood that is a high-dimensional equivalent of the multivariate Normal density, where the dimensionality is driven by the dimensionality of the output variable $\boldsymbol{Y}$, (or equivalently, of the function $\boldsymbol{f}(\cdot)$). For example, if $\boldsymbol{Y}$ is a matrix – and therefore $\boldsymbol{f}(\cdot)$ is a matrix-valued function – then a matrix-variate GP is invoked to model this function. Such a GP will then induce a likelihood that is a 3rd-order tensor Normal density, parametrised by

DOI: 10.1201/9781003296331-4

a mean tensor and three covariance matrices. For a training set of size n, the "inter-realisation" covariance matrix of this likelihood is then $n \times n$-dimensional. Here the "inter-realisation" covariance matrix is s.t. its ij-th element is the covariance between the outputs realised at $\boldsymbol{X} = \boldsymbol{x}_i$ and $\boldsymbol{X} = \boldsymbol{x}_j$.

By replacing the ij-th element of this $n \times n$-dimensional inter-realisation covariance matrix, with the kernel function computed at values of $\boldsymbol{X} = \boldsymbol{x}_i$ and $\boldsymbol{X} = \boldsymbol{x}_j$, the burden of learning reduces from each of the $(n^2 - n)/2$ distinct elements of this symmetric covariance matrix, to the kernel hyperparameters. Here $i, j = 1, \ldots, n$. This is in general a big reduction in the number of parameters that need to be learnt, (as long as the number of kernel hyperparameters is much lower than $(n^2 - n)/2$), but *is it correct to replace each element of the covariance matrix by the appropriately computed kernel function?* In other words, is such kernel parametrisation of the covariance matrix, correct?

Before we proceed, a comment needs to be made regarding notation. Though the kernel function is depicted above as one with the input that is a difference between two values of $\boldsymbol{X} \in \mathcal{X}$, (e.g. $\boldsymbol{x}_i$ and $\boldsymbol{x}_j$), in the following, we will often represent "K(difference between $\cdot$ and $\cdot$)" as $K(\cdot, \cdot)$.

When kernel parametrisation is likely to fail

We can see immediately that if the correlation between $\boldsymbol{Y}_i$ and $\boldsymbol{Y}_j$ is not reproduced as $K(\boldsymbol{x}_i, \boldsymbol{x}_j)$, then the undertaken kernel parametrisation will render the GP-based learning of the sought function $\boldsymbol{f}(\cdot)$ erroneous, leading to wrong prediction (or learning) of $\boldsymbol{Y}$ (or $\boldsymbol{X}$) at $\boldsymbol{X} = \boldsymbol{x}$ (or $\boldsymbol{Y} = \boldsymbol{y}$). So, to transform the burden of learning from the infeasible (learning of $n(n-1)/2$ parameters) to the achievable (learning of the few kernel hyperparameters) – as promised by the parametrisation of a covariance structure with a kernel – it is important that we establish the relevance of the considered kernel function, to the data at hand. Then, when will the kernel incorrectly reproduce a general element of the covariance matrix?

The answer to that can be traced to the shortcomings of the hyperparameters that are intrinsic to the chosen (typically parametric) form of the kernel. Indeed, the hyperparameter values could have been wrongly learnt, given the data at hand. Additionally, retaining the same values of the hyperparameters, at all values of $\boldsymbol{X} \in \mathcal{X}$, may not be the correct approach. This perceived risk is synonymous to the worry that correlation between all pairs of output variables may not fade in the same way – after all, the kernel hyperparameters determine the nature of the depreciation of the correlation between two output variables realised at respective inputs, as distance between these inputs increases. Thus, the variation of the correlation between output variables $\boldsymbol{Y}_i$ and $\boldsymbol{Y}_j$ over values of $\boldsymbol{X}$, may be different to how the correlation between $\boldsymbol{Y}_{i'}$ and $\boldsymbol{Y}_{j'}$ varies with $\boldsymbol{x}$, where $i \neq i', j \neq j', i, j, i', j' \in \{1, \ldots, n\}$.

In other words, the correlation structure in one subset of the training data, may turn out to be different from that in another subset of the dataset, i.e. the correlation structure of the training data might be uneven. This could be rendered possible if the intrinsic process that the data is sampled from, is not invariant to shift in input space, i.e. is non-stationary s.t. outputs realised over different inputs, are differently correlated. Alternatively, the used training data could result from the collation of multiple datasets obtained from assorted sources that are not homogeneous in one or multiple ways, (including different error structures).

So to summarise, it is possible that real-world data is marked by an inhomogeneous correlation structure, and if that is true of the available data – that we employ to learn the function $\boldsymbol{f}(\cdot)$ – then a kernel function that is assigned (unknown) constant hyperparameters, will wrongly model the covariance function of the GP that models $\boldsymbol{f}(\cdot)$. On the other hand, kernel hyperparameters that vary from one input location to another, give rise to a kernel that is enabled to correctly model a correlation structure that varies from one subset of input values to another. So in general, kernels that vary with the input location – and not simply depend on the difference between a pair of input values – can correctly model inhomogeneous correlation structures. Such kernels are referred to as non-stationary and (as indicated above), these are not invariant to a shift in values of the input parameter $\boldsymbol{X}$, s.t. the correlation function of the GP that such a kernel parametrises, is not shift-invariant, i.e. the underlying GP is non-stationary, [114, 95, 91, 105, 60, 100, 4, 127].

4.2 Kernels given real-world data

So from the above, we intuitively follow that we are better-advised to invoke a non-stationary GP – the covariance function of which is parametrised using a non-stationary kernel – to model the relationship between an output and an input variable, given data on the output that bears inhomogeneities in its correlation structure. Then the kernel that we would want to work with, given such inhomogeneously-correlated data, will be such, that this kernel computed at input values of $\boldsymbol{x}_i$ and $\boldsymbol{x}_j$, will be different from that at $\boldsymbol{x}_{i'}$ and $\boldsymbol{x}_{j'}$, even for $|\boldsymbol{x}_i - \boldsymbol{x}_j| = |\boldsymbol{x}_{i'} - \boldsymbol{x}_{j'}|$, but $\boldsymbol{x}_i \neq \boldsymbol{x}_{i'}$ and/or $\boldsymbol{x}_j \neq \boldsymbol{x}_{j'}$. Here $|\cdot|$ is a norm in $\mathcal{X}$. Thus, the kernel computed at $\boldsymbol{x}_i$ and $\boldsymbol{x}_j$ depends not just on $|\boldsymbol{x}_i - \boldsymbol{x}_j|$, but is unique to the index pair i and j. Here, this kernel $K(\boldsymbol{x}_i, \boldsymbol{x}_j)$ models the correlation $corr(\boldsymbol{Y}_i, \boldsymbol{Y}_j)$ between the outputs $\boldsymbol{Y}_i$ and $\boldsymbol{Y}_j$ that are realised at the input values of $\boldsymbol{x}_i$ and $\boldsymbol{x}_j$ respectively, $\forall i, j = 1, \ldots, n$.

The kernel function is likely to bear hyperparameters that are collated into the vector $\boldsymbol{\Theta}$; hence we update the notation of the kernel to $K(\cdot, \cdot; \boldsymbol{\theta})$, where $\boldsymbol{\theta}$ is a realisation of $\boldsymbol{\Theta}$.

Thus, it may appear that a ready way of ensuring that the kernel $K(\cdot, \cdot; \boldsymbol{\theta})$ is specific to a pair of locations in $\mathcal{X}$, is to motivate a kernel that takes as its input, the norm of the vector that is the difference between these input locations, and has hyperparameters that are components of vector $\boldsymbol{\theta}$ where these hyperparameters vary from one input pair to another. Such a kernel computed at inputs of $\boldsymbol{x}_i$ and $\boldsymbol{x}_j$, will be represented as the function

$$K(|\boldsymbol{x}_i - \boldsymbol{x}_j|; \boldsymbol{\theta}_{i,j}),$$

i.e. notation of the non-stationary kernel is updated to suggest an i, j-dependent realisation $\boldsymbol{\theta}_{i,j}$ of the hyperparameter vector

$\boldsymbol{\Theta}$ of this kernel, at these inputs. However, there is an obvious downside to this choice of a model of kernel hyperparameters that will lead us to discard this model.

Next we discuss this shortcoming of this model.

4.2.1 Why $K(|\boldsymbol{x}_i - \boldsymbol{x}_j|; \boldsymbol{\theta}_{i,j})$ is a bad model for a kernel

As motivated above, the whole purpose of parametrising $corr(\boldsymbol{Y}_i, \boldsymbol{Y}_j)$ with $K(\boldsymbol{x}_i, \boldsymbol{x}_j)$ is the reduction of the burden of learning to the few hyperparameters of the kernel, from the distinct $(n^2 - n)/2$ elements of the inter-realisation covariance matrix $\boldsymbol{\Sigma}_p = [corr(\boldsymbol{Y}_i, \boldsymbol{Y}_j)]$. But in the model of the kernel proposed in the grey box above, $\boldsymbol{\theta}_{i,j}$ will need to be learnt for each ij-pair, so that there are at least as many hyperparameters, as relevant i, j pairs – which as we have just recalled, is $(n^2 - n)/2$. For example, for a training set with 10 data points, we would then need to learn at least 45 hyperparameters – which is already a large learning exercise, even for this small a training set. Then in this choice of the kernel design, the purpose of kernel parametrisation is defeated. This is why we refer to the model proposed in the last grey box to be "a bad model for a kernel".

In its place, there are a few suggestions for models of non-stationary kernels in the literature. We will discuss these below, and supplement this discussion with a new non-stationary kernel that emerges from a new non-parametric kernel that we will first consider. Indeed, a driving factor in our model choice for a useful kernel, is the need to learn very few parameters – namely the few components of the hyperparameter vector $\boldsymbol{\Theta}$ of the kernel under consideration, where by "few", we imply fewer than $(n^2 - n)/2$ for sure, and as few as possible, given the nature of the input and output variables. We will find the upcoming non-parametric kernel to be such that, for a scalar-valued output Y and a scalar-valued input X, we need to learn only two hyperparameters.

4.2.2 Motivating a non-parametric kernel

It is imperative that the model used for the kernel is not responsible for obstructing our attempt at replicating the correlation structure of the data, as the correlation structure of the sought function $\boldsymbol{f}(\cdot)$ – that outputs $\boldsymbol{Y}$ at input $\boldsymbol{X} = \boldsymbol{x}$. The very choice of a parametric form for the kernel, imposes a correlation structure on the sought function $\boldsymbol{f}(\cdot)$, and it is to mitigate this *ad hoc* imposition, that input-dependence of kernel hyperparameters is welcomed. Indeed, when the kernel hyperparameters are made to vary from one pair of design inputs to another, the form of the kernel is rendered different at the different points within the input space $\mathcal{X}$.

Designing the kernel to be non-parametrically learnt – in addition to learning $\boldsymbol{f}(\cdot)$ non-parametrically – ensures efficient countering of any *ad hoc* representation of the empirical correlation structure. This serves the purpose of efficient replication of the empirical correlation structure, as the correlation structure of the sought function $\boldsymbol{f}(\cdot)$, leading to less inaccurate predictions than what is possible with other kernel designs.

In fact, it was the need to circumvent model-dependence, that drove us to seek non-parametric learning of the functional relation between an input and an output variable, in the first place. It is then not satisfactory to substitute the dependence on the parametric nature of the functional relation between $\boldsymbol{X}$ and $\boldsymbol{Y}$, by the dependence on the parametric form of the kernel – that parametrises the covariance structure of the GP that the sought functional relation is modelled to be a sample function of. Our learning of $\boldsymbol{f}(\cdot)$ is truly non-parametric, if our learning of the kernel is non-parametric as well. Then there is no model-driven distortion imposed on the empirical correlation structure, as we attempt duplicating this correlation as the correlation structure of the sought functional relation.

With the above in mind, the learning strategy that we will discuss later in this chapter, will encompass the dual desire:

1. of pursuing a non-parametric kernel, and

2. of the need to learn as few kernel hyperparameters as possible.

At the same time, every ingredient of this desired learning needs to be mathematically sound and the implementation needs to be streamlined.

4.2.3 The protocol for fully Bayesian supervised learning

Let us begin our discussions with a recap of the notation that we carry over from previous chapters. Our basic aim is to learn the functional relationship between an input variable $\boldsymbol{X} \in \mathcal{X} \subseteq \mathbb{R}^d$ and an output variable $\boldsymbol{Y}$ which in general is an m-th ordered tensor of dimension $k_1 \times k_2 \times \ldots \times k_m$, and lives in the space $\mathcal{Y}$. The relation between $\boldsymbol{X}$ and $\boldsymbol{Y}$ is motivated as in the previous chapters, to be formulated as $\boldsymbol{Y} = \boldsymbol{f}(\boldsymbol{X})$ where the unknown function $\boldsymbol{f} : \mathcal{X} \longrightarrow \mathcal{Y}$ is an m-th ordered tensor-valued, random function, which we model as a sample function from a tensor-variate (of order m) GP. The parameters of this GP are learnt given the training data that comprises pairs of designed values of $\boldsymbol{X}$, and value of the output $\boldsymbol{Y}$ at each such design point, i.e. the training data is $\mathbf{D} = \{(\boldsymbol{x}_i, \boldsymbol{y}_i)\}_{i=1}^{n}$, while we define the $m+1$-th ordered tensor $\mathbf{D}_Y = \{\boldsymbol{y}_1, \ldots, \boldsymbol{y}_n\}$. Thus, $\mathbf{D}_Y$ is constructed by collating the n realisations $(\boldsymbol{y}_1, \ldots, \boldsymbol{y}_n)$ of the output at the respective design points $\boldsymbol{x}_1, \ldots, \boldsymbol{x}_n$.

In fact it is the standardised values of outputs that are used in the construction of $\mathbf{D}_Y$ and $\mathbf{D}$. By this we mean that a sample of the observed values of any component of the tensor-valued output variable is first collated, for each such component. The sample is then used to compute the sample mean and standard deviation of the component, for each such component. Using the sample mean and standard deviation, the observed values of this component are standardised. Thus, the mean of the standardised data on the output is a null tensor and a zero-mean GP is invoked to model $\boldsymbol{f}(\cdot)$.

The following is a gist of the protocol that facilitates a fully Bayesian supervised learning of $\boldsymbol{f}(\cdot)$, where such learning is reliable and robust.

- The likelihood of the model parameters is rendered $m+1$-th ordered tensor Normal in form, with a mean tensor (that is null in our consideration) and $m+1$ number of covariance matrices, each of which informs on the covariance between pairs of m-th ordered tensor-valued slices of the tensor $\mathbf{D}_{\boldsymbol{Y}}$.

- Since the data comprises standardised values of the outputs realised at the different inputs, the covariance matrices are in fact, correlation matrices.

- Then the parameters of this model include:

  — hyperparameters of the kernels that are invoked to parametrise any of these correlation matrices, and

  — secondly, individual elements of some other correlation matrices that are learnt directly using the inference scheme – as long as these matrices are small in size.

  Sample estimates of relevant components of outputs are used to populate the remaining correlation matrices.

- We will invoke priors on the model parameters, and write down the joint posterior probability density of all these parameters given the data $\mathbf{D}_{\boldsymbol{Y}}$ as well as the design points $\boldsymbol{x}_1, \ldots, \boldsymbol{x}_n$.

- Thereafter, inference is performed on the parameters. In the fully Bayesian setting that we motivate throughout the book, inference is performed by sampling from the joint posterior density using Markov Chain Monte Carlo techniques (MCMC techniques).

- Once the parameters of the correlation structure of the underlying GP are learnt, closed-form forward prediction is possible on the expected value – and its dispersion – that the output attains, at a test input.

4.3 Inhomogeneous correlation structure of data

Before we embark on the development of the desirable kernel, we define inhomogeneities in the correlation structure of the data, and clarify why such inhomogeneities discourage the usage of stationary kernels. Naturally, whatever we mean by a heterogeneous correlation structure of the data, has to be defined in terms of the data alone. However, the data at hand – i.e. the available training data – comprises a set of discrete points $(\boldsymbol{x}_1, \boldsymbol{y}_1), \ldots, (\boldsymbol{x}_n, \boldsymbol{y}_n)$, with $\boldsymbol{y}_i$ finite $\forall i = 1, \ldots, n$. *How could we use such a set of n points to define a "correlation structure" – let alone qualify that correlation structure as plagued by inhomogeneities, (or not).* On examining the connotation, it appears that one is in fact referring to the correlation structure of the random (sample) function (drawn from a GP) that is fit to this data.

To be more specific, it is the correlation function of the underlying GP, inhomogeneities in which are being sought – we would think. However, it is the learning of this very correlation function of the underlying GP, that is considered here – conditional of course, on the available training data $\{(\boldsymbol{x}_i, \boldsymbol{y}_i)\}_{i=1}^n$. Thus it appears that we need to invoke sample functions of the underlying GP to motivate inhomogeneities in the empirical correlation structure, which however contradicts the aforesaid requirement that such inhomogeneities should be motivated/investigated using the data alone – *what then is the way out of this circularity?*

Let us then identify a benchmark functional representation of the sought function $\boldsymbol{f}(\cdot)$, and in terms of this, categorise "correlation of the data" as inhomogeneous or homogeneous – as well as parametrise the inhomogeneity in the correlation of the data. We will refer to this representation as the "*trivial*" function, and it is chosen to be the piecewise-linear function that joins every successive pair of training points, with a linear piece.

4.3.1 Searching for heterogeneity in correlation structure of each component of the output variable

The suggested "linear" pieces that define the "*trivial*" function given the data $\mathbf{D} = \{(\boldsymbol{x}_i, \boldsymbol{y}_i)\}_{i=1}^N$ will however need to be qualified given that the input to the sought function $\boldsymbol{f}(\cdot)$, has been declared to be a vector, and the output is a tensor in general. We apreciate that for a scalar-valued and univariate $f(\cdot)$, the "*trivial*" function will be the piecewise linear function that joins successive training points with straight lines. Indeed, for a sought tensor-valued function, we investigate homogeneity – or the lack of it – in the correlation between each scalar component of the tensor-valued output (of this sought function). Additionally, to address the nature of the input of a sought function, the following is relevant.

- For a scalar-valued input, if the tensor-valued output variable is $\boldsymbol{Y} \in \mathcal{Y} \subseteq \mathbb{R}^{k_1 \times k_2 \times \ldots \times k_m}$, one of its components is $Y^{(i_1,\ldots,i_t,\ldots,i_m)} \in \mathbb{R}$, $\forall i_t = 1, \ldots, k_t; t = 1, \ldots, m$. So after standardising the data on $Y^{(i_1,\ldots,i_t,\ldots,i_m)}$, we construct the n-sized dataset $\mathbf{D}_{Y^{(i_1,\ldots,i_t,\ldots,i_m)}}$ that comprises the n values of this component. Then the piecewise linear function that we refer to as "*trivial*", is constructed by joining successive values of $Y^{(i_1,\ldots,i_t,\ldots,i_m)}$ - realised at the corresponding design points $x_1, \ldots, x_n$ – with straight lines. We do this for each component of the tensor-valued output variable in $\mathcal{Y}$. Thus, in the case of the high-dimensional output, we search for inhomogeneities in the correlation structure of each individual component of the output.

- When the input is vector-valued, we fit a piecewise linear function to the set of pairs of values of the datum index (j) and the value of each component of the generally tensor-valued output variable, for $j = 1, \ldots, n$. Thus, this piecewise linear function connects the point (j, j-th value of $Y^{(i_1,\ldots,i_t,\ldots,i_m)}$) with the point ($j+1$, $j+1$-th value of $Y^{(i_1,\ldots,i_t,\ldots,i_m)}$) with straight lines, for $j = 1, \ldots, n-1$. Thus, for the case of a high-dimensional output and a vector-variate input, we search for inhomogeneities in the correlation structure in each individual component of the output realised as n values in the data $\mathbf{D}_{Y^{(i_1,\ldots,i_t,\ldots,i_m)}}$, irrespective of the nature of the input.

Our interest transcends mere identification of the presence of inhomogeneities in the correlation of the data – we will compute an inhomogeneity parameter for a given data. Towards this, we will integrate our treatment of vector-valued and scalar-valued inputs by treating the data, as the set of pairs of datum index (j) and the observed value of the output variable at this given j.

4.3.2 Current focus: scalar-valued output and input

To begin our current discussions, let us consider the output to be scalar-valued. We use the notation Y for the output. The functional relation $f(\cdot)$ between this scalar-valued output $Y \in \mathcal{Y}$ and the scalar-valued input X is discussed here; said relation is treated as the sample function from a scalar-variate GP in this section. The data on the output is then $\mathbf{D}_Y = \{y_1, \ldots, y_n\}$. Let the design points $x_1, \ldots, x_n$ that contribute to the training data $\mathbf{D} = \{(x_i, y_i)\}_{i=1}^n$ be s.t. $x_1, \ldots, x_n \in [x_{in}, x_{out}] \subset \mathcal{X}$. Here $\mathcal{Y}$ and $\mathcal{X}$ are metric spaces. In the case considered in this section, the *trivial* or piecewise linear function $f_t(\cdot)$, is the one that joins the adjacent observed values of $Y \in \mathbf{D}_Y$ by straight lines.

We recall from above, that for the purpose of computing the parameter that measures inhomogeniety in the correlation structure of $\mathbf{D}$, we will consider the observed values of the output Y at the corresponding data index j, instead of at the corresponding design point, where $j = 1, \ldots, n-1$.

We will introduce this parameter after discussing identification of inhomogeneity in the correlation structure of the data.

4.3.2.1 Identifying inhomogeneity: the case of scalar-valued, univariate function

In order to define a criterion for identifying the presence of "inhomogeneities in the correlation structure" of data $\mathbf{D}$, we first introduce a way to compute the Lipschitz constant (L_i) of the *trivial* sample function $f_t(\cdot)$ over the i-th partition ($[x_i, x_{i+1})$) of $[x_{in}, x_{out}]$, and define a δ_i-neighbourhood of L_i, $\forall i = 1, \ldots, n-1$.

- Lipschitz continuity of the *trivial* function that is fit to data **D** in interval $[x_i, x_{i+1})$, is expressed via L_i, s.t.

$$d_Y(f_t(a), f_t(a')) \leq L_i d_X(a, a'),$$

$$\forall a, a' \in [x_i, x_{i+1}) \subset \mathcal{X};\ f_t(a), f_t(a') \in \mathcal{Y},$$

for $i = 1, \ldots, n-1$, with $d_Y(\cdot, \cdot)$ denoting the distance function of the metric space $\mathcal{Y}$ and $d_X(\cdot, \cdot)$ is the distance function of space $\mathcal{X}$.
Then setting:

$$L_i = \frac{d_Y(y_i, y_{i+1})}{d_X(x_i, x_{i+1})},$$

does not counter the demand of the Lipschitz nature of the piece of this function that joins the data points (x_i, y_i) and (x_{i+1}, y_{i+1}). We identify this data-driven Lipschitz constant L_i for the i-th partition $[x_i, x_{i+1})$, of $[x_{in}, x_{out}]$. This is relevant $\forall i = 1, \ldots, n-1$.

- After computing $L_1, \ldots, L_n$, we identify a "similarity-neighbourhood" of width δ_i around L_i, to define "similarly-Lipschitz" partitions of $[x_{in}, x_{out}]$.

Definition 4.1 *A "similarity-neighbourhood" of width $2\delta_i$ around L_i is given as the interval*

$$[L_i - \delta_i, L_i + \delta_i],$$

$\forall i = 1, \ldots, n-1$ *and a chosen* $\delta_i > 0$.

Here δ_i is a small positive number that intuitively represents a scale of noise in the observation of output Y_i that is realised at $X = x_i$. It could be given by the error in observations of Y_i. In case there is no error available on Y_i, we choose $\delta_i = \delta$, $\forall i = 1, \ldots, n-1$ and report heterogeneity in the correlation structure of the data on the output, for this chosen δ.

Indeed we could set δ as 5% or 10% of the sample mean; or any fraction of another sample summary statistic of the sample $\{L_1, \ldots, L_{n-1}\}$; etc. Attributable values of δ are not set in stone, but upto the learner to choose, as they consider the evaluation of "similarity" amongst values of Lipshitz constants that they compute, given the data $\mathbf{D}$ at hand.

Definition 4.2 *The partitions* $[x_i, x_{i+1})$ *and* $[x_j, x_{j+1})$ *are "similarly-Lipschitz" -* $\forall i, j = 1\ldots, n-1$ *- if the similarity neighbourhoods of* L_i *and* L_j *overlap, given* δ_i *and* δ_j*. In other words, partitions* $[x_i, x_{i+1})$ *and* $[x_j, x_{j+1})$ *are "similarly-Lipschitz" if*

$$[L_i - \delta_i, L_i + \delta_i] \cap [L_j - \delta_j, L_j + \delta_j] \neq \phi,$$

for chosen $\delta_i, \delta_j \geq 0$.

Typically, δ_i is set to be equal to δ_j $(= \delta)$, $\forall i, j = 1, \ldots, n-1; i \neq j$.

- We will use the absolute difference between a pair of points in $\mathcal{Y}$ as the distance $d_Y(\cdot, \cdot)$ in the output space $\mathcal{Y}$, and use the absolute difference between a pair of design points as well, (as $d_X(\cdot, \cdot)$).

Then a criterion for the presence of "inhomogeneities in the correlation structure" of data $\mathbf{D}$ can be defined as follows.

Definition 4.3 *If* $\exists i \in \{1, \ldots, n\}$ *s.t.*

$$L_i \notin [L_j - \delta, L_j + \delta], \; \forall j \neq i; \; j = 1, \ldots, n-1,$$

then inhomogeneities exist in the correlation structure of $\mathbf{D}$.

Thus, for this *trivial* function, if there exists even one partition of $[x_{in}, x_{out}]$, that is not "*similarly-Lipschitz*" to at least another partition, we declare that data $\mathbf{D}$ bears inhomogeneities in its correlation structure. Thus, there is no inhomogeneity in the correlation of the data $\mathbf{D}$, if all partitions of design values are "*similarly-Lipschitz*". The intuition is that if inhomogeneities are noted to exist in the correlation structure of $\mathbf{D}$, in reference to this *trivial* function, then inhomogeneities in $\mathbf{D}$ will be noted in generic sample functions, that are comparatively more non-linear.

Example 4.1 *We recall the training data* $\mathbf{D}$ *comprising about 200 pairs of temperature and energy consumption values, within Tetouan city, where* $\mathbf{D}$ *is built by randomly sampling from the publicly-available Tetouan city data that we discussed in Chapter 3, [113]. The trivial – or piecewise linear – function is plotted on the left of Figure 4.1. This trivial sample function leads to Lipschitz constants defined as the ratio of the absolute differences in the output, to that of the design points in* $\mathbf{D}$. *Some of the resulting Lipschitz constants do not fall within the* δ*-neighbourhood of any of the other computed Lipschitz constants. So this dataset does bear inhomogeneities in its correlation structure. As we will soon see, given such a training set, the learning of the functional relation between the input and the output will call for a non-stationary GP.*

Example 4.2 *However, when we consider the monthly Brent crude oil price data, we find that there are more similarly-Lipschitz partitions of the range of values of the input – than in the Tetouan city example. In fact, had it not been for intervals realised at approximately the 150-th and 290-th partitions of the data, the sample of computed Lipschitz constants would not pass the criterion delineated in Definition 4.3. Had it not been for these two data points, we would identify this oil price time series dataset to be absent of inhomogeneities in its correlation structure. However, the Lipschitz constants relevant to the intervals in input space, around the 150-th and 290-th time points imply that even this data bears inhomogeneities in its correlation structure.*

4.3.3 Defining a parametrisation of inhomogeneity in the correlation of an observed dataset

In principle it is possible to define a parametrisation of how inhomogeneous the correlation structure of a given dataset is. We could

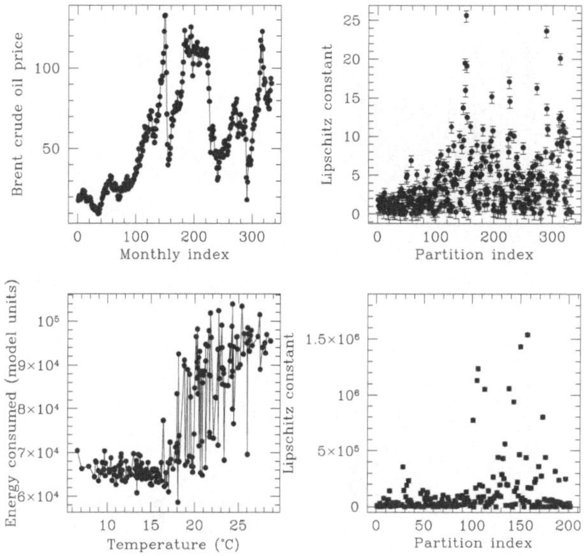

FIGURE 4.1
Bottom left: plot of the *trivial* (or linearly-piecewise) function that joins successive training data points in the Tetouan city example. Black filled circles represent points of the training data **D** – constructed by randomly sampling from the publicly available Tetouan city data discussed in Chapter 3 – used for the learning of the temperature-dependence of the energy consumed across all three zones of Tetouan city. *Bottom right:* in black filled circles are plotted the Lipschitz constant relevant to a partition of the range of temperatures included inside **D**, computed using the *trivial* function. A δ-neighbourhood of 15% of the sample mean ($\approx$ 120863) of all Lipschitz constants computed over all partitions is defined, and superimposed as an error bar on either side of the computed Lipschitz constant. *Top left:* plot of the *trivial* function that connects successive monthly variation values in Brent crude oil price from May 1987, to October 2023, taken from the site: `https://www.eia.gov/dnav/pet/hist/LeafHandler.ashx?n=PET&s=RBRTE&f=M`. *Top right:* δ-neighbourhood of Lipschitz constants given by 15% of the sample mean ($\approx$ 4.0596) of all Lipschitz constants computed for this data.

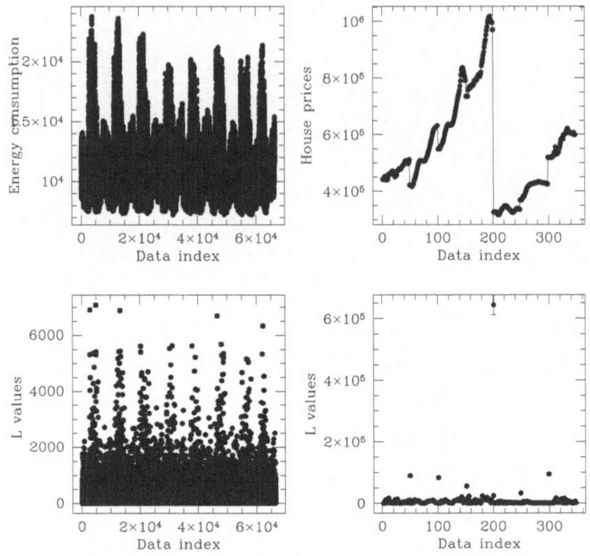

FIGURE 4.2
Top left: plot of the *trivial* (or linearly-piecewise) function that joins successive data points of the publicly-available data COMED_hourly.csv comprising time series of energy consumption, available at `https://www.kaggle.com/datasets/robikscube/hourly-energy-consumption/data?select=COMED_hourly.csv`. The dataset comprises 66,497 data points. *Bottom left:* the absolute value of the difference between successive output observations is plotted (as "L values"), against the index of the data point. For δ set as 5% fractional error of L_i defined in Definition 4.4, $\forall i = 1, \ldots, 66497$, inhomogeneity parameter $p = 1/66497$, while for $\delta = 1\%$, $p = 6/66497$. *Top right:* plot of the *trivial* function relevant to the time series dataset ma_lga.csv available at `https://www.kaggle.com/datasets/htagholdings/property-sales` comprising information on moving average of median house sale price. *Bottom right:* similar to the plot on the bottom left panel, except the current panel uses the ma_lga.csv dataset. For a fractional error of 1% on each datum, inhomogeneity parameter $p = 88/347 \approx 0.26$, while for a fractional error of 5%, $p = 23/347$.

compute this inhomogeneity parameter as the fraction of the number of partitions that are not *similarly-Lipschitz*, out of the total, i.e. $n-1$ number of partitions in a training set with n data points in it. Such an inhomogeneity parameter for the Tetouan city data will be higher than the Brent crude oil price time series data – at the same choice of δ.

Definition 4.4 *In order to compute the inhomogeneity parameter of the given dataset* $\mathbf{D} = \{(x_i, y_i)\}_{i=1}^n$, *consider the dataset* $\mathbf{D}^{/} := \{(i, y_i)\}_{i=1}^n$ *and compute* $L_1, L_2, \ldots, L_{n-1}$ *where* $L_i := |y_{i+1} - y_i|$, $\forall i = 1, \ldots, n-1$. *Then inhomogeneity parameter*

$$p := \sum_{i=1}^{n} \sum_{j=1, j\neq i}^{n} \frac{|([L_i - \delta, L_i + \delta] \cap [L_j - \delta, L_j + \delta])|}{n},$$

where $|A|$ *is the cardinality of set* A.

We do not interpret L_i as a Lipschitz constant in this definition.

We compute the inhomogeneity parameter p for distinct publicly-available datasets and in Figures 4.1 and 4.2, plot the *trivial* functions that linearly join adjacent data points of each of these datasets, as well as the L_i values computed as delineated in Definition 4.4 above. We have assigned arbitrary values of the fractional error on these data points and computed p.

It is interesting to note that while both datasets used to construct Figure 4.2 are non-stationary, the inhomogeneity in the correlation of one of the datasets - namely, the energy consumption data – is nearly zero, compared to the other. Indeed, non-stationarity of the process that generates the data on the output, will give rise to input-dependence in the correlation structure of the data on this output.

Again, in reference to the Tetouan city data, we find that when δ of 15% of the sample mean is used as the width of the similarity neighbourhood, the inhomogeneity parameter of the correlation in this data is 2/203, while for a δ of 1% of the sample mean, leads to a p of $49/203 \approx 0.24$. For the Brent crude oil data, for δ of 15% of the sample mean, the inhomogeneity parameter of the correlation in this data is $12/133$, while a δ of 1% of the sample mean, leads to a p of $21/133 \approx 0.16$. So, the width of the similarity neighbourhood is 1% of the sample mean, the Brent crude oil data bears less inhomogeneities in its correlation than the subset of the Tetouan city data that we introduced in Chapter 3.

Correlation structure in an observed data could potentially vary across input values bounded by the range $[x_{in}, x_{out}]$ that is relevant to the given training set. Such variation could emerge

- if the data comprises observations that are sampled at different frequencies over different subsets of this range, and/or
- if data on the output realised over such different subsets of $[x_{in}, x_{out}]$ were collected with distinct error distributions, from different experiments/sources, and/or
- if the collected data on the output results from drawing multiple samples from an inherently non-stationary process.

The discussed inhomogeneity in the correlation of the data aims at identifying - as well as parametrising – differences in the correlation between outputs (of the function that are fit to said data), realised at variously-located input pairs.

4.3.4 Connection between kernel hyperparameters & inhomogeneities in the correlation structure of data

We have motivated a data-driven way for identifying inhomogeneities in the correlation structure of the training data $\mathbf{D}$, but next we ask if such detected inhomogeneities can inform on the kernel structure of the GP that is invoked to generate the function $f(\cdot)$ that represents the relation between the input and the output, where $\mathbf{D}$ serves as the training data that permits the learning of this sought function.

Let us consider the training data $\mathbf{D}$ to be such that inhomogeneities in its correlation structure have been detected on the basis of the criterion that was discussed in the previous subsection using the *trivial* function. Then we are aware of the variation in values of the Lipschitz constant across the $n-1$ partitions of the interval $[x_{in}, x_{out}]$ that accommodates all the design points of $\mathbf{D}$, where the computation of the Lipschitz constants used the distance functions $d_Y(\cdot,\cdot)$ and $d_X(\cdot,\cdot)$ in the output and input space respectively.

We seek the correlation – or rather the absolute correlation – between any pair of output variables – say $Y^\star, Y^{\star\star}$ – that are respectively realised at input values $x^\star, x^{\star\star}$. Let us consider $|corr(Y^\star, Y^{\star\star})|$ parametrised using a stationary kernel $K(\cdot,\cdot,\boldsymbol{\theta})$, i.e. set

$$|corr(Y^\star, Y^{\star\star})| = K(d_X(x^\star, x^{\star\star}), \boldsymbol{\theta}),$$

where the constant kernel hyperparameters are collated into vector $\boldsymbol{\Theta} \in \mathbb{R}^d$, that takes a value of $\boldsymbol{\theta}$ in the available data. Then irrespective of whether $x^\star, x^{\star\star} \in [x_i, x_{i+1})$, or $x^\star, x^{\star\star} \in [x_j, x_{j+1})$, the correlation between these two considered output variables is a constant. This result owes to our undertaking of the parametrisation of the correlation with a stationary kernel.

However, if the training data $\mathbf{D}$ *bears inhomogeneous correlation structure,* $corr(Y^\star, Y^{\star\star})$ *varies with the location in input space* $\mathcal{X}$, *of the input values at which these output variables are realised, i.e. on whether* $x^\star, x^{\star\star} \in [x_i, x_{i+1})$, *or* $x^\star, x^{\star\star} \in [x_j, x_{j+1})$. We can prove this by first defining a distance function $d_Y(\cdot,\cdot)$ in the metric space $\mathcal{Y}$, as the square root of the logarithm of the reciprocal of the absolute of the correlation between $Y^\star$ and $Y^{\star\star}$.

Definition 4.5 *Let us define a distance function* $d_Y(\cdot,\cdot)$ *in* $\mathcal{Y}$ *s.t.*

$$d_Y(y^\star, y^{\star\star}) := \sqrt{-\log|corr(Y^\star, Y^{\star\star})|}.$$

That this function $d_Y(\cdot,\cdot)$ is a distance function in $\mathcal{Y}$, is stated in Theorem 4.2 below and proved subsequently.

From the definition of this distance function in $\mathcal{Y}$, we see that $|corr(Y^\star, Y^{\star\star})| = \exp(-d_Y(y^\star, y^{\star\star})^2)$.

Theorem 4.1 *If there are inhomogeneities in the correlation* *structure of the available training data* $\mathbf{D}$, *then the input location dependence of the correlation between the outputs* Y_i *and* Y_j *realised at* $X = x_i$ *and* $X = x_j$ *respectively, cannot be modelled only via the distance* $d_X(x_i, x_j)$ *between* x_i *and* x_j. *This holds* $\forall i \neq j;\ i, j = 1, \ldots, n-1$, *with* $\mathbf{D} = \{(x_i, y_i)\}_{i=1}^{n}$.

Proof *We assume that* $corr(Y_i, Y_j)$ *is dependent on inputs* x_i *and* x_j *only via their mutual distance* $d_X(x_i, x_j)$. *Thus, we assume that* $corr(Y_i, Y_j)$ *can be parametrised with the kernel function* $K(d_X(x_i, x_j); \boldsymbol{\theta})$, *where the vector* $\boldsymbol{\theta}$ *of hyperparameter values is a constant vector.*

In the model distance function $|corr(Y_i, Y_j)| = \exp(-d_Y(y_i, y_j)^2)$, *this assumption is equivalent to assuming that*

$$|corr(Y_i, Y_j)| = \exp(-d_Y(y_i, y_j)^2) = K(d_X(x_i, x_j); \boldsymbol{\theta}),$$

where $\boldsymbol{\theta}$ *is a constant.*

Here the training data is s.t. inhomogeneities exist in its correlation structure. Then from the criterion that defines the presence of inhomogeneities in a given training set, we state:

$$\exists i, j \in \{1, \ldots, n-1\},\ \textit{with } i \neq j,\ \textit{s.t. } L_i \notin [L_j - \delta, L_j + \delta],$$

for a chosen δ.

Let $Y^\star$ *be realised as* $y^\star$ *at an input of* $x^\star$ *that is in the i-th partition of the range of values of design points, i.e.* $x^\star \in [x_{i-1}, x_i)$. *Again, let* $Y^{\star\star} = y^{\star\star}$ *be realised at* $x^{\star\star} \in [x_{i-1}, x_i)$.

This implies, $d_Y(y^\star, y^{\star\star}) = L_i d_X(x^\star, x^{\star\star})$.

Then, it follows that

$$|corr(Y^\star, Y^{\star\star})| = \exp(-d_Y(y^\star, y^{\star\star})^2) = \exp(-L_i^2 d_X(x^\star, x^{\star\star})^2) = K(d_X(x^\star, x^{\star\star}); \boldsymbol{\theta}). \tag{4.1}$$

Next, consider $x^\circ \in [x_{j-1}, x_j)$ *and* $x^{\circ\circ} \in [x_{j-1}, x_j)$, *for* $i \neq j$, *s.t.* $d_X(x^\star, x^{\star\star}) = d_X(x^\circ, x^{\circ\circ})$.

Then, for the outputs Y° *and* $Y^{\circ\circ}$ *realised at* x° *and* $x^{\circ\circ}$, *as* y° *and* $y^{\circ\circ}$, *we get* $d_Y(y^\circ, y^{\circ\circ}) = L_j d_X(x^\circ, x^{\circ\circ}) = L_j d_X(x^\star, x^{\star\star})$, *s.t.*

$$|corr(Y^\circ, Y^{\circ\circ})| = \exp(-d_Y(y^\circ, y^{\circ\circ})^2) = \exp(-L_j^2 d_X(x^\star, x^{\star\star})^2) = K(d_X(x^\star, x^{\star\star}); \boldsymbol{\theta}). \tag{4.2}$$

Then by $|corr(Y^\circ, Y^{\circ\circ})| = K(d_X(x^\star, x^{\star\star}); \boldsymbol{\theta})$ *from Equation 4.2, and* $|corr(Y^\star, Y^{\star\star})| = K(d_X(x^\star, x^{\star\star}); \boldsymbol{\theta})$ *from Equation 4.1,* $|corr(Y^\circ, Y^{\circ\circ})| = |corr(Y^\star, Y^{\star\star})|$*. This owes to the constancy of* $\boldsymbol{\theta}$*.*

However, by $|corr(Y^\circ, Y^{\circ\circ})| = \exp(-L_j^2 d_X(x^\star, x^{\star\star})^2)$ *from Equation 4.2, and* $|corr(Y^\star, Y^{\star\star})| = \exp(-L_i^2 d_X(x^\star, x^{\star\star})^2)$ *from Equation 4.1,* $|corr(Y^\circ, Y^{\circ\circ})| \neq |corr(Y^\star, Y^{\star\star})|$*. This owes to the difference between* L_i *and* L_j *in general, for* $i \neq j$*, given that dataset* $\mathbf{D}$ *bears inhomogeneities in its correlation structure.*

Thus, the conclusions from the last two paragraphs suggest a contradiction that can only be removed if – in this case of $L_i \neq L_j$ *–* $corr(Y_i, Y_j)$ *depends on* x_i *and* x_j*, not just via their mutual distance* $d_X(x_i, x_j)$*. In other words, it is true that* $\boldsymbol{\theta}$ *is not a constant, but in instead different in* $[x_{i-1}, x_i)$ *from what it is in* $[x_{j-1}, x_j)$*, (for* $i \neq j; i, j = 1, \ldots, n-1$*). In other words,* $\boldsymbol{\theta}$ *should be dependent on the input locations at which the outputs are generated.*

Hence, correlation between a pair of outputs is dependent on the inputs (at which the outputs are realised), not only via the inter-input distance, but also via the input locations themselves.

Had the data **D** not borne inhomogeneities in its correlation structure, input-output correlation could be modelled as dependent only on the distance between the inputs at which the considered outputs are realised. Then the kernel-parametrised inpter-output correlation could be modelled with a kernel, each hyperparameter of which is a constant.

However, when the empirical correlation structure is inhomogeneous, the hyperparameters cannot in general be constants, but will each be input-dependent.

One remark is that with increasing data size, the identification of inhomogeneity in the correlation of the data can alter.

Example 4.3 *If we use the SQE kernel and data on* Y *that is standardised, there is only one kernel hyperparameter that we need to invoke, towards the learning of the univariate, scalar-valued* *function $f(\cdot)$, where $Y = f(x)$. This one kernel hyperparameter is the length scale hyperparameter ℓ. In other words, $\boldsymbol{\theta}$ reduces in this case, to the scalar-valued hyperparameter ℓ. When the training data* $\mathbf{D}$ *is detected to bear inhomogeneities in its correlation structure, the kernel-parametrised correlation computed between outputs realised at input values $x^\star, x^{\star\star} \in [x_i, x_{i+1})$, reduces to* $\exp(-(d_X(x^\star, x^{\star\star}))^2/\ell^2)$, *which should be different from kernel-parametrised covariance between outputs realised at $x^\star, x^{\star\star} \in [x_j, x_{j+1})$ – for a given $d_X(x^\star, x^{\star\star})$. So ℓ cannot be a constant. [Here $y^\star$ is the value attained – at input value $x^\star$ – by a sample function that is drawn from the underlying GP. Similarly $y^{\star\star}$ is the output value realised at $X = x^{\star\star}$].*

4.3.4.1 $d_Y(\cdot,\cdot)$ used above is a distance function in $\mathcal{Y}$

We present this distance function for a high-dimensional output variable $\boldsymbol{Y}$ in $\mathcal{Y}$.

Theorem 4.2 *The function $d_\mathcal{Y} : \mathcal{Y} \times \mathcal{Y} \longrightarrow \mathbb{R}_{\geq 0}$, defined as*

$$d_\mathcal{Y}(\boldsymbol{Y}, \boldsymbol{Y}^/) := \sqrt{-\log|corr(\boldsymbol{Y}, \boldsymbol{Y}^/)|} \equiv \sqrt{-\log|corr(\tilde{\boldsymbol{f}}(\boldsymbol{x}), \tilde{\boldsymbol{f}}(\boldsymbol{x}^/))|},$$

is a distance function in $\mathcal{Y}$, for non-independent $\boldsymbol{Y}, \boldsymbol{Y}^/ \in \mathcal{Y}$, with $\boldsymbol{Y} = \tilde{\boldsymbol{f}}(\boldsymbol{x})$, and $\boldsymbol{Y}^/ = \tilde{\boldsymbol{f}}(\boldsymbol{x}^/)$.

Proof

1. *$d_\mathcal{Y}$ is invariant to the order of $\boldsymbol{Y}$ and $\boldsymbol{Y}^/$ in its definition.*
2. $\sqrt{-\log|corr(\boldsymbol{Y}, \boldsymbol{Y}^/)|} \geq 0$ *since* $|corr(\boldsymbol{Y}, \boldsymbol{Y}^/)| \in (0, 1]$.
3. $\sqrt{-\log|corr(\boldsymbol{Y}, \boldsymbol{Y}^/)|} = 0 \implies corr(\boldsymbol{Y}, \boldsymbol{Y}^/) = \pm 1$, *i.e. $\boldsymbol{Y}$ is perfectly correlated or perfectly anti-correlated with $\boldsymbol{Y}^/$.*
   If $\boldsymbol{Y}$ is perfectly correlated or perfectly anti-correlated with $\boldsymbol{Y}^/$, $\sqrt{-\log|corr(\boldsymbol{Y}, \boldsymbol{Y}^/)|} = 0$.
4. *Now we check for the triangle rule.*
   Let us assume that:

$$\sqrt{-\log|corr(\boldsymbol{Y}_1, \boldsymbol{Y}_2)|} + \sqrt{-\log|corr(\boldsymbol{Y}_1, \boldsymbol{Y}_3)|} <$$

$$\sqrt{-\log|corr(\boldsymbol{Y}_3, \boldsymbol{Y}_2)|},$$

where $\boldsymbol{Y}_1, \boldsymbol{Y}_2, \boldsymbol{Y}_3 \in \mathcal{Y}$.
Since $|corr(\boldsymbol{Y}_3, \boldsymbol{Y}_2)| \in (0, 1]$, *both sides of this assumed inequality is positive. Then squaring both sides, we get:*

$$2\sqrt{\log|corr(\boldsymbol{Y}_1, \boldsymbol{Y}_2)| \log|corr(\boldsymbol{Y}_1, \boldsymbol{Y}_3)|} <$$

$$\log\left(\frac{|corr(\boldsymbol{Y}_1, \boldsymbol{Y}_2)||corr(\boldsymbol{Y}_1, \boldsymbol{Y}_3)|}{|corr(\boldsymbol{Y}_2, \boldsymbol{Y}_3)|}\right),$$

Then if $|corr(\boldsymbol{Y}_1, \boldsymbol{Y}_2)||corr(\boldsymbol{Y}_1, \boldsymbol{Y}_3)| < |corr(\boldsymbol{Y}_2, \boldsymbol{Y}_3)|$, *the RHS is negative. However, the LHS is non-negative.*

So this is a contradiction. Therefore, the initial assumption is false.

Hence, it holds that

$$\sqrt{-\log|corr(\boldsymbol{Y}_1, \boldsymbol{Y}_2)|} + \sqrt{-\log|corr(\boldsymbol{Y}_1, \boldsymbol{Y}_3)|} \geq$$

$$\sqrt{-\log|corr(\boldsymbol{Y}_3, bY_2)|},$$

$\forall \boldsymbol{Y}_1, \boldsymbol{Y}_2, \boldsymbol{Y}_3 \in \mathcal{Y}$.

Thus, the triangle rule holds.

Then on the basis of points 1-4, we forward $d_{\mathcal{Y}}(\cdot, \cdot)$ *as a distance function in* $\mathcal{Y}$.

□

4.3.4.2 Wrapping up the case of a scalar-valued output

If the given training data bears a homogeneous correlation structure, then we have not encountered any reason for the kernel to be chosen as non-stationary, i.e. then the kernel that is invoked to parametrise the covariance function of the GP, can bear hyperparameters that attain the same values at all inputs in the range $[x_{in}, x_{out}]$. Now, a stationary kernel is one that depends on a pair of values of the input variable, via a distance $d_X(\cdot, \cdot)$, (or more generally, a difference), between these input values. So the hyperparameters of such a stationary kernel do not bear any location-specific dependence, where by "location", we imply location in the input space $\mathcal{X}$. Rather, these hyperparameters attain globally-relevant values. Thus, if the training data does not bear inhomogeneities, then it suffices to model the functional relation between the input and output with a *GP*, the covariance function of which is parametrised by a stationary kernel.

On the other hand, when the training data bears an inhomogenous correlation structure, then the kernel that we use, should bear hyperparameters that vary from one subset of $[x_{in}, x_{out}]$ to another, i.e. should in general depend on the input values at which the kernel is being computed, Then using a stationary kernel will wrongly parametrise the covariance function of the *GP* that is invoked to model the sought functional relation between the input and output variables. After all, it is the correlation structure of the data that needs to be correctly translated into information about the correlation function of this invoked *GP*, and thereby into the correlation between pairs of realisations of the sought function at different input values. So if the correlation structure of the data displays inhomogeneities, using a stationary kernel will give rise to erroneous learning of this function, leading to wrong predictions about the output at test inputs. In such a data situation, a kernel that is parametrised by hyperparameters that are location-specific in general, i.e. vary from one input value to another, will better model the inhomogeneous correlation structure of the available data, (than a kernel with constant hyperparameters).

4.3.5 Generalising to higher dimensions

Let us generalise the above definition of an inhomogeneously-correlated dataset, to one that bears measurements of a generic tensor-valued output variable $\boldsymbol{Y}$ that is associated with a vector-valued input variable $\boldsymbol{X}$. Our ulterior aim is to seek the (tensor-valued) function that represents the relationship between these input and output variables. We do not attempt interpreting L_i in our higher-dimensional generalisation.

Definition 4.6 *Let us consider a random variable* $\boldsymbol{Y} \in \mathcal{Y} \subseteq \mathbb{R}^{m_1 \times m_2 \times \ldots \times m_k}$, *where* $\boldsymbol{Y} = \boldsymbol{y}_i$ *at* $\boldsymbol{X} = \boldsymbol{x}_i$, *with* $\boldsymbol{X} \in \mathcal{X} \subseteq \mathbb{R}^d$. *Here*

$\boldsymbol{Y} = \boldsymbol{f}(\boldsymbol{x})$. *The (k-th ordered) tensor-valued function* $\boldsymbol{f}(\cdot)$ *is considered a sample function of a k-th ordered tensor-variate GP. Let a distance function of* $\mathcal{Y}$ *be* $d_Y(\cdot,\cdot)$, *and a distance function of* $\mathcal{X}$ *be* $d_X(\cdot,\cdot)$. *The training data set* $\mathbf{D} = \{(\boldsymbol{x}_i, \boldsymbol{y}_i)\}_{i=1}^n$ *is identified to bear inhomogeneities in its correlation structure if* $\exists i \in \{1,\ldots,n-1\}$ *s.t.*

$$L_i \notin [L_1 - \delta, L_1 + \delta] \text{ and } L_i \notin [L_2 - \delta, L_2 + \delta] \text{ and } \ldots$$

$$L_i \notin [L_{i-1} - \delta, L_{i-1} + \delta] \text{ and } L_i \notin [L_{i+1} - \delta, L_{i+1} + \delta] \text{ and } \ldots$$

$$\text{and } L_i \notin [L_{n-1} - \delta, L_{n-1} + \delta],$$

where

$$L_i := \frac{d_Y(\boldsymbol{y}_i, \boldsymbol{y}_{i+1})}{d_X(\boldsymbol{x}_i, \boldsymbol{x}_{i+1})}.$$

Here, δ *is the semi-width of an interval of values centred at* L_i *s.t. if* $L_i \in [L_j - \delta, L_j + \delta]$, L_i *and* L_j *are referred to as "similar"; for* $i \neq j$, $i, j = 1, \ldots, n-1$.

We can choose δ variously. Sometimes, we choose δ as $\alpha\bar{L}$, where $\bar{L}$ is the mean of the sample $\{L_1, L_2, \ldots, L_{n-1}\}$, and α is chosen. For instance, in the examples provided above with data on scalar-valued outputs, we set α to be 0.15, and in Figure 4.3 we have displayed results using $\alpha = 0.15$ as well. However, in Figure 4.2 we chose δ as a constant fraction (0.01 or 0.15) of the value of L_i, $\forall i = 1, \ldots, n-1$.

In Figure 4.3, the value of L_i computed given the training pairs $(\boldsymbol{x}_{i-1}, y_{i-1})$ and $(\boldsymbol{x}_i, y_i)$ is plotted for $i = 1, \ldots, n-1$, for the training data $\mathbf{D}$ of size ($n =$)203 that was randomly selected from the publicly-available Tetouan city data that we discussed in Section 3.7 of Chapter 3. In this dataset, the energy consumed over all three zones of Tetouan city is measured, in addition to 5 metereological variables, though in our work we use data on only 4 of these variables. We treat the scalar-valued energy as the output Y. We standardise the 4 atmospheric conditions – temperature (X_1), humidity (X_2), general diffuse flows (X_4), diffuse flows (X_5) – and collate them into the 4-dimensional input vector $\boldsymbol{X} := (X_1, X_2, X_4, X_5)^T$. In our computation of L_i values presented

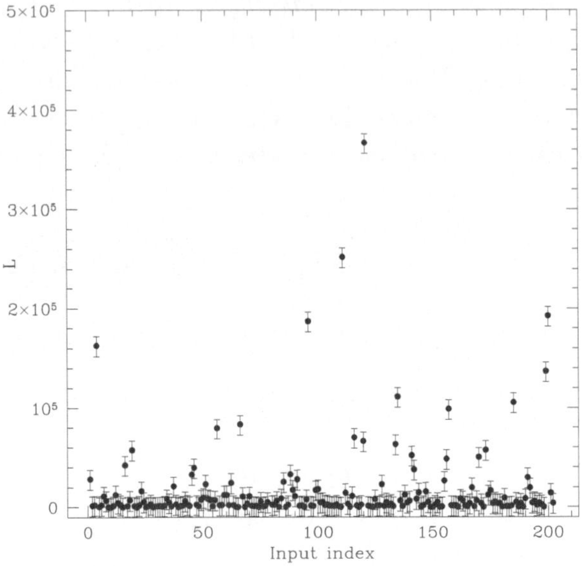

FIGURE 4.3
Bottom left: plot of $L_i := d_Y(y_{i-1}, y_i)/d_X(\boldsymbol{x}_{i-1}, \boldsymbol{x}_i)$ against i for the data on the variation in energy consumed (Y) within Tetouan city with changing multivariate atmospheric conditions, including temperature, humidity, general diffuse flows and diffuse flows – with the four standardised atmospheric condition variables collated into the 4-dimensional vector-valued input $\boldsymbol{X}$. This multivariate dataset was used to learn the function that represents the relationship between this output Y and input $\boldsymbol{X}$ in Section 3.7.2 of Chapter 3. This dataset contains 203 training points, so $i = 1, \ldots, 202$. A δ-neighbourhood of 15% of the sample mean $\bar{L}$ (of ≈ 60301.01) has been superimposed as error-bar on either side of the computed L at a given input index.

in Figure 4.3, we have chosen $\| \cdot - \cdot \|^2$ as $d^2_X(\cdot, \cdot)$, and $| \cdot - \cdot |$ as $d_Y(\cdot, \cdot)$. The L values computed in this sample are s.t. this data is identified to bear inhomogeneities in its correlation structure, and its modelling performed with a stationary kernel – as discussed in Section 3.7.2 – is therefore short of what will provide correct predictions at a new or test value of the input $\boldsymbol{X}$.

4.4 Closed-form predictions

The case of scalar-valued output

For a scalar-valued output – and therefore a scalar-valued function $f(\cdot)$ treated as a sample function of a scalar-variate GP – the closed form prediction of the expectation of the function at any test input $\boldsymbol{a} \in \mathcal{X}$, is given in terms of the covariance structure of this scalar-variate GP, as noted in Section 4.2.3. In fact, the prediction of the mean value of the function at a test input is then given in terms of the kernel that parametrises the covariance function of this GP. The kernel computed at the i-th and j-th design inputs is then the ij-th element of the covariance matrix of the multivariate Normal likelihood that ensues from our modelling of $f(\cdot)$ as a sample function of this GP.

When the output is higher-dimensional

But when the sought function $\boldsymbol{f}(\cdot)$ is higher-dimensional, the likelihood is then tensor Normal in general, and such a tensor Normal density is parametrised by multiple covariance matrices, of which, there is only one inter-realisation covariance matrix. This covariance matrix is s.t. its ij-th element is $Cov(\boldsymbol{f}(\boldsymbol{x}_i), \boldsymbol{f}(\boldsymbol{x}_j))$, i.e. the covariance between the realisation of the random function $\boldsymbol{f}(\cdot)$ generated at the i-th design input, and the realisation generated at the j-th input. This inter-realisation covariance matrix can be parametrised using a kernel. Certain applications might permit kernel-parametrisation of a few of the other covariance matrices of this tensor Normal likelihood, but in general, there are covariance matrices of this likelihood that will remain unparametrised, [28, 24]. The expectation $\mathbb{E}(\boldsymbol{f}(\boldsymbol{x}_{test}))$ of the variable that is the random function $\boldsymbol{f}(\cdot)$ computed at a test input $\boldsymbol{x}_{test}$ in $\mathcal{X}$, is is contributed to, by the kernel that parametrises the inter-realisation matrix, as well as the other covariance matrices. Therefore, such an expectation is erroneous, if the inter-realisation covariance matrix is parametrised with a stationary kernel, when the data on the high-dimensional output $\boldsymbol{Y}$, bears inhomogeneities as stated in Definition 4.6.

Example: when the output is vector-valued

We consider the example of a vector-valued output $\boldsymbol{Y}$, with dimension m, s.t. the function $\boldsymbol{f}(\cdot)$ – where $\boldsymbol{Y} = \boldsymbol{f}(\boldsymbol{x})$ – is likewise vector-valued. The expectation of $\boldsymbol{f}(\boldsymbol{x}_{test})$ is then given as:

$$\mathbb{E}(\boldsymbol{f}(\boldsymbol{x}_{test})) = \mathbf{L}^{(\boldsymbol{\theta})}_{x,train}(\mathbf{L}^{(\boldsymbol{\theta})})^{-1}\mathbf{vec}(\boldsymbol{y}).$$

— Here, $\mathbf{vec}(\boldsymbol{y})$ is the mn-dimensional vector that is obtained by vectorising the $n \times m$-dimensional matrix, the i-th row of which is the realisation of the m-dimensional vector-valued output variable $\boldsymbol{Y}$, at the i-th design point $\boldsymbol{x}_i$ in the training data $\{(\boldsymbol{x}_i, \boldsymbol{y}_i)\}_{i=1}^n$,

— with the $nm \times nm$-dimensional matrix $\mathbf{L}^{(\boldsymbol{\theta})} = \mathbf{V} \otimes \mathbf{K}^{(\boldsymbol{\theta})}$, where

  ⋆ the $n \times n$-dimensional inter-realisation matrix $\mathbf{K}^{(\boldsymbol{\theta})} = [K(\boldsymbol{x}_i, \boldsymbol{x}_j; \boldsymbol{\theta})]$, defined at the value $\boldsymbol{\theta}$ of the hyperparameter vector, with $\boldsymbol{x}_i$ and $\boldsymbol{x}_j$ the i-th and j-th of the n design inputs in the training data $\{(\boldsymbol{x}_i, \boldsymbol{y}_i)\}_{i=1}^n$. This training data is used to learn the vector-valued function $\boldsymbol{f}(\cdot)$;

  ⋆ the $m \times m$-dimensional inter-component matrix $\mathbf{V}$ is s.t. its ij-th element is the covariance between the i-th and j-th components of the vector-valued output variable $\boldsymbol{Y}$; $i, j = 1, \ldots, m$.

— Lastly, $\mathbf{L}^{(\boldsymbol{\theta})}_{x,train}$ is the $m \times mn$-dimensional matrix that results from $\boldsymbol{V}^{(m\times m)} \otimes [K(\boldsymbol{x}_{test}, \boldsymbol{x}_i; \boldsymbol{\theta})]^{(1\times n)}$, where the n-dimensional vector $[K(\boldsymbol{x}_{test}, \boldsymbol{x}_i)]$ is s.t. its i-th element is the value of the kernel computed at the relevant test input $\boldsymbol{x}_{test}$ and the i-th design input $\boldsymbol{x}_i$, $i = 1, \ldots, n$.

For higher-dimensional outputs still, the expectation of the function-valued random variable $\boldsymbol{f}(\boldsymbol{x}_{test})$ is more complicated, with more than two covariance matrices then combining. But in each case, the mean (and the variance) of $\boldsymbol{f}(\boldsymbol{x}_{test})$ is predicted in a closed-form way that is identifiable.

4.5 A new non-parametric kernel

Let us summarise the ambition declared above. We attempt the non-parametric learning of the sought functional relationship between the output $\boldsymbol{Y}$ and input $\boldsymbol{X}$, by treating this function as a random realisation from a GP with a covariance function that is parametrised by a non-parametrically learnt covariance kernel. We appreciate that the kernel needs to acknowledge "inhomogeneities in the correlation structure of the data". In Section 4.3.3 we discussed one way of quantifying such "inhomogeneities in the correlation structure of the data", by defining L_i as the ratio $d_{\mathcal{Y}}(\boldsymbol{y}_i, \boldsymbol{y}_{i+1})/d_{\mathcal{X}}(\boldsymbol{x}_i, \boldsymbol{x}_{i+1})$, where $(\boldsymbol{x}_i, \boldsymbol{y}_i)$ is the i-th training point inside training data $\mathbf{D}$ that is available for the learning of $\boldsymbol{f}(\cdot)$, where $\boldsymbol{Y} = \boldsymbol{f}(\boldsymbol{x})$. We said that if $\exists i \in \{1, 2, \ldots, n-1\}$, s.t. L_i does not lie within a δ-neighbourhood of any $L_j \in \{L_1, \ldots, L_{i-1}, L_{i+1}, \ldots, L_{n-1}\}$, then $\mathbf{D}$ is identified to manifest inhomogeneities in its correlation structure. L_i is "unique" in this sense, compared to any L_j, $\forall j \neq i; i, j \in \{1, \ldots, n-1\}$.

One way of ensuring that the kernel is learnt non-parametrically and is able to address the inhomogeneities in the correlation structure of the data, is by working with the model suggested in Section 4.3.4. To recall, under such a model, at each pair of inputs that the kernel is computed at, the hyperparameters of the kernel are different. As stated above, the learning load in such a model is huge, thus rendering such a proposed model infeasible to implement.

4.5.1 Way out of input-pair dependence of hyperparameters – let hyperparameter be a function of the sample function

While the input-pair dependence of each kernel hyperparameter – as proposed in Section 4.3.4 – clearly acknowledges the variation in the empirical correlation from one location in input space to another, it is the dependence on the input-pairs that pushes up the learning load to infeasible levels. Then we are motivated to substitute the input-pair dependence of each kernel hyperparameter, with dependence of kernel hyperparameters on such a variable, that the computational load is mitigated, while the information content in that alternative treatment remains as comprehensive in the explicitly non-stationary model – in which the covariance kernel varies from one input-pair to another.

From basic principles, we can motivate any kernel hyperparameter θ_c to be dependent on the sample function $\tilde{\boldsymbol{f}}(\cdot)$ that is drawn from the GP that is invoked to model the function $\boldsymbol{f}(\cdot)$ that links $\boldsymbol{X}$ to $\boldsymbol{Y}$. After all, θ_c informs on the nature of variation of the correlation between $\boldsymbol{Y}_i$ (realised at $\boldsymbol{X} = \boldsymbol{x}_i$) and $\boldsymbol{Y}_j$ (realised at $\boldsymbol{X} = \boldsymbol{x}_j$), as we move along a direction of the input space $\mathcal{X}$ that hosts the d-dimensional input variable $\boldsymbol{X}$. Here $c = 1, \ldots, h$. Details of such correlation variation will itself vary, with every new sample function that is drawn from the relevant GP. Indeed, a sample function might be drawn from the GP such that correlation between an earmarked pair of outputs, (say $\boldsymbol{Y}_i$ and $\boldsymbol{Y}_j$) is noted to depreciate in a certain manner, along the c-th direction in $\mathcal{X}$, with increasing distance from $\boldsymbol{x}_i$. However, another sample function might be s.t. it is comparatively less rough in the neighbourhood of $\boldsymbol{x}_i$, and $corr(\boldsymbol{Y}_i, \boldsymbol{Y}_j)$ then fades less rapidly. Then the kernel hyperparameter θ_c that controls the nature of the fading of the correlation between a pair of outputs, along the c-th direction in input space $\mathcal{X}$, can be generically modelled as a function – albeit an unknown function $g_c(\cdot)$ - of the sample function, i.e.

$$\theta_c = g_c(\tilde{\boldsymbol{f}}(\cdot)), \quad \forall c = 1, \ldots, h.$$

When the shape of the sample function drawn from the GP changes – i.e. a new sample function is drawn from the GP – there are changes to how $corr(\boldsymbol{Y}_i, \boldsymbol{Y}_j)$ varies with $|\boldsymbol{x}_i - \boldsymbol{x}_j|$, and thereby to θ_c. This holds $\forall c = 1, \ldots, h$.

Then to kernel parametrise $corr(\boldsymbol{Y}_i, \boldsymbol{Y}_j)$ correctly, in the face of heterogeneities in the data on the output, we will draw a sample function from the GP, at every input pair. In other words, in computing the kernel $K(|\boldsymbol{x}_i - \boldsymbol{x}_j|; \theta_1, \ldots, \theta_h)$ at this ij-th input pair, we will in fact need to draw (from the GP) a sample function $\tilde{\boldsymbol{f}}_{ij}(\cdot)$, s.t. the kernel is in fact correctly represented as:

$$K(|\boldsymbol{x}_i - \boldsymbol{x}_j|; \theta_1 = g_1(\tilde{\boldsymbol{f}}_{ij}(\boldsymbol{x})), \ldots, \theta_h = g_h(\tilde{\boldsymbol{f}}_{ij}(\boldsymbol{x}))),$$

As motivated above, $g_1(\cdot), \ldots, g_h(\cdot)$ are unknown functions. Such a kernel is manifestly non-stationary, as its hyperparameters vary from one input pair to another.

Since we are within a Bayesian framework, anything that is an unknown, is to be considered a random variable, s.t. the unknown function $g_c(\cdot)$ is a function-valued random variable, or a random function – which in principle can be treated as a sample function of an adequately-chosen Stochastic Process. This holds $\forall c = 1, \ldots, h$. A Gaussian Process performs well in generating sought functions, given the flexibility that GPs offer on the functions that are learnt as their sample functions; generalisability across dimensions; ease of computation. So we will treat $g_c(\cdot)$ as generated by a GP as well, as we will soon see. We summarise our above considerations now.

We can introduce non-stationarity in the model if we render each kernel hyperparameter unique to a given input pair. That way, the kernel will be allowed to adopt to the unevenness of the correlation structure of the GP - i.e unevenness of the correlation structure of the data – since in our attempt, the GP correlation is designed to mimic the empirical correlation. Such input pair-dependence of any kernel hyperparameter can follow, if we set each such hyperparameter to depend on the sample function variable, with a new sample function plucked from the underlying GP at each input pair.

However, to proceed to consider the learning of $g_c(\cdot)$, we need to properly understand this model. *What does it even mean to say that the hyperparameter θ_c is a function of the sample function of the underlying GP?* All this and more, we discuss below.

4.5.2 Why $\theta_c = g_c(\tilde{\boldsymbol{f}}(\cdot))$ is not implementable

The current subsection – and a few subsequent subsections – are technical by nature, and ultimately motivate the utter futility of the suggestion that a kernel hyperparameter is an unknown function of the sample function drawn from the GP under consideration. So if the restless reader would like to move directly to Section 4.5.3, they will find a feasible model that ensures input-dependence for kernel hyperparameters, and is still highly parsimonious. The patient reader is welcome to the following.

The learning of the function $g_c(\cdot)$ is called into question, within the model $\theta_c = g_c(\tilde{\boldsymbol{f}}(\cdot))$, where this random function is computed using the sample function $\tilde{\boldsymbol{f}}_{ij}(\cdot)$ that is drawn from the GP, at the input pair $\boldsymbol{x}_i$ and $\boldsymbol{x}_j$ that the kernel is computed at, when parametrising $corr(\boldsymbol{Y}_i, \boldsymbol{Y}_j)$. Learning this $g_c(\cdot)$ function would demand a training set that comprises pairs of: a sample function drawn from the mother GP at the ij-th input pair, and a value of θ_c used in the kernel computed at $\boldsymbol{x}_i$ and $\boldsymbol{x}_j$ – with the training points spread across the relevant ij pairs, i.e. $\forall i, j = 1, \ldots, n;\ i \neq j$. *This model is manifestly non-stationary.* However, learning the function $g_c(\cdot)$ – that takes a function-valued input – appears difficult. We understand why this is so in the following three subsubsections, but prior to that, we clarify notation.

Let the tensor-valued random function $\boldsymbol{f} : \mathcal{X} \longrightarrow \mathcal{Y}$, be s.t. the output variable $\boldsymbol{Y} = \boldsymbol{f}(\boldsymbol{x})$, where input $\boldsymbol{X} = \boldsymbol{x}$, and $\boldsymbol{X} \in \mathcal{X}$ is a d-dimensional vector while $\boldsymbol{Y} \in \mathcal{Y}$ is an m-th ordered tensor. Let the unknown function $\boldsymbol{f}(\cdot)$ – that links the input variable $\boldsymbol{X}$ and the output $\boldsymbol{Y}$ – be modelled as a sample function $\tilde{\boldsymbol{f}}(\cdot)$ from an m-th ordered tensor-variate GP that we refer to as $\mathcal{GP}_{non-stat}$. We set each of the hyperparameters $\theta_1, \ldots, \theta_h$ of the kernel (that parametrises the covariance function of $\mathcal{GP}_{non-stat}$), to be a function of the sample function of $\mathcal{GP}_{non-stat}$, i.e. we set

$$\theta_c = g_c(\tilde{\boldsymbol{f}}(\boldsymbol{x})), \quad \forall c = 1, \ldots, h.$$

4.5.2.1 Learning the composition of functions?

- One interpretation of this model is that we learn the composition of the functions $g_c(\cdot)$ and $\tilde{\boldsymbol{f}}(\cdot)$ as

$$\gamma_c(\cdot) = g_c \circ \tilde{\boldsymbol{f}}(\cdot),$$

  i.e. $\gamma_c : \mathcal{X} \longrightarrow \mathbb{R}$, and compute the output to this function $\gamma_c(\cdot)$ at any input $\boldsymbol{x} \in \mathcal{X}$, to provide the value of the c-th hyperparameter θ_c. In this treatment, it appears that

(a) we are setting ourselves the task of learning the composite function $\gamma_c(\cdot)$, along each direction in input space $\mathcal{X}$, $\forall c = 1, \ldots, h$;

(b) and as if that is not difficult enough, we now have to learn this composite function $\gamma_c(\cdot)$ anew, every time we pluck a new sample function $\tilde{\boldsymbol{f}}(\cdot)$ from the Gaussian Process $\mathcal{GP}_{non-stat}$ – where said sample function models the function $\boldsymbol{f}(\cdot)$ that represents the sought relationship between the output $\boldsymbol{Y}$ and the input $\boldsymbol{X}$. So every time a new sample function is drawn from $\mathcal{GP}_{non-stat}$, a new training set will need to be generated and the new composite function $\gamma_c(\cdot)$ will need to be learnt. Such functional learning at every iteration of the undertaken inference, is infeasible.

4.5.2.2 Difficulties with training data required for learning $g_c(\cdot)$

The interpretation of using the composition of an unknown function and the sample function of the underlying process $\mathcal{GP}_{non-stat}$, does not lend to a feasible implementation. Instead we seek the training data that will be required to learn the function $g_c(\cdot)$. This training set is:

$$\{(\{\tilde{\boldsymbol{f}}(\boldsymbol{x})_{\boldsymbol{x}\in\mathcal{X}}\})_k, (\theta_c)_k\}_{k=1}^{N_{max}},$$

where $\left(\{\tilde{\boldsymbol{f}}(\boldsymbol{x})\}_{\boldsymbol{x}\in\mathcal{X}}\right)_k$ denotes the tensor-valued sample function generated from the tensor-variate process $\mathcal{GP}_{non-stat}$ at the k-th "instance" – out of the total number of N_{max} "instance"s – and $(\theta_c)_k$ is the c-th kernel hyperparameter generated at the k-th realisation of the sample function of this process $\mathcal{GP}_{non-stat}$. To clarify, by "instance" above, we imply any instance during which a new sample function is drawn from $\mathcal{GP}_{non-stat}$. Here, at any iteration index or time-step of the iteration – during which the mother $\mathcal{GP}_{non-stat}$ is updated – there are as many sample functions drawn from the current $\mathcal{GP}_{non-stat}$, as there are pairs of design points.

We appreciate that even if such a training data were to become available, the learning of the sought function $g_c(\cdot)$ - *modelled as a sample function from a scalar-variate GP (referred to as $\mathcal{GP}_c$)* –

would be difficult. Indeed, we would then require the computation of the covariance between:

- the random variable $\theta_c^{(i)}$ that is the c-th kernel hyperparameter realised for the i-th chosen shape of the sample function $\tilde{\boldsymbol{f}}(\cdot)$, and
- $\theta_c^{(j)}$ realised for the j-th shape of this sample function – where a new shape is chosen at every "instance".

Learning every distinct element of the $N_{max} \times N_{max}$-dimensional inter-"instance" covariance matrix, directly from an inferential scheme (such as MCMC) would not be possible for $N_{max} \gtrsim 6$. So we would need to resort to parametrisation of this inter-"instance" covariance matrix using a kernel, that would need to be computed at the difference between the tensor-valued sample functions – of the tensor-variate process $\mathcal{GP}_{non-stat}$ – picked at two "instance"s. This difference computation will be difficult. Basically, it is difficult to perform kernel parametrisation with high-dimensional function-valued kernels.

So to summarise, the model stated above – that sets each kernel hyperparameter as an unknown function of the sample function of $\mathcal{GP}_{non-stat}$ – seems useless!

4.5.2.3 Helps to approximate a sample function as a vector?

It may then be considered possible to avoid difficulties in dealing with a (sample) function-valued input to the sought function $g_c(\cdot)$, by considering an amenable approximation of the sample function $\tilde{\boldsymbol{f}}_{ij}(\cdot)$. Let the output values realised at a set of chosen inputs: $\boldsymbol{X} = \boldsymbol{x}_1^{(try)}, \boldsymbol{X} = \boldsymbol{x}_2^{(try)}, \ldots, \boldsymbol{X} = \boldsymbol{x}_M^{(try)}$, respectively be $\boldsymbol{y}_1^{(try)}, \ldots, \boldsymbol{y}_M^{(try)}$. Then, the input $\tilde{\boldsymbol{f}}_{ij}(\cdot)$ to the function $g_c(\cdot)$, can be approximated by the high-dimensional vector $(\boldsymbol{y}_1^{(try)}, \ldots, \boldsymbol{y}_M^{(try)})^T$. We refer to this vector as "high-dimensional" since we anticipate that M should be high, in order for the shape of the sample function to be adequately sampled.

Then θ_c – that is modelled as the output of the function $g_c(\cdot)$ computed at $\tilde{\boldsymbol{f}}_{ij}(\cdot)$ – can be approximated as $g_c(\cdot)$ computed at $(\boldsymbol{y}_1^{(try)}, \ldots, \boldsymbol{y}_M^{(try)})^T$. The learning of $g_c(\cdot)$ would then require the training set $\{((\boldsymbol{y}_1^{(try,k)}, \ldots, \boldsymbol{y}_M^{(try,k)})^T, \theta_c^{(k)})\}_{k=1}^{N_{max}}$. Here, $\boldsymbol{y}_q^{(try,k)}$ is the output of the (originally-sought) function $\boldsymbol{f}(\cdot)$ computed at $\boldsymbol{X} = \boldsymbol{x}_q^{(try)}$, given the sample function that is drawn from $\mathcal{GP}_{non-stat}$ on the k-th instance. (Here $q = 1, \ldots, M$). This however leads us to question: how is this $\mathcal{GP}_{non-stat}$ specified in the k-th iteration of the inference, so that we can draw a sample function from it, (in the k-th instance)? Said sample function will then be approximated by $(\boldsymbol{y}_1^{(try,k)}, \ldots, \boldsymbol{y}_M^{(try,k)})^T$.

In fact, the k-th design set of values $\theta_1^{(k)}, \ldots, \theta_h^{(k)}$ of the kernel hyperparameters will need to be invoked, and used to kernel parametrise the covariance function of the GP in this k-th iteration/instance, s.t. a sample function is then drawn from the thus specified GP. Then $\boldsymbol{y}_q^{(try,k)}$ could be suggested to be the expectation of the output of $\boldsymbol{f}(\cdot)$, realised at $\boldsymbol{X} = \boldsymbol{x}_q^{(try)}$, $q = 1, \ldots, M$, where the expectation is predicted in a closed-form way, using $\theta_1^{(k)}, \ldots, \theta_h^{(k)}$. Thus, it may appear possible to create a viable training set relevant for the learning of $g_c(\cdot)$.

However, we require the current values of *all* the h hyperparameters – i.e. of $\theta_1, \ldots, \theta_h$ - to predict the output values at each of the M chosen inputs $\boldsymbol{x}_1^{(try)}, \ldots, \boldsymbol{x}_M^{(try)}$. Thus, the training data that we will conjure up to learn $g_c(\cdot)$, is rendered the same as the one that we will use to learn $g_{c'}(\cdot)$ with, for $c \neq c'$; $c, c' = 1, \ldots, h$. This is a fundamental problem with this model for the multivariate case, even under this stretched approximation of the sample function as a vector of values of outputs realised at M number of chosen values of the input $\boldsymbol{X}$.

Even in the univariate case in which we need to learn only a single kernel hyperparameter θ_1, reducing the model $\theta_1 = g_1(\tilde{\boldsymbol{f}}(\cdot))$ to one in which $g_1(\cdot)$ takes $(\boldsymbol{y}_1^{(try)}, \ldots, \boldsymbol{y}_M^{(try)})^T$ as its input implies a very difficult learning exercise, given that M is large. This is because, the M-dimensional vector input to $g_1(\cdot)$ implies that we will then have to learn the large number (M) of hyperparameters of the kernel that parametrises the covariance function of the GP that $g_1(\cdot)$ is modelled using.

From the above we realise that we need to do better than to suggest that θ_c is a function of a sample function that is drawn anew from the mother GP ($\mathcal{GP}_{non-stat}$) when the kernel is computed at the input pair $\boldsymbol{x}_i, \boldsymbol{x}_j$, as distinguished from the input pair $\boldsymbol{x}_{i/}, \boldsymbol{x}_{j/}$. *Instead of suggesting that each kernel hyperparameter is an unknown function of the sample function, we could state that θ_c is a function of a parameter that we identify to be such, that when this parameter attains a new value, a new sample function is drawn from the mother GP.*

4.5.3 A hyperparameter is a function of that parameter, at which, a new sample function $\tilde{\boldsymbol{f}}(\cdot)$ is generated.

The difficulties with the previously-suggested model for θ_c arise from the fact that in this model, the input to the function $g_c(\cdot)$ is another, (generally tensor-valued) function, (that takes a vector as its input). As we saw above, this would lead to the need for working with a high-dimensional function-valued kernel. Then every time the shape of this (tensor-valued) input (function) changes, a different value of a kernel hyperparameter is output.

Instead, we attempt identifying a parameter $T \in \mathcal{T}$, that marks the picking of a new shape of this (tensor-valued) sample function of the mother GP s.t. a different value of a kernel hyperparameter θ_c will be output, when this parameter T changes value from t_1 to t_2. Then we can confidently use T as the input to the function that outputs θ_c, as long as the information content in modelling each kernel hyperparameter as T-dependent is similar to that in learning θ_c as sample function-dependent. This is achieved if on T attaining a new value, a new sample function is drawn from the process $\mathcal{GP}_{non-stat}$.

In other words, at every new value of T, a new shape of the sample function $\tilde{\boldsymbol{f}}(\cdot)$ is picked, i.e. the shape of the sample function, is a function $q(\cdot)$ of T, where $q(\cdot)$ is unknown. Then in this newly proposed model - as distinguished from the previously proposed model in which the hyperparameter θ_c is modelled as $g_c(\boldsymbol{f}(\boldsymbol{x}))$ – we propose θ_c to be $g_c(q(t))$ at $T = t$.

This variable – at each distinct value of which we can generate a new sample function – is the time step variable T of the iteratve inferential scheme that is undertaken to conduct the learning of the sought function $\boldsymbol{f}(\cdot)$. Thus, in this newly proposed model, we consider the input of $g_c(\cdot)$ to be an unknown function $q(\cdot)$ of the time step variable $T \in \mathcal{T} \subset \mathbb{N}$, so that $q : \mathcal{T} \longrightarrow \mathcal{Y}$.

Then comparing the information content in the two models that we have proposed, we say the following.

- Modelling the c-th kernel hyperparameter θ_c as $g_c(\boldsymbol{f}(\boldsymbol{x}))$ implies that we are sampling the input of function $g_c(\cdot)$ by the (shape of the) sample function of $\mathcal{GP}_{non-stat}$.
- On the other hand, modelling θ_c as $g_c(q(t))$ implies sampling the input of function $g_c(\cdot)$ by the time step variable T.

Assuming that the system that generates the training data $\mathbf{D}$ is ergodic, in the inferential scheme used to undertake the learning – of which T is the time-step (or iteration index) variable – the average taken over time-steps will converge to the average taken over sample functions drawn from $\mathcal{GP}_{non-stat}$.

In other words, for an ergodic system,

- sampling (during a given time-step), multiple times (i.e. $n(n-1)/2$ times) from a tensor-variate GP, is equivalent to
- sampling once during each of N time-steps, from a distinct tensor-variate GP - that is distinguished from another, by its covariance function, as the GPs are updated from one time-step to another, within the undertaken iterative inference.

This is true for a large $n(n-1)/2$ and a large N, where there are n design inputs in the training set and N is the number of time-steps of the undertaken inference, with the qualification that the time step is of the equilibrium (i.e. post-burnin) stage of said undertaken inferential procedure.

Indeed, in this book we will use MCMC as the chosen inferential scheme, so that each learnt parameter converges in an average sense, after the chain runs for a large number of iterations.

At the same time, considering $q(t)$ to be the input to $g_c(\cdot)$ is an easier model to work with – as we will now appreciate – over the previously suggested model, while the essence of non-stationarity that was carried by the suggestion that θ_c is $g_c(\boldsymbol{f}(\boldsymbol{x}))$, will be extended to the model in which we set θ_c as $g_c(q(t))$. We will in fact scrutinise this, by constructing a non-stationary equivalence of the kernel comprising hyperparameter θ_c defined as $g_c(q(t))$, $\forall c = 1, \ldots, h$, in Section 4.8.

4.5.4 Proceeding with model $\theta_c = g_c(q(t))$

Having started with the model $\theta_c = g_c(\tilde{\boldsymbol{f}}(\cdot))$, we proceed hereafter with the model

$$\theta_c = g_c(q(t)) \equiv \eta_c(t), \quad c = 1, \ldots, h$$

where the composition of functions $g_c(\cdot)$ and $q(\cdot)$ is $\eta_c(\cdot)$. Here

- $g_c : \mathcal{Y} \longrightarrow \mathbb{R}$, since the sample function of the tensor-variate process $\mathcal{GP}_{non-stat}$ models the relationship $\boldsymbol{f}(\cdot)$ where the tensor-valued output $\boldsymbol{Y} = \boldsymbol{f}(\boldsymbol{x})$ and $\boldsymbol{Y} \in \mathcal{Y}$.

- Also, $q : \mathcal{T} \longrightarrow \mathcal{Y}$.

- So $\eta_c : \mathcal{T} \longrightarrow \mathbb{R}$.

Then $\theta_c = \eta_c(t)$ holds $\forall c = 1, \ldots, h$. This interpretation of the model for θ_c appears implementable, as we will soon see.

4.5.5 Press pause – clarifying notation & pulling everything together

We recall the following.

- We refer to the generally tensor-variate and non-stationary GP – sample functions $\tilde{\boldsymbol{f}}(\cdot)$ of which are considered models for the sought function $\boldsymbol{f}(\cdot)$ – as $\mathcal{GP}_{non-stat}$.
- We refer to the scalar-variate GP, sample functions of which model the functional relationship $\eta_c(\cdot)$ – between the kernel hyperparameter θ_c and the time step variable T – as $\mathcal{GP}_c$, $\forall c = 1, \ldots, h$.

Thus, we view these GPs in two layers – the *outer layer* is built by $\mathcal{GP}_{non-stat}$ and the *inner layer* comprises h scalar-variate GPs $\mathcal{GP}_1, \ldots, \mathcal{GP}_h$.

The full learning of the covariance functions of $\mathcal{GP}_{non-stat}$ as well as $\mathcal{GP}_1, \ldots, \mathcal{GP}_h$, should be undertaken, to do justice to the messiness of the correlation structure of real-world datasets.

So let us summarise all relevant notation now, before we start in-depth discussions on this kernel.

- We seek the functional relation between the (generally) m-th ordered tensor-valued output variable $\boldsymbol{Y}$ and the d-dimensional vector-valued input parameter $\boldsymbol{X}$, i.e. we seek the tensor-valued function $\boldsymbol{f}(\cdot)$ where $\boldsymbol{Y} = \boldsymbol{f}(\boldsymbol{x})$.
- We model this sought function $\boldsymbol{f}(\cdot)$ as a sample function $\tilde{\boldsymbol{f}}(\cdot)$ that is drawn from the (generally) tensor-variate and non-stationary $\mathcal{GP}_{non-stat}$.
- To be precise, the data on $\boldsymbol{Y}$ that is considered in this discussion, is standardised, so that we invoke a zero-mean tensor-variate $\mathcal{GP}_{non-stat}$.

The correlation function of this zero-mean non-stationary tensor-variate process $\mathcal{GP}_{non-stat}$ is parametrised by the kernel

$K(\cdot, \cdot; \boldsymbol{\theta})$, where $\boldsymbol{\theta} \in \mathbb{R}^h$ is the value of the vector-valued variable $\boldsymbol{\Theta}$ that comprises all the hyperparameters of this kernel. We choose to render this kernel non-parametric, by modelling each component of $\boldsymbol{\theta}$ – i.e. each kernel hyperparameter – as a *random function of the sample function from* $\mathcal{GP}_{non-stat}$. The motivation for this model – as condensed using material discussed above – is the following.

The kernel serves the purpose of fitting the inter-realisation correlation matrix of the tensor Normal likelihood – that is induced by $\mathcal{GP}_{non-stat}$ – to that of the training data. (By inter-realisation correlation, we imply correlation between a pair of output variables that are realised at a given pair of design inputs, where the design inputs are included in the training set). It follows that if the kernel has a fixed shape – given by fixed values of hyperparameters of this kernel – then such a parametric kernel will be incapable of fitting the correlation structure of the GP to the generally inhomogeneity-bearing correlaton structure of the data. Then, the GP-generated sample functions are incompatible to the data at hand.

On the other hand, if the kernel smartly adapts its morphology to the roughness of the GP correlation structure, it will be more capable of matching the empirical correlation structure, irrespective of inhomogeneities in the latter. A generic model of such a desired kernel is one, the hyperparameters of which adapt to the sample function of the GP, i.e. a kernel hyperparameter should ideally vary with the sample function, for the kernel to allow a good fit to the empirical data. This is the motivation behind setting each kernel hyperparameter to be an unknown function of the sample function of $\mathcal{GP}_{non-stat}$.

If we allow for the c-th *kernel hyperparameter to be a random function* $g_c(\cdot)$*, the output of which varies from one sample function (of the underlying* $\mathcal{GP}_{non-stat}$*) to another sample function*, then the resulting kernel will produce the correct mapping of the distance $d_{\boldsymbol{X}}(\boldsymbol{x}, \boldsymbol{x}')$ to the correlation between the outputs generated at these inputs, namely, $\boldsymbol{Y}$ and $\boldsymbol{Y}'$ respectively, $\forall \boldsymbol{x}, \boldsymbol{x}'$. (Here

$c = 1, \ldots, h$). Indeed, this construct will do the job, as long as we ensure that the shape of the kernel is driven by that of the sample function of $\mathcal{GP}_{non-stat}$ in such a way that the kernel computed at $\boldsymbol{X} = \boldsymbol{x}$ and $\boldsymbol{X} = \boldsymbol{x}^{/}$ correctly produces the correlation between $\boldsymbol{Y}$ and $\boldsymbol{Y}^{/}$.

Fallibility of sample function dependent model & resolution using ergodicity

At this point we reiterate (from above), the fallibility of the sample function dependent model for a hyperparameter of the kernel used to specify $\mathcal{GP}_{non-stat}$. We recall from above that learning the function $g_c(\cdot)$ that has a function-valued input – namely, the sample function of $\mathcal{GP}_{non-stat}$ – is hard. This input to $g_c(\cdot)$ could in principle be approximated by the learning of the vector of outputs predicted at (a large number of) chosen values of $\boldsymbol{X}$, but for $h > 1$, the training data required to learn each of the functions $g_1(\cdot), \ldots, g_h(\cdot)$, will basically concur. Additionally, even for $h = 1$, the number of parameters that we will have to learn – in order to specify the covariance function of the GP that we will need to invoke to model $g_1(\cdot)$ – is very large indeed, since only a large number of outputs (predicted at the chosen $\boldsymbol{X}$ values), can approximate a sample function drawn from $\mathcal{GP}_{non-stat}$.

The alternative is then to invoke ergodicity [14, 89, 53, 85] of the system for which $\boldsymbol{f}(\cdot)$ is a relevant function, to assert the following. The effect of drawing different sample functions from $\mathcal{GP}_{non-stat}$ – once at each pair of design points at which the kernel $K(\cdot, \cdot; \boldsymbol{\theta})$ is computed – is the same as drawing one sample function from each of a sequence of GPs, the covariance functions of which are slightly different from each other, as rendered during the equilibrium stage of an inferential chain that is run to learn the covariance function of $\mathcal{GP}_{non-stat}$. *Then for the kernel computed at design points $\boldsymbol{x}_i$ and $\boldsymbol{x}_j$, instead of setting θ_c a function $g_c(\cdot)$ computed at the ij-th draw of a sample function from $\mathcal{GP}_{non-stat}$, we can sample once from the $\mathcal{GP}_{non-stat}$ that is current in each post-burnin iteration of our inferential exercise, namely, MCMC.*

Thus, the drawn sample function changes every time the value of the iteration index or time-step $T \in \mathcal{T}$ changes, i.e. the sample function of $\mathcal{GP}_{non-stat}$ in the t-th iteration is a random function $q : \mathcal{T} \longrightarrow \mathcal{Y}$, computed at $T = t$. Given this, we suggest the following model of the hyperparameter:

$$\theta_c = g_c(q(t)) \equiv \eta_c(t), \quad c = 1, \ldots, h$$

where the composition of functions $g_c(\cdot)$ and $q(\cdot)$ is $\eta_c(\cdot)$. Here

- $g_c : \mathcal{Y} \longrightarrow \mathbb{R}$, since the sample function of the tensor-variate process $\mathcal{GP}_{non-stat}$ models the relationship $\boldsymbol{f}(\cdot)$ where the tensor-valued output $\boldsymbol{Y} = \boldsymbol{f}(\boldsymbol{x})$ and $\boldsymbol{Y} \in \mathcal{Y}$.
- Also, $q : \mathcal{T} \longrightarrow \mathcal{Y}$.
- So $\eta_c : \mathcal{T} \longrightarrow \mathbb{R}$.

Then $\theta_c = \eta_c(t)$ holds $\forall c = 1, \ldots, h$.

Given that θ_c is going to be modelled as an unknown/random function of T, we foresee no loss of generality in using a simple overall parametric form for the kernel $K(\cdot, \cdot)$, such as the Square Exponential or the SQE form. Thus, the kernel that parametrises the covariance function of $\mathcal{GP}_{non-stat}$ is SQE with length scale hyperparameters $\ell_1, \ell_2, \ldots, \ell_d$, each of which is a distinct (unknown) function of the iteration index T. So we acknowledge that in this paradigm, $d = h$. Then in our consideration,

$$K(\boldsymbol{x}, \boldsymbol{x}'; \ell_1, \ldots, \ell_d) = \exp\left(-\frac{(\boldsymbol{x} - \boldsymbol{x}')^T \boldsymbol{Q}(\boldsymbol{x} - \boldsymbol{x}')}{2}\right),$$

where the diagonal matrix $\boldsymbol{Q}$ is s.t.

$$\boldsymbol{Q} = [q_{cc'}^2], \text{ with } q_{cc'} = 0, \forall c \neq c', \text{ and } q_{cc'} \equiv q_c = \frac{1}{\ell_c}, \forall c = c',$$

for $c, c' \in \{1, \ldots, d\}$.

So will refer to the length scale hyperparameter ℓ_c as $1/q_c$, where $\ell_c = \eta_c(t)$, with $\eta_c(\cdot)$ a random function. In our Bayesian mindset, anything that is an unknown, is a random variable. In other words, the unknown function $\eta_c(\cdot)$ is a function-valued random variable, i.e. a random function. Then this function is in turn, modelled as a sample function from a Stochastic Process. We choose this process to be a GP that we refer to as $\mathcal{GP}_c$.

In general, the process – the sample function of which models $\eta_c(\cdot)$ – will be distinct from the process that is invoked to model the function $\eta_{c'}(\cdot)$ that outputs the c'-th length scale variable $\ell_{c'}$. Here, $c \neq c'$, and $c, c' \in \{1, \ldots, d\}$. So we will denote the latter process $\mathcal{GP}_{c'}$, while the process that is invoked to model the c-th length scale hyperparameter ℓ_c, is referred to as $\mathcal{GP}_c$.

Now, $\mathcal{GP}_c$ is scalar-variate, since it models the scalar-valued random variable ℓ_c that is the c-th length scale hyperparameter. We recall the interpretation of ℓ_c from the previous chapter; ℓ_c is a scale along the c-th direction of the input space $\mathcal{X}$, s.t. it informs on the distance in $\mathcal{X}$ over which correlation along this direction fades by a stipulated factor. So we can see the structure of the construct that we are aiming to establish, in order to undertake the learning of this high-dimensional functional relation $\boldsymbol{f}(\cdot)$ between two generic variables, given data that is messy in as many ways as one can perceive. For now it appears that this construct comprises d number of scalar-valued $\mathcal{GP}$s:

$$\mathcal{GP}_1, \ldots, \mathcal{GP}_d,$$

with these d scalar-valued $\mathcal{GP}$s nested within the mother non-stationary, tensor-valued process $\mathcal{GP}_{non-stat}$. Figure 4.4 represents this learning strategy that comprises the d number of scalar-variate $\mathcal{GP}$s in the inner layer and a non-stationary tensor-variate $\mathcal{GP}$ in the outer layer.

4.5.6 Why not model hyperparameters of kernels of $\mathcal{GP}_1, \ldots, \mathcal{GP}_d$ with further GPs?

At this stage, the engaged reader is perhaps curious about the sufficiency of two layers of processes in this learning strategy. Why stop at modelling the hyperparameters of $\mathcal{GP}_{non-stat}$ as random functions of the iteration index T? In other words, what about modelling each hyperparameter of the covariance function of $\mathcal{GP}_c$ as a random function as well, $(\forall c = 1, \ldots, d)$? The answer to these questions is that there is a very convincing reason for us to stop this nested architecture for the learning at the aforementioned "inner" stage.

> We will prove that each process $\mathcal{GP}_1, \ldots, \mathcal{GP}_d$, is a stationary process, and prove that the hyperparameters of such stationary processes do not need to be modelled as random functions. Hyperparameters of such processes can be treated as global (unknown) constants, which are then learnt given the *relevant* data. This negates any perceived need for adding further layers to our dual-staged learning strategy that is motivated above.

4.5.7 Training data for learning $\eta_c(\cdot)$

In the last shaded box, there was a convincing reason for emphasising the data that is relevant to the learning of the "global (constant)" hyperparameters of the GPs in the inner layer – as distinguished from the available training set. What is the training data required for the learning of $\eta_c(\cdot)$ for any $c \in \{1, 2, \ldots, d\}$?

Since we will model this sought function $\eta_c(\cdot)$ as a sample function from the scalar-variate $\mathcal{GP}_c$, we will learn $\eta_c(\cdot)$ using the training data set $\{(t_k, \theta_c^{(k)})\}_{k=1}^{N_{lookback}}$, where $\theta_c^{(k)}$ is the value of the random variable θ_c, as observed at $T = t_k$. We access $N_{lookback}$ number of such observations of θ_c at the current value of T, i.e. at the current value of T, we recall $N_{lookback}$ number of sample functions that have been drawn from the process $\mathcal{GP}_{non-stat}$. But how do we understand $N_{lookback}$?

4.5.7.1 What is "$N_{lookback}$"?

The training data required to learn $\eta_c(\cdot)$ is $\mathbf{D}_c := \{(t_k, \theta_c^{(k)})\}_{k=1}^{N_{lookback}}$. These $N_{lookback}$ instances - during which the training set is generated – should be s.t. the value of the hyperparameter θ_c is known to us at each such instance. Then these instances could equivalently be time-steps or iteration indices of the *past* $N_{lookback}$ *iterations of the inferential scheme*, during which we know the value of the random variable θ_c, as updated during such an iteration. The undertaken inference could be MCMC, or a procedure to optimise the posterior of the unknowns given the data, etc. But whatever the inferential strategy, the value of θ_c is updated to $\theta_c^{(k)}$ in the k-th iteration, $\forall k = 1, \ldots, N_{iter}$, where the inference is undertaken over N_{iter} iterations, with $N_{burnin}+1$ the iteration index that marks the beginning of the equilibrium stage of the inference, (or the post-burnin stage, if using MCMC for inference). *Then training data that is available at the t-th iteration of the undertaken inferential scheme – from updated values of θ_c collated over the past $N_{lookback}$ iterations - will be different from the training data made available at the t'-th iteration, for $t \neq t'$. In other words, the set of values of θ_c collected in the "previous $N_{lookback}$" iterations at one time-step (or iteration), is distinguished from that collected at another time-step.*

Thus, the training data used for the learning of $\eta_c(\cdot)$ in the k-th iteration, is different from training data used for such learning in the k'-th iteration. *Yet, the learning of $\eta_c(\cdot)$ should converge – such should be the design of the learning strategy.* We will leave the confirmation of such a desirable later, but using this recently-gathered understanding of the dynamic nature of this training data, the data that we use to learn $\eta_c(\cdot)$ in the k-th iteration, is denoted to bear a dependence on k – denoted as $\mathbf{D}_c^{(k)}$. Such a dataset will be available, $\forall c = 1, \ldots, h$, with

$$\mathbf{D}_c^{(k)} := \{(t, \theta_c^{(t)})\}_{t=k-N_{lookback}}^{k-1},$$

where $\theta_c^{(t)}$ is the value of kernel hyperparameter θ_c, updated in the t-th iteration of the inference, for $c = 1, \ldots, h$.

4.5.8 The dual–layered learning scheme

The learning strategy suggested above, is summarised below.

1. In our search for the relationship between tensor-valued

$\boldsymbol{Y}$ and vector-valued $\boldsymbol{X}$, we set $\boldsymbol{Y} = \boldsymbol{f}(\boldsymbol{x})$, with

$$\boldsymbol{f}(\cdot) \sim \mathcal{GP}_{non-stat}(\underline{0}, \boldsymbol{K}(\cdot, \cdot)),$$

where $\mathcal{GP}_{non-stat}$ is tensor-variate and is non-stationary in general, to address inhomogeneities in the available training data $\mathbf{D}$.

2. Of the resulting tensor-Normal likelihood, the induced covariance matrix – the ij-th element of which is $Cov(\boldsymbol{Y}_i, \boldsymbol{Y}_j)$ – is parametrised using a kernel, where this kernel is in turn parametrised by hyperparameters $\theta_1, \ldots, \theta_h$.
3. For the first n_0 number of iterations, $(n_0 > N_{lookback})$, θ_c is held an unknown constant that we learn using $\mathbf{D}$, within an inferential scheme, $\forall c = 1, \ldots, h$.
4. We use MCMC for our inference, so that in the first n_0 iterations, we update $\theta_1, \ldots, \theta_h$ within each iteration, by proposing θ_c from an appropriate proposal density, and by using appropriate priors $\forall c = 1, \ldots, h$, where the joint posterior of all these parameters given the data $\mathbf{D}$ is known as the scaled product of the multivariate Normal likelihood and the chosen priors.
5. The $n_0 + 1$-th iteration onward, at every iteration, we update θ_c in a 2-staged way, that is mimicked within two successive blocks of each iteration of the inference scheme.

   - In the *first-block* of the k-th iteration, $(n_0 < k \leq N_{iter})$, $\theta_1, \ldots, \theta_h$ are all updated using the inference scheme, s.t. the respective current values of these hyperparameters are $\theta_1^{(k1)}, \ldots, \theta_h^{(k1)}$. The joint posterior of these hyperparameters, given $\mathbf{D}$, is known. Here $n_0 > N_{lookback}$.
   - In the *second block* of the k-th iteration, we use the training data $\mathbf{D}_c^{(k)}$ to learn $\eta_c(\cdot)$, by treating the random function $\eta_c(\cdot)$, as modelled by a scalar-variate process $\mathcal{GP}_c$. The last value of θ_c within this dynamically-changing training data $\mathbf{D}_c^{(k)}$, is the value

of θ_c that was current at the end of the $k-1$-th iteration. This holds $\forall c = 1, \ldots, h$.

6. How do we specify $\mathcal{GP}_c$, for any $c = 1, \ldots, h$? From the definition of $\eta_c(\cdot)$, i.e. from

$$\theta_c = \eta_c(t),$$

the function $\eta_c(\cdot)$ is a mapping from $\mathbb{N}$ to $\mathbb{R}$, because the iteration step, (or time step) $T \in \mathcal{T} \subset \mathbb{N}$, and $\theta_c \in \mathbb{R}$. But all functions $\eta_c : \mathbb{N} \longrightarrow \mathbb{R}$ are continuous. This is a result that is known from Analysis 101. We will see the proof soon. For now, knowing that $\eta_c(\cdot)$ is continuous, we will be able to show that $\mathcal{GP}_c$ can be stationary. The proof for this is also what we will soon see. Thus, it implies that the covariance of $\mathcal{GP}_c$ can be parametrised using a stationary kernel, the hyperparameters of which can be held (unknown) constants, that we will learn from the data; this holds for any $c = 1, \ldots, h$.

7. In the second block of the k-th iteration (where $k > n_0$), we will undertake such learning of the hyperparameters of the stationary kernel that parametrises the covariance function of $\mathcal{GP}_c$, $\forall c = 1, \ldots, h$. For this, we have to update each hyperparameter of the covariance structure of each of $\mathcal{GP}_1, \ldots, \mathcal{GP}_h$ that comprise the inner layer – of our dual-layered learning strategy – the sample functions of which model the respective function $\eta_1(t), \ldots, \eta_h(t)$. Such learning is performed using the standardised look-back data $\mathbf{D}_1^{(k)}, \ldots, \mathbf{D}_h^{(k)}$ respectively.

8. In this 2nd block of the k-th iteration, $\mathcal{GP}_c$ induces a multivariate Normal likelihood with a covariance matrix, the ij-th element of which is $corr[\theta_c^{(i)}, \theta_c^{(j)}] \equiv corr[\eta_c(i), \eta_c(j)]$ – which, as we will soon see, is parametrised by a stationary (SQE) kernel with (unknown) constant hyperparameter δ_c. Here, $i, j = k-1, k-2, \ldots, k - N_{lookback}$; $c = 1, \ldots, h$. We will propose each sought hyperparameter $(\delta_1, \ldots, \delta_h)$ and compute the prior probability density of each such hyperparameter at this proposed value,

to generate the joint posterior of these parameters given data $\mathbf{D}_c^{(k)}$. We will then accept/reject the proposed values of the hyperparameters $\delta_1, \ldots, \delta_h$, as determined by the acceptance criterion of the used inference scheme. Then within this block of the k-th iteration, we will use the updated values of the kernel hyperparameters $\delta_1, \ldots, \delta_h$ (of the covariance matrices induced by $\mathcal{GP}_1, \ldots, \mathcal{GP}_h$ respectively), to perform a closed-form prediction of the mean and variance of the outputs $\theta_1, \ldots, \theta_h$ of the respective functions $\eta_1(\cdot), \ldots, \eta_h(\cdot)$, at input $T = k$. Let the mean of θ_c predicted (in a closed-form way) in the 2nd block of the k-th iteration be $\bar{\theta}_c^{(k)}$. Then:

— $\bar{\theta}_c^{(k)}$ becomes the current value of θ_c, when the 1st block of the $k+1$-th iteration begins, $(\forall c = 1, \ldots, h)$;

— the last element of $\mathbf{D}_c^{(k+1)}$ is $\bar{\theta}_c^{(k)}$, with its first element given by the second element of $\mathbf{D}_c^{(k)}$.

In other words, at every new iteration, the new lookback data is constructed using the elements of the lookback data in the previous iteration, except that the first element of the previous dataset is dropped and the newly predicted value ($\bar{\theta}_c^{(k)}$) of θ_c is included. We record $\bar{\theta}_c^{(k)}$ and the predicted standard deviation of θ_c in the k-th iteration, and after the completion of the chain, use the sample of such values recorded over all iterations, to generate the trace of θ_c – and thereby the marginal density of θ_c – given the data $\mathbf{D}$. We check trendlessness of the trace of θ_c $\forall c = 1, \ldots, h$, to confirm convergence of the chain.

9. We continue this two-staged learning for N_{iter} number of iterations.

4.5.8.1 Existing dual-layered GP-based strategies

Multiple non-stationary kernels exist in the literature, where these strategies model the dependence of each kernel hyperparameter on the input location, as a sample function of a GP, [106, 96]. Thus, these methods invoke two sets of GPs – one GP is at the outer layer, and it is non-stationary in general. This GP at the outer layer then models the sought function $\boldsymbol{f}(\cdot)$ which can be tensor-valued in general. Then in the inner layer, multiple GPs are invoked – each to model the dependence of a kernel hyperparameter on each component of the input variable $\boldsymbol{X}$.

— However, a generic tensor-variate and non-stationary GP in the outer layer, has not been reported – or implemented – so far in the literature.

— More importantly, the existing methods assume that the GPs in the inner layer are all stationary – though, there exists no apriori reason to believe that the GPs in the inner layer are stationary. In other words, stationarity at the inner layer is tacitly assumed in multiple existing work (referred to above) without a proof for this assumption motivated.

— Here, we will soon prove the stationarity of each inner layer GP in the proof to Theorem 4.5 (in Section 4.5.11).

— However, this presented proof cannot be retrospectively invoked to explain the assumption of stationarity in the existing two-layered GP-based learning strategies, for the following reason. This proof stems from our modelling an inner layer GP as the generating process of a $\mathbb{N} \longrightarrow \mathbb{R}$ mapping – to be precise, the inner layer process $\mathcal{GP}_c$ generates the function $\eta_c : \mathbb{N} \longrightarrow \mathbb{R}$, that is learnt using the dynamical lookback data described above. If an existing strategy were designed s.t. an inner layer GP generated such a mapping, we could have extended the presented proof to justify the assumption of stationarity at the inner layer. This is however not the case with the existing methods that use two layers of GPs. In these methods, the function generated by an inner layer GP does not model a kernel hyperparameter as an unknown function of $T \in \mathcal{T} \subset \mathbb{N}$. (In fact, often, existing strategies do not learn the covariance function of the inner layer GP, but choose this covariance function by hand). So the available proof cannot be extended to justify the assumption of the inner layer stationarity.

— Even under the assumption of stationarity, the constant value of hyperparameters of a stationary kernel are still unknowns, and therefore, need to be learnt. Such is however often not undertaken in existing work. Instead, the GPs in the inner layer are specified using covariance functions that are *ad hoc* choices, as hinted above. While the relevance of these chosen covariance functions may be considered optimal in certain datasets, in other data, such choices do not stand justified. Indeed, if the motivation is to forward a robust methodology that can work ubiquitously across datasets of various dimensions, types and sizes, then we need to do better than what the current situation is in the literature.

Indeed, the motivation behind the correct learning of the covariance function of the mother GP – that generates the originally sought function $\boldsymbol{f}(\cdot)$ – is to perform correct prediction of the output at a test input, even when the data is high-dimensional and/or inhomogeneously-correlated.

To summarise, currently a few models have been forwarded, that model each hyperparameter of the kernel – that parametrises the covariance function of the GP that models the sought function $\boldsymbol{f}(\cdot)$ – as dependent on the input, with the input-dependence modelled as a sample function of another GP. In the literature, such a GP is typically – and without justification - treated as stationary, and importantly, it is specified by hand, i.e. there is no learning undertaken at the inner stage of this strategy. These shortcomings are addressed in a genuinely dual-staged learning strategy that we will now come across.

It is clear, that to learn the covariance structure of a GP that models the dependence of a kernel hyperparameter on the input variable $\boldsymbol{X}$, we will require the relevant training data. It is emphasised that the empirically-observed data $\{(\boldsymbol{x}_i, \boldsymbol{y}_i)\}_{i-1}^N$ is not the data that can inform on the function that represents the input-dependence of the hyperparameter.

So the training datasets sought for the learning at the inner layer, need to be motivated – as we have done above – and will consider this below in detail.

An aside: in high-dimensions

When the output variable $\boldsymbol{Y}$ is high-dimensional, not all the covariance matrices can be kernel parametrised. This is another aspect of supervised learning that we will discuss in this chapter. We will illustrate this with an empirical example in which the output is a vector-valued variable.

4.5.9 What if sample functions of a GP are always continuous?

In our attempt at establishing the formulation of the dual-staged learning strategy that we hinted at above, we explore the availability of added information that all sample functions drawn from a GP – such as from $\mathcal{GP}_c$ – are Lipschitz continuous. In the past, a lot of work has been done on continuity – and the nature of continuity – of sample functions of stationary GPs, [12, 45, 19, 1]. In fact, Theorem 1 of [12] states that for a stationary GP with a continuous correlation function, "either with probability one the sample functions are continuous or with probability one they are unbounded in every finite interval".

To monitor continuity of the GPs in the learning strategy at hand, we first recall from [98], that the correlation function of a weakly stationary Stochastic Process is continuous if the correlation is stationary at the input of 0.

Theorem 4.3 *Correlation function of a weakly stationary process* $\{X_t\}_{t\in\mathbb{R}}$ *– and therefore of a stationary GP – is continuous if the correlation function is stationary at* $t = 0$*.*

Proof *We consider* $\{X_t\}_{t\in\mathbb{R}}$ *to be a zero-mean process. So* $\mathbb{E}(X_t) = 0$*,* $\forall t \in \mathbb{R}$*.*

For $\delta \in \mathbb{R}$*,*

$$|corr(X_{t+\delta}, X_0) - corr(X_t, X_0)|^2 = |\mathbb{E}(X_{t+\delta}X_0) - \mathbb{E}(X_tX_0)|^2 =$$

$$|\mathbb{E}((X_{t+\delta} - X_t)X_0)|^2 \leq \mathbb{E}[X_{t+\delta} - X_t]^2 \mathbb{E}[X_0]^2,$$

using the Cauchy Schwartz inequality. But

$$\mathbb{E}[X_{t+\delta} - X_t]^2 = \mathbb{E}(X_{t+\delta})^2 + \mathbb{E}(X_t)^2 - 2\mathbb{E}(X_{t+\delta}X_t).$$

Also,

$$\mathbb{E}(X_i)^2 \equiv Var(X_i) = \mathbb{E}(X_0)^2 \; \forall i \in \mathbb{R}$$

for this stationary process. Then it follows that

$$|corr(X_{t+\delta}, X_0) - corr(X_t, X_0)|^2 \leq \mathbb{E}(X_{t+\delta})^2 + \mathbb{E}(X_t)^2 - 2\mathbb{E}(X_{t+\delta}X_t) =$$

$$\mathbb{E}(X_\delta)^2 + \mathbb{E}(X_0)^2 - 2\mathbb{E}(X_0 X_\delta),$$

where the first two terms result from stationarity of the process and the last term from the stationarity of the correlation function at $t = 0$*, that allows for* $\mathbb{E}(X_{t+\delta}X_t)$ *to be equated to* $\mathbb{E}(X_\delta X_0) \equiv corr(X_\delta, X_0)$*. Thus,*

$$|corr(X_{t+\delta}, X_0) - corr(X_t, X_0)|^2 \leq (\mathbb{E}(X_\delta) - \mathbb{E}(X_0))^2.$$

Then as $\delta \longrightarrow 0$*,* $|corr(X_{t+\delta}, X_0) - corr(X_t, X_0)|^2 \longrightarrow 0$*. Thus, the correlation function of this stationary process (with stationary correlation at* $t = 0$*) is continuous* $\forall t \in \mathcal{T}$*.*

- Now, we would soon want to prove that continuity of sample functions drawn from a GP implies that this GP can be a stationary process – where by arranging the correlation function of this GP to be stationary at the input value of 0, we set the GP correlation function to be continuous, as per Theorem 4.5.
- Indeed, we can see from Theorem 1 of [12] that universally continuous sample functions can be generated by a stationary GP with a continuous correlation function - where such continuity of the GP correlation function can be ascertained by holding this correlation function to be stationary at input value of 0.
- Following this, we argue that a stationary GP (with shift-invariant correlation function at an input of 0) can be invoked to generate the continuous sample functions at hand.

4.5.10 $\mathcal{GP}_c$ is a stationary process

Theorem 4.4 *Given* $\ell_c = \eta_c(t)$*, with* $T \in \mathcal{T} \subset \mathbb{N}$ *and* $\ell_c \in \mathbb{R}$*, the map* $\eta_c : \mathcal{T} \subset \mathbb{N} \longrightarrow \mathbb{R}_{\geq 0}$ *is a Lipschitz-continuous map,* $\forall c = 1, \ldots, h$.

The proof of this theorem is a standard proof of the basic result that the map from the set of natural numbers to reals, is continuous.

Proof *Distance* $d_{t_{1,2}}$ *between* $t_1, t_2 \in \mathcal{T}$ *is* $|t_1 - t_2|$. *Similarly, distance* $d_{\eta_{c;1,2}}$ *between* $\eta_c(t_1), \eta_c(t_2) \in \mathbb{R}_{\geq 0}$ *is* $|\eta_c(t_1) - \eta_c(t_2)|$.

$$\textit{Assume} \quad \frac{d_{\eta_{c;1,2}}}{M} > d_{t_{1,2}}, \quad M > 0, \ M \ \textit{is finite} \ \forall t_1, t_2 \in \mathcal{T}.$$

$$\textit{Consider} \quad \frac{d_{\eta_{c;1,2}}}{M} = 1/2 \Longrightarrow |t_1 - t_2| < 1/2 \quad \textit{by our assumption,}$$

i.e. for this choice of the LHS of the assumed inequality, the only solution for $|t_1 - t_2| < 1/2$ *is* $t_1 = t_2$*, as* $t_1, t_2 \in \mathbb{N}$.

But $t_1 = t_2 \Longrightarrow \eta_c(t_1) = \eta_c(t_2)$ *for injective* $\eta_c(\cdot)$*, i.e. LHS of the assumed inequality is then 0.*
This is a contradiction (contradicts our choice of 1/2 for the LHS).
$\therefore$ *Our assumption is wrong,*
$\Longrightarrow$ *the correct inequation is:*

$$\frac{d_{\eta_{c;1,2}}}{M} \leq d_{t_{1,2}}, \quad M > 0, \ M \ \textit{is finite} \ \forall t_1, t_2 \in \mathcal{T},$$

i.e.

$$|\eta_c(t_1) - \eta_c(t_2)| \leq M|t_1 - t_2|, \quad M > 0, \ M \ \textit{is finite} \, \forall t_1, t_2 \in \mathcal{T} \subset \mathbb{N}$$

i.e. $\eta_c(\cdot)$ *is Lipschitz continuous.*

□

4.5.11 $\mathcal{GP}_c$ is stationary

Theorem 4.5 *The process $\mathcal{GP}_c$ – any sample function of which models the function $\eta_c(\cdot)$ – can be a stationary process with a continuous correlation function.*

Proof *We only need to show that it suffices for $\mathcal{GP}_c$ to be stationary, s.t. $\eta_c(\cdot)$ abides by the constraint of being Lipschitz, as stated in Theorem 4.4.*

By Theorem 4.4 $\eta_c : \mathcal{T} \in \mathbb{N} \longrightarrow \mathbb{R}_{\geq 0}$ is a Lipschitz continuous map.
Hence no sample function of $\mathcal{GP}_c$ - that models the function $\eta_c(\cdot)$ – is unbound within any finite interval.

— *Now, if $\mathcal{GP}_c$ is a stationary process s.t. its covariance is stationary at $t = 0$, then by Theorem 4.3, the correlation function of $\mathcal{GP}_c$ is continuous $\forall t \in \mathcal{T}$.*

— *We choose the covariance function of $\mathcal{GP}_c$ to be stationary at $T = 0$, $\forall c = 1, \ldots, h$.*

— *Then it follows that if $\mathcal{GP}_c$ is modelled as a stationary process, it will be a stationary process with a continuous correlation function.*

Also, Theorem 1 of [12], states that for a stationary GP with a continuous correlation function, "either with probability one the sample functions are continuous or with probability one they are unbounded in every finite interval".

Then by Theorem 1 of [12], if $\mathcal{GP}_c$ is modelled as a stationary process, it will generate universally continuous sample functions.
$\Longrightarrow$ *it is sufficient for $\mathcal{GP}_c$ to be stationary.*

□

As each GP in the inner layer of the dual-layered learning strategy is stationary, its covariance function can be parametrised with a stationary kernel, the hyperparameters of which are constants that we can learn using the lookback data that we know how to generate. Then there is no need to nest further GPs inside $\mathcal{GP}_c$ for any $c = 1, \ldots, h$, where $\mathcal{GP}_1, \ldots, \mathcal{GP}_h$ were nested within $\mathcal{GP}_{non-stat}$.

Definition 4.7 *We model the correlation function of $\mathcal{GP}_c$ as kernel parametrised with an SQE kernel, as:*

$$Corr(\ell_{c,i}, \ell_{c,j}) = \left(\exp\left[-\frac{(t_i - t_j)^2}{\delta_c^2}\right]\right),$$

where the scale hyperparameter of this kernel is $\delta_c \neq 0$ that is a constant. This holds $\forall c = 1, \ldots, h$.

4.6 Implementational details and application

Let us discuss the details of implementing this dual-layered learning strategy, via the simple univariate real-world application to the learning of the thermal variation of the total energy consumed in Tetouan city data, [113]. While the full dataset is available at `https://archive.ics.uci.edu/ml/datasets/Power+consumption+of+Tetouan+city`, in this section, we will discuss the learning of the relationship between the total energy consumed (Y) on different days in Tetouan city in 2017, and the ambient temperature (X_1), using the subset data $\mathbf{D}_1 := \{(x_{1,i}, y_i)\}_{i=1}^{N_{data}}$, of this full publicly-available dataset $\mathbf{D}$, where we had discussed this subset earlier in Chapter 3. Here $N_{data} = 203$ and the number N_{test} of test temperatures is about 50, as used in Chapter 3. Hereon, in the context of this application, we refer to X_1 as X; then the training data is $\mathbf{D} := \{(x_i, y_i)\}_{i=1}^{N_{data}}$.

Unlike in Chapter 3, here we employ the dual-layered learning scheme to perform the learning of the relationship $f(\cdot)$ between Y ad X, to then make predictions of consumed energy, at the same set of test temperatures $x_1^{(test)}, x_2^{(test)}, \ldots, x_{N_{test}}^{(test)}$ that we made

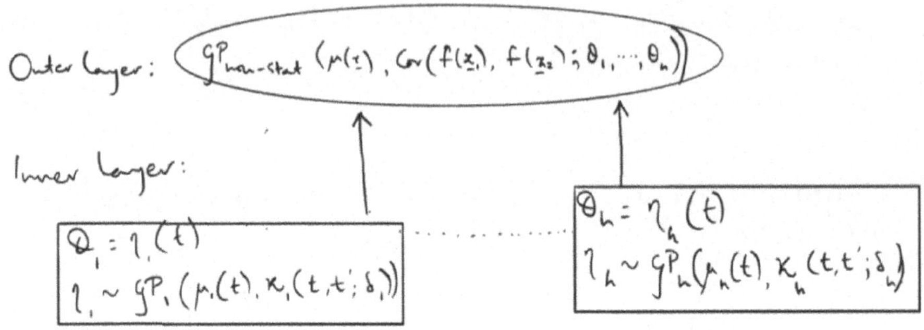

FIGURE 4.4
Cartoon to depict the dual-layered learning strategy that comprises the non-stationary and generally tensor-variate $\mathcal{GP}_{non-stat}$ in the outer layer, where each of the h hyperparameters $(\theta_1, \ldots, \theta_h)$ of the kernel – that parametrises the covariance function of $\mathcal{GP}_{non-stat}$ – is modelled as the output of an unknown function. This unknown (and therefore random) function is in turn modelled as a sample function from another GP. Each such GP $(\mathcal{GP}_1, \ldots, \mathcal{GP}_h)$ then comprises the inner layer. Each such GP is a scalar-variate and – as proven above – stationary process. We construct a dynamic lookback data to undertake the learning of the constant hyperparameter δ_c of the kernel that parametrises the covariance function of $\mathcal{GP}_c$, $\forall c = 1, \ldots, h$.

predictions at, in Chapter 3. In Chapter 3, we had undertaken the learning using a suite of stationary kernels. In this chapter, we will first undertake the learning with the dual staged learning in which the kernel is itself non-parametric. We will then show that this non-parametric kernel is in fact equivalent to a non-stationary kernel, and lastly, we will develop a blending of the non-parametric kernel and its non-stationary equivalence.

4.6.1 1st block

So in this application we learn the function $f(\cdot)$, where $Y = f(X)$. As in the "1st exercise" of the previous chapter, let the N_{data} standardised observations on Y that live within the training set $\mathbf{D}$, comprise the set $\mathbf{d}_{output}$. In anticipation of inhomogeneities in the data, we model $f(\cdot)$ as a sample function $\tilde{f}(\cdot)$ of a zero-mean non-stationary Gaussian Process $\mathcal{GP}_{non-stat}$, the mean function and covariance function of which induce the mean vector $\boldsymbol{\mu}$ and the covariance matrix $\boldsymbol{\Sigma}$ - of the resulting multivariate Normal likelihood, (see Definition 3.1). Given that the data is standardised, $\boldsymbol{\mu} = \underset{\sim}{0}$ since we model the function-valued random variable $f(\cdot)$ as a realisation from a zero-mean process $\mathcal{GP}_{non-stat}$. Again, for this standardised data, the covariance matrix $\boldsymbol{\Sigma}$ reduces to a correlation matrix, the ij-th element of which is the correlation between the output variable corresponding to $X = x_i$ and that at $X = x_j$, $\forall i, j = 1, \ldots, N_{data}$.

4.6.1.1 Likelihood in the 1st block

For the sought function

$$f(\cdot) \sim \mathcal{GP}_{non-stat}(\mu(\cdot), K(\cdot, \cdot)),$$

where $Y = f(x)$, the likelihood of the model parameters is multivariate Normal:

$$\mathcal{L}(\boldsymbol{\mu}, \boldsymbol{\Sigma}; \mathbf{d}_{output}, x_1, \ldots, x_{N_{data}}) =$$
$$(2\pi)^{-n/2}|\boldsymbol{\Sigma}|^{-1/2} \exp\left(-\left((\boldsymbol{y} - \boldsymbol{\mu})^T \boldsymbol{\Sigma}^{-1} (\boldsymbol{y} - \boldsymbol{\mu})\right)\right).$$

where $\boldsymbol{y} = (y_1, \ldots, y_{N_{data}})^T$, with $y_i \in \mathbf{d}_{output}$, for $i = 1, \ldots, N_{data}$. We had in fact chosen to model $f(\cdot)$ with a zero-mean process, s.t. $\boldsymbol{\mu} = \underset{\sim}{0}$.

4.6.1.2 Kernel parametrisation of $\boldsymbol{\Sigma}$

— We undertake kernel parametrisation of the correlation matrix $\boldsymbol{\Sigma}$ as $\boldsymbol{\Sigma} = [Corr(f(x_i), f(x_j)] = [K(x_i, x_j); \theta_1, \ldots, \theta_h]$, with the kernel $K(\cdot, \cdot; \theta_1, \ldots, \theta_h)$ given by a "stationary-looking" form that bears h hyperparameters, where each such hyperparameter is modelled as an unknown function of T, which is the time step variable – or iteration index – of the (MCMC-based) inferential scheme that we use.

— In this application, we in fact choose this "stationary-looking" kernel to be "SQE-looking", the length scale hyperparameter, (ℓ) of which, is modelled as an unknown function $\eta(\cdot)$ that takes a value of T as its input, s.t. $K(x_i, x_j; \ell) = \exp(-(x_i - x_j)^2/\ell^2)$, with

$$\ell = \eta(t), \text{ where } \eta(\cdot) \sim \mathcal{GP}(\mu_{inner}(\cdot), K_{inner}(\cdot, \cdot; \delta)),$$

$i, j \in \{1, \ldots, N_{data}\}$. So in this application, h=1 and $\theta_1 \equiv \ell$.

4.6.1.3 Inference in the 1st block

So within the MCMC chain that we run, in the 1st block of the k-th iteration, we propose a value for ℓ, as randomly selected from a proposal density; in fact, this proposal is Normal by our choice. Variance of this Normal proposal is chosen experimentally.

- The mean of this Normal proposal density is the value of ℓ that is current at the end of the 2nd block of the $k-1$-th iteration, which is given as a closed-form prediction made after updating the value of hyperparameter δ of kernel $K_{inner}(\cdot, \cdot)$ in the 2nd block of the $k-1$-th iteration. Updating of the hyperparameter of this stationary kernel used for specifying the stationary process $\mathcal{GP}$ in the inner layer, allows for the closed-form prediction of the mean output ℓ of the function $\eta(\cdot)$ computed at $T = k-1$, where $\eta(\cdot) \sim \mathcal{GP}(\mu_{inner}(\cdot), K_{inner}(\cdot, \cdot; \delta))$. The predicted ℓ at the end of the 2nd block of an iteration (such as the $k-1$-th iteration) is considered to be the *current* value of ℓ at the end of that iteration. The above holds for $k > n_0$.

- If $k \leq n_0$, the mean of the Normal proposal density in the 1st block of this iteration is the value of ℓ that was current as of the end of the $k-1$-th iteration, where for the first n_0 iterations, Random Walk Metropolis Hastings is undertaken, to learn ℓ in the single block of each such iteration.

The posterior probability density of the proposed and current

ℓ in the k-th iteration are computed given training set $\mathbf{D}$ with the standardised $\boldsymbol{Y}$ values, where the posterior is itself given as proportional to the product of the chosen prior on ℓ, and the multivariate likelihood $\mathcal{L}(\boldsymbol{\mu}, \boldsymbol{\Sigma}; \mathbf{d}_{output}, x_1, \ldots, x_{N_{data}})$ ($\equiv \mathcal{L}(\ell; \mathbf{d}_{output}, x_1, \ldots, x_{N_{data}})$);

4.6.1.4 Priors

If we have useful information on ℓ, we use it to formulate informative priors; else, we work with weak priors. In this application we had no prior information on values of ℓ, and have therefore used weak priors. Using the posterior of the proposed and current ℓ, given the training data, we accept/reject the proposed value of ℓ. Thus, ℓ is updated in the 1st block of the k-th iteration, $\forall k > n_0$.

4.6.2 2nd block

Again, at the beginning of the 2nd block of the k-th iteration, $\forall k > n_0$, we will collect the lookback data on the previous $N_{lookback}$ values of the kernel (length scale) hyperparameter ℓ, that were current as at the end of the previous $N_{lookback}$ 2nd blocks, where $k > n_0 + N_{lookback}$. In our application, we choose n_0 – at a given $N_{lookback}$ – to be s.t. $n_0 > 2N_{lookback}$.

— Using the estimated mean and standard deviation of this collected sample of values of ℓ, we will standardise each element of this sample and then learn the parameters of the process $\mathcal{GP}$ in the inner layer (that models the unknown function $\eta(\cdot)$ that outputs the length scale hyperparameter ℓ).

— That we work with standardised outputs, motivates $\mathcal{GP}$ to be a zero-mean process. The likelihood induced by this $\mathcal{GP}$ is multivariate Normal, with a null vector for the mean.

— The correlation matrix of this likelihood is Δ, which we kernel parametrise with a stationary SQE kernel – with constant hyperparameter δ that we will learn using the data. This choice of the kernel is supported by the aforementioned mathematical result that each scalar-variate GP in the inner layer of our dual-layered learning strategy, is stationary.

4.6.2.1 Lookback data & the inner layer

Since the correlation matrix $\boldsymbol{\Delta}$ of the scalar-variate inner layer process $\mathcal{GP}$ is parametrised with an SQE kernel, bearing the constant length scale $\delta \neq 0$, the only model parameter of the multivariate Normal likelihood (relevant at the 2nd block of the k-th iteration) is δ – which we will update using the lookback data $\mathbf{D}^{(k)}_{lookback}$. As stated above, this lookback data is collated using the current value of ℓ at the end of the 2nd block of the $N_{lookback}$ previous iterations, (for $k > n_0$, when $n_0 > 2N_{lookback}$, as used in this application).

Thus the lookback data available in the 2nd block of the k-th iteration of the inferential scheme, (i.e. for $T = k$) – to update δ - comprises the $N_{lookback}$-sized set:

$$\mathbf{D}^{(k)}_{lookback} = \{(i, \ell^{(i)}\}_{i=k-N_{lookback}}^{k-1}.$$

At the beginning of the 2nd block of the k-th iteration, we standardise the values of ℓ using the sample $\{\ell^{(k-1)}, \ldots, \ell^{(k-N_{lookback})}\}$, and collate the standardised values into the dataset $\mathbf{d}^{(k)}_{lookback} := \{\ell^{(i)}\}_{i=k-N_{lookback}}^{k-1}$.

4.6.2.2 Likelihood in the 2nd block & inference

Then in the 2nd block of the k-th iteration, the multivariate Normal likelihood of the model parameters is

$$\mathcal{L}(\delta; \mathbf{d}^{(k)}_{lookback}, k-1, \ldots, k-N_{lookback}) =$$

$$(2\pi)^{-N_{lookback}/2} |\boldsymbol{\Delta}|^{-1/2} \exp\left(-\left((\boldsymbol{\ell}^{(k)})^T \boldsymbol{\Delta}^{-1} (\boldsymbol{\ell}^{(k)})\right)\right),$$

where $\boldsymbol{\ell}^{(k)} := (\ell^{(k-1)}, \ldots, \ell^{(k-N_{lookback})})^T \delta_k$.

In the 2nd block of this k-th iteration, we will propose a new value of δ from a Normal proposal density with an experimentally-chosen variance, and a mean that is the currently accepted value of δ, from the 2nd block of the $k-1$-th iteration, for $k > n_0$. For $k = n_0$, this mean is the seed for δ. For $k < n_0$, we do not have two blocks in our MCMC chain, so this subsection is not relevant for such values of k.

The posterior probability density of δ given the lookback data is computed separately at the proposed – and the current – value of δ, where this posterior is given as proportional to the product of the likelihood $\mathcal{L}(\delta; \mathbf{d}^{(k)}_{lookback}, k-1, \ldots, k-N_{lookback})$ mentioned above, and (weak) priors, where this weakness typically arises out of lack of information on δ at the outset. Then using the posterior computed at the proposed and current values of δ, given the lookback data relevant to the k-th iteration, we accept/reject the proposed value of δ. Thus, δ is updated at the end of the 2nd block of the k-th iteration.

4.6.2.3 Predicting ℓ using the updated GP in the inner layer

Using this updated δ, the mean and variance of the (standardised) output – namely, the length scale ℓ that is output by the function $\eta(\cdot)$ – are predicted in a closed form way, and these predicted mean and standard deviation of ℓ are transformed back to the unstandardised values, using the standardisation details imposed on the lookback data at the beginning of the 2nd block of the k-th iteration. This mean ℓ, predicted at the end of the 2nd block of the k-th iteration, using $\mathbf{D}^{(k)}_{lookback}$ and the value δ_k (of δ) that is updated in this 2nd block of the k-th iteration, is:

$$\begin{aligned} \ell^{(k)} = \quad & (e^{-((k-1)-k)^2/2(\delta_k)^2}, \ldots, e^{-(k-N_{lookback}-k)^2/2(\delta_k)^2})\boldsymbol{\Delta} \\ & (\ell^{(k-1)}, \ldots, \ell^{(k-N_{lookback})})^T, \end{aligned}$$

$\boldsymbol{\Delta} = [\exp(-(i-j)/2(\delta_k)^2]$, for $i, j = k-N_{lookback}, \ldots, k-1$. Thus $\forall n_0 < k < N_{iter}$, the matrix $\boldsymbol{\Delta}$ retains its form - only the value of δ varies as we move from one iteration to another.

So to summarise, in this univariate and uni-dimensional application, likelihood of the covariance matrix induced by the zero-mean outer GP $\mathcal{GP}_{non-stat}$ is

$$\begin{aligned} \mathcal{L}(\ell; \mathbf{d}_{output}, x_1, \ldots, x_{N_{data}}) = \\ (2\pi)^{-n/2}|\boldsymbol{\Sigma}|^{-1/2}\exp\left(-\left((\boldsymbol{y})^T\boldsymbol{\Sigma}^{-1}(\boldsymbol{y})\right)\right), \end{aligned} \tag{4.3}$$

where ℓ is included within $\mathbf{\Sigma}$ as:

$$\mathbf{\Sigma} = \left[\exp\left(-\frac{(x_i - x_j)^2}{2\ell^2}\right)\right], \tag{4.4}$$

with

$$\ell := \eta(k) \text{ at } T = k,$$

and the function

$$\eta(\cdot) \sim \mathcal{GP}(\mu_{inner}(\cdot), Cov_{inner}(\cdot, \cdot)),$$

s.t. likelihood of the covariance matrix $\mathbf{\Delta}$ at $T = k$ – induced by $\mathcal{GP}$ in the inner layer – is:

$$\mathcal{L}(\mathbf{\Delta}; \mathbf{d}^{(k)}_{lookback}, k-1, \ldots, k - N_{lookback}) = (2\pi)^{-N_{lookback}/2} |\mathbf{\Delta}|^{-1/2} \exp\left(-\left((\boldsymbol{\ell}^{(k)})^T \mathbf{\Delta}^{-1} \boldsymbol{\ell}^{(k)}\right)\right), \tag{4.5}$$

while $\mathbf{\Delta}$ is kernel parametrised in the k-th step of the inference, using the value δ_k of the hyperparameter δ updated in the k-th iteration. Thus in this k-th iteration:

$$\mathbf{\Delta} = \left[\exp\left(-\frac{(i-j)^2}{(\delta_k)^2}\right)\right], \tag{4.6}$$

$\forall i, j = k-1, \ldots, k - N_{lookback}$.

4.6.3 Results of modelling thermal dependence of energy consumption using nested model

To learn the relation between temperature and the energy consumed in the whole of Tetouan city – using a subset of the publicly available Tetouan city data that we discuss in Section 3.7 of Chapter 3 and display in Figure 3.3 – we implement the nested model that we have discussed in this section. It is only a subset of 203 training data points that are pairs of values of design temperature and the energy consumption realised at this temperature, that constitute the training set $\mathbf{D}$ relevant to this current application. So the sought relation is treated as a sample function drawn from a non-stationary process $\mathcal{GP}_{non-stat}$, the covariance function of which is non-parametric – underlined by another GP (called $\mathcal{GP}$), where the nature of $\mathcal{GP}_{non-stat}$ and $\mathcal{GP}$ are discussed above. The length scale hyperparameter ℓ of the kernel that parametrises the covariance function $\mathcal{GP}_{non-stat}$, and the length scale δ of the kernel that parametrises the covariance function of $\mathcal{GP}$, are learnt within a Metropolis-within-Gibbs algorithm that is discussed in Chapter 3, [9, 30]. Traces of these length scale hyperparameters are depicted in Figure 4.5. Histograms of the output values predicted at 16 (out of the 48 test inputs at which prediction is undertaken) are plotted in Figure 4.6. The true value of the output at the given test input is depicted in red, (or gray in the monochromatic version).

Predictions made at 48 test inputs are shown in the left-most panel of Figure 4.7. The predictions are over-plotted on the true values of the output (energy consumption), where the latter are plotted in red/grey. The error bar plotted on either side of the predicted mean at any test input, is 2 times the predicted standard deviation of the output at this input. The middle and right-most panels of this figure refer to predictions performed with other models of the kernel that we discuss soon below.

4.7 An aside: correlation matrix of inner GP is symmetric circulant

We understand that the kernel that parametrises the correlation function $\kappa_c(\cdot,\cdot;\delta_c)$ of the $\mathcal{GP}_c$ – in the inner layer of the dual-layered learning strategy – lends itself to a periodic form. Here $c = 1,\ldots,h$. Let us explore this using a simple SQE form for this kernel used to specify $\mathcal{GP}_c$. In the t-th iteration, the lookback data is $\mathbf{D}^{(c,t)}_{lookback} = \{(i, \ell^{(c)}_i)\}^{t}_{i=t-N_{lookback}+1}$. Then at this t-th iteration,

$$\kappa_c(t_i, t_j; \delta_k) = \exp\left(-\frac{(i-j)^2}{2\delta_k^2}\right),$$

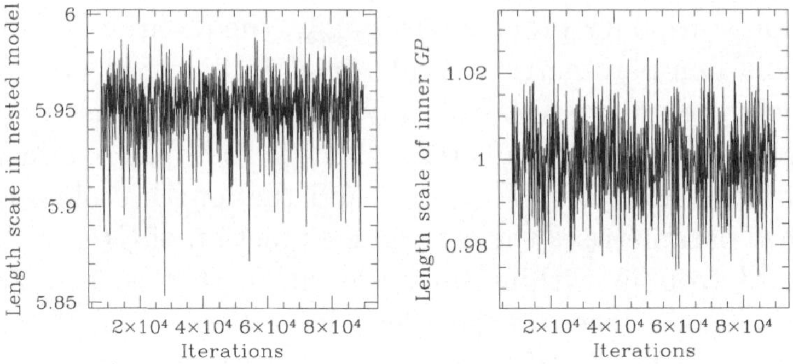

FIGURE 4.5
Left: trace of the length scale hyperparameter of $\mathcal{GP}_{non-stat}$, updated in the first block of the MCMC chain run to learn the relation between temperature and energy consumption in Tetouan city. *Right:* trace of the length scale hyperparameter of the inner GP that generates the function that outputs the length scale of $\mathcal{GP}_{non-stat}$.

$$\forall i, j = t - N_{lookback} + 1, t - N_{lookback} + 2, \ldots .t.$$

So we can see that at any iteration, the argument of the (exponential term in the) form of the kernel lies in the set $\{N_{lookback} - 1, N_{lookback} - 2, ..., 0\}$. Indeed, the correlation matrix $\mathbf{\Delta}_k = [\kappa_k(t_i, t_j; \delta_k)]$ is

$$\mathbf{\Delta}_k =$$

$$\begin{bmatrix} 1 & e^{-\frac{(1)^2}{2\delta_k^2}} & \ldots & \ldots & \ldots & e^{-\frac{(N_{lookback}-1)^2}{2\delta_k^2}} \\ e^{-\frac{(1)^2}{2\delta_k^2}} & 1 & e^{-\frac{(1)^2}{2\delta_k^2}} & \ldots & \ldots & e^{-\frac{(N_{lookback}-2)^2}{2\delta_k^2}} \\ e^{-\frac{(2)^2}{2\delta_k^2}} & e^{-\frac{(1)^2}{2\delta_k^2}} & 1 & e^{-\frac{(1)^2}{2\delta_k^2}} & \ldots & e^{-\frac{(N_{lookback}-3)^2}{2\delta_k^2}} \\ \ldots & \ldots & \ldots & \ldots & \ldots & \ldots \\ e^{-\frac{(N_{lookback}-1)^2}{2\delta_k^2}} & e^{-\frac{(N_{lookbak}-2)^2}{2\delta_k^2}} & \ldots & \ldots & e^{-\frac{(1)^2}{2\delta_k^2}} & 1 \end{bmatrix}$$

Thus, this correlation matrix is a symmetric circulant matrix [111]. A circulant matrix is one in which the i-th row is the same as

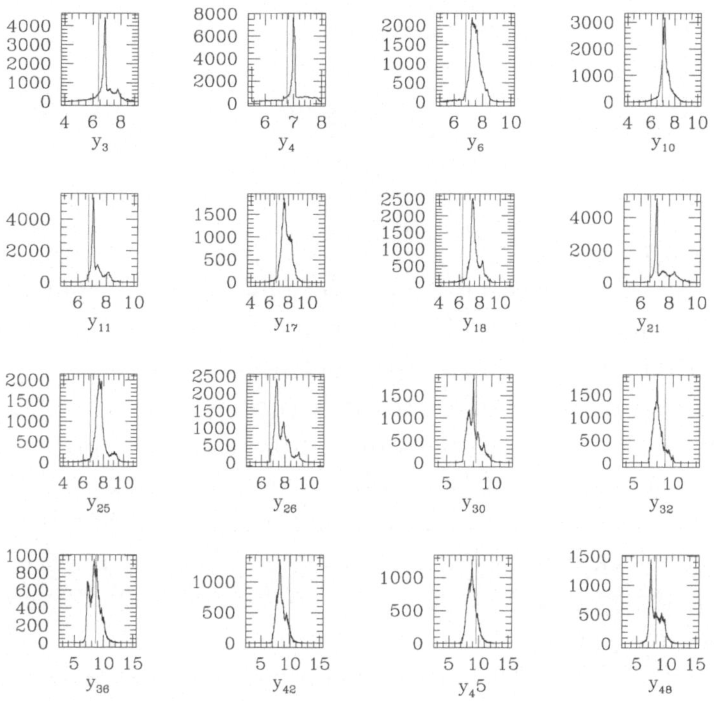

FIGURE 4.6
Plots of histograms of predicted values of the output at 16 out of the 48 considered test inputs – with a prediction done at each iteration of the Metropolis-within-Gibbs chain undertake to learn the (length scale) hyperparameter of the kernel that parametrises the covariance function of $\mathcal{GP}_{outer}$ and that of the kernel that parametrises the covariance function of $\mathcal{GP}_{inner}$. The true value of the output at the given test input is depicted in the red vertical line superimposed on each histogram – or in grey line in the monochromatic version.

the $i+1$-th row, except that every element of the $i+1$-th row is shifted by one column, over the i-th row. A symmetric matrix with such a structure is a symmetric circulant matrix. The interesting property of circulant matrices is that operating a vector with a circulant matrix is equivalent to circular convolution [43].

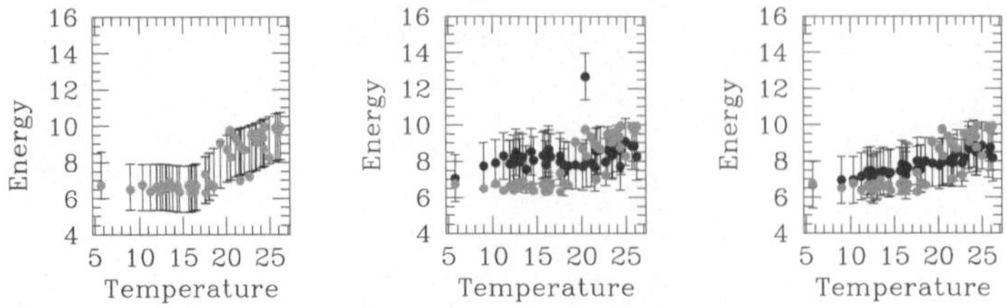

FIGURE 4.7
Left: in black is represented predictions of the output made at 48 test inputs, by modelling the sought function $f(\cdot)$ as a sample function of $\mathcal{GP}_{non-stat}$, the correlation function of which is parametrised by a non-parametric kernel – where $Y = f(x)$, with input X representing temperature, and output Y representing energy consumed. 95% HPD credible regions on each prediction are shown in the black error bars, about the mean that is depicted as a filled black circle. The true value of the consumed energy is in green, (or grey in the monochromatic version). *Middle:* same as in the left panel, except the correlation function of $\mathcal{GP}_{non-stat}$ is parametrised by the non-stationary kernel that is constructed using the non-parametric kernel. *Right:* same as in the left panel, except the correlation function of $\mathcal{GP}_{non-stat}$ is parametrised by the blending of the non-parametric and non-stationary kernels.

We are interested in the multiplication of a vector by this circulant matrix $\mathbf{\Delta}_k$, because such a multiplication is part of the prediction of the output (ℓ_k) at a test input (value of iteration index T), where such an input and output are of the function $\eta_c(\cdot) \sim GP_c(\mu_c(\cdot), \kappa_c(\cdot,\cdot))$, where $\ell_c = \eta_c(t)$. Indeed, then the output $\ell_c^{(test)}$ predicted at a test input $t^{(test)}$ is

$$\ell_c^{(test)} =$$

$$(e^{(t-t^{(test)})/2\delta_c^2} \ldots e^{(t-N_{lookback}+1-t^{(test)})/2\delta_c^2})\mathbf{\Delta}_c(\ell_t \;\ldots\; \ell_{t-N_{lookback}+1})^T.$$

Thus, $\mathbf{\Delta}_c(\ell_t, \ldots, \ell_{t-N_{lookback}+1})^T$ is indeed a circular convolution. Therefore, the predicted value of the length scale hyperparameter at iteration index $t^{(test)}$ is then a (transposed) vector multiplied by this convolution. Above, $c = 1, \ldots, h$.

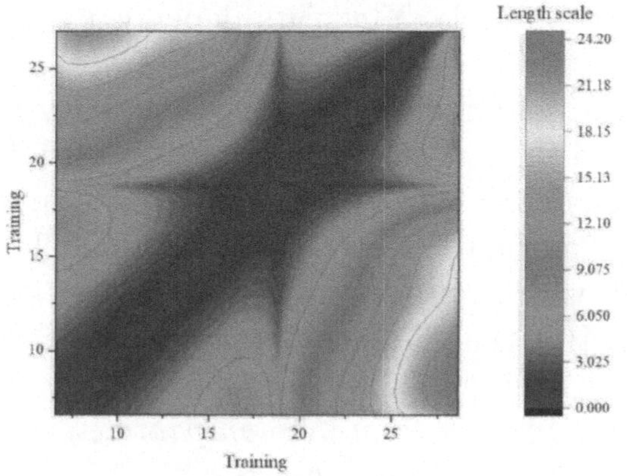

FIGURE 4.8
Contour plot of length scale hyperparameters computed for the non-stationary kernel, as a function of design input pairs, in the illustration using the data on thermal variation of energy consumption in Tetouan city. This non-stationary kernel is equivalent to the non-parametric kernel that parametrises the covariance function of $GP_{non-stat}$ on the outer layer of our 2-layered learning strategy. Length scale hyperparameters are also computed for input pairs, one of which is a design point, and the other a test point. Similarly, length scales at each pair of test inputs are also computed.

4.8 Non-stationary equivalence of the above non-parametric model

In the model development clarified above, the kernel that is used to parametrise the covariance function of $\mathcal{GP}_{non-stat}$ – that models the generally tensor-variate function $\boldsymbol{f}(\cdot)$ – is allowed hyperparameters that vary with every iteration of the undertaken inference. It is after all a dependence on the shape of the sample function drawn from $\mathcal{GP}_{non-stat}$, that marks each kernel hyperparameter in our model – and a sample function with a new shape is drawn from $\mathcal{GP}_{non-stat}$ at every time-step of the undertaken inference. Also, this model inherently renders each hyperparameter dependent on the input variable $\boldsymbol{X}$ via its dependence on the sample function of $\mathcal{GP}_{non-stat}$.

To paraphrase, the model $\theta_c = \eta_c(t)$, is a proxy for the model $\theta_c = g_c(\tilde{\boldsymbol{f}}(\boldsymbol{x}))$, and the former is comparatively easy to implement than the latter model. If it is a proxy of the latter model, then we expect the same information content in this model as there is in the latter model, which imposes an $\boldsymbol{X}$-dependence on each kernel hyperparameter. But the $\boldsymbol{X}$-dependence appears absent from the model $\theta_c = \eta_c(t)$. *Then how can we extract the input-variation of the kernel hyperparameters from the non-parametric model that we have motivated in the previous subsection?*

4.8.1 A non-stationary non-parametric kernel

We state $\theta_c = \eta_c(t)$, where $\eta_c(\cdot) \sim \mathcal{GP}_c(\mu_c(\cdot), K_c(\cdot, \cdot; \delta_c)))$, s.t. at the k-th time-step – i.e. at the k-th iteration of the inferential scheme undertaken to learn unknowns – the likelihood of δ_c given the lookback data $\mathbf{d}_c^{(k)}$ is multivariate Normal with a null mean vector and correlation matrix $\boldsymbol{\Delta}_c$. Here $\boldsymbol{\Delta}_c = [K_c(t, t^{/}; \delta_c)]$; $t, t^{/} = 1, \ldots .N_{lookback}$, $c = 1, \ldots, h$. (Since $\mathcal{GP}_c$ is stationary, $K_c(\cdot, \cdot, \delta_c)$ is invariant to a shift of the input variable T; hence these bounds for $t, t^{/}$). Here $c = 1, 2, \ldots, h$.

We would like to compute the correlation between the output realised at the i-th design input $\boldsymbol{x}_i$, and that realised at the j-th design input $\boldsymbol{x}_j$ in this dual-staged learning strategy, and compare this to $corr(\boldsymbol{Y}_i, \boldsymbol{Y}_j)$ modelled using a non-stationary kernel that parametrises the covariance function of $\mathcal{GP}_{non-stat}$. Said comparison will inform on the link between the hyperparameters of the proposed non-stationary kernel and the inputs $\boldsymbol{x}_i, \boldsymbol{x}_j$. Such could be made possible with an estimate of $corr(\boldsymbol{Y}_i, \boldsymbol{Y}_j)$ obtained using a sample of realisations $\{\boldsymbol{y}_i^{(1)}, \boldsymbol{y}_i^{(2)}, \ldots, \boldsymbol{y}_i^{(N_\epsilon)}\}$ of the random variable $\boldsymbol{Y}_i$, and a similarly-sized sample of realisations of $\boldsymbol{Y}_j$, where these samples are generated using the non-parametric kernel model in which the hyperparameter θ_c is modelled as $\eta_c(t)$. How could we generate such samples of $\boldsymbol{Y}_i$ and $\boldsymbol{Y}_j$? Here $c = 1, \ldots, h$ and $i \neq j$, $i, j \in \{1, \ldots, n\}$.

We understand at the very outset, that it is not possible to attain (via prediction) different values of the random variable $\boldsymbol{Y}_i$ and $\boldsymbol{Y}_j$ realised at the design points $\boldsymbol{x}_i$ and $\boldsymbol{x}_j$ respectively, since all sample functions will pass through the values of the output *observed* at each of the design points in the available training data.

Then instead of sampling at $\boldsymbol{X} = \boldsymbol{x}_i$ *across different shapes of sample functions of* $\mathcal{GP}_{non-stat}$*, we choose to sample at different values of* $\boldsymbol{X}$ *within an* ϵ*-neighbourhood of* $\boldsymbol{x}_i$*, given a sample function (of* $\mathcal{GP}_{non-stat}$*).* Now, we recall that in each time-step of the iterative inferential scheme that we undertake, a sample function $\tilde{\boldsymbol{f}}(\cdot)$ – of a given shape – is drawn from $\mathcal{GP}_{non-stat}$.

Using this, we rephrase our approach: at a given time-step - or iteration – of the undertaken inference, we predict the output at different values of the input $\boldsymbol{X}$, from within an ϵ-neighbourhood of $\boldsymbol{x}_i$.

— So we set up a neighbourhood of width ϵ along each of the d directions in input space $\mathcal{X}$, centred around the i-th design point $\boldsymbol{x}_i = (x_1^{(i)}, x_2^{(i)}, \ldots, x_d^{(i)})^T$. (Recall that $\boldsymbol{X} = (X_1, \ldots, X_d)^T \in \mathcal{X}$).

— Thus, along the c-th direction, we choose N_ϵ number of values of X_c lying in the interval $(x_c^{(i)}, x_c^{(i)} + \epsilon]$, $\forall c = 1, \ldots, d$.

— Such chosen values of X_c are denoted

$$x_{c,1}^{(i)}, x_{c,2}^{(i)}, \ldots, x_{c,N_\epsilon}^{(i)},$$

s.t. we use these to construct the sample $S_c^{(i)}$ of N_ϵ values of the vector $\boldsymbol{X}$, close to the i-th design point.

— This sample $S_c^{(i)}$ includes the N_ϵ values of the d-dimensional vector $\boldsymbol{X} = (X_1, \ldots, X_d)^T$, with the c-th component X_c (of $\boldsymbol{X}$) attaining values in the ϵ-neighbourhood of $x_c^{(i)}$, while all the other $d-1$ components attain their designed values. Thus,

$$S_c^{(i)} := \{(x_1^{(i)}, x_2^{(i)}, \ldots, x_{c-1}^{(i)}, x_{c,k}^{(i)}, x_{c+1}^{(i)}, \ldots, x_d^{(i)})^T\}_{k=1}^{N_\epsilon}.$$

— We understand that

$$|x_c^{(i)} - x_{c,k}^{(i)}| \leq \epsilon, \quad \forall k = 1, \ldots, N_\epsilon,$$

and therefore $x_{c,k}^{(i)}$ is close to $x_c^{(i)}$ since ϵ is small by choice.

— Thus, we generate $d \times N_\epsilon$ number of neighbours of the i-th design point $\boldsymbol{x}_i$, within this ϵ-neighbourhood that we have generated. The sample of such generated neighbours of $\boldsymbol{x}_i$ are in the set:

$$\{S_1^{(i)}, S_2^{(i)}, \ldots, S_d^{(i)}\}.$$

All such neighbours of $\boldsymbol{x}_i$ are close to $\boldsymbol{x}_i$, since ϵ is small.

— We then predict expected values of the output at each of such $N_\epsilon \times d$ number of neighbours of $\boldsymbol{x}_i$ that live within its ϵ-neighbourhood. Let the expected output predicted at

$$\boldsymbol{x}_i^{(c,k)} := (x_1^{(i)}, x_2^{(i)}, \ldots, x_{c-1}^{(i)}, x_{c,k}^{(i)}, x_{c+1}^{(i)}, \ldots, x_d^{(i)})^T$$

be $\boldsymbol{y}_i^{(c,k)}$.

— Then values of the output variable, predicted at those neighbours of $\boldsymbol{x}_i$ that live in the sample $S_c^{(i)}$, are: $\boldsymbol{y}_i^{(c,1)}, \boldsymbol{y}_i^{(c,2)}, \ldots, \boldsymbol{y}_i^{(c,N_\epsilon)}$. We set the sample $\psi_c^{(i)}$ to be:

$$\psi_c^{(i)} := \{\boldsymbol{y}_i^{(c,1)}, \boldsymbol{y}_i^{(c,2)}, \ldots, \boldsymbol{y}_i^{(c,N_\epsilon)}\}.$$

So samples $\psi_1^{(i)}, \ldots, \psi_d^{(i)}$ are all generated.

— Similarly, we identify $d \times N_\epsilon$ number of neighbours of $\boldsymbol{x}_j$ that live within an ϵ-neighbourhood of $\boldsymbol{x}_j$, and predict the outputs at each such generated neighbours, to generate the samples $\psi_1^{(j)}, \ldots, \psi_d^{(j)}$.

— We note that all elements of $\psi_c^{(\cdot)}$ are predicted using the kernel that parametrises the covariance function of $\mathcal{GP}_{non-stat}$, with hyperparameters of this kernel modelled as $\theta_c = \eta_c(t)$, $\forall c = 1, \ldots, h \equiv d$. Thus, as suggested by this notation, we simplify our discussions by choosing to construct the non-stationary version of the non-parametric kernel under the constraint that there are as many hyperparameters in this kernel, (each modelled as a random function of T), as there are dimensions to the input variable $\boldsymbol{X}$. It is not infeasible to do this for $h > d$, since we *compute* and do not learn the hyperparameters of this sought non-stationary kernel, as we will soon appreciate. However, we choose to work with $h = d$ merely to ease the current discussions. We recall that h is the number of kernel hyperparameters describing the kernel that parametrises the covariance function of $\mathcal{GP}_{non-stat}$. On the other hand, d is the dimension of the input variable $\boldsymbol{X}$. The hyperparameters could be length scales along each of the d directions in input space $\mathcal{X}$, where the length scale $\ell_c = 1/q_c \equiv 1/\theta_c$ is the distance we need to move along the c-th direction in space $\mathcal{X}$, for the inter-output correlation to drop by an identified factor.

— So to recap, elements of the samples $\psi_1^{(i)}, \ldots, \psi_d^{(i)}$ are the realisations of the output at neighbours of $\boldsymbol{x}_i$ that differ from $\boldsymbol{x}_i$ in only a small difference to the 1-th, 2-th, ..., d-th component respectively. Similarly, elements of $\psi_1^{(j)}, \ldots, \psi_d^{(j)}$ are realisations of outputs at neighbours of $\boldsymbol{x}_j$ that bear a small difference in their 1-st, 2-nd, ..., d-th components respectively, from the corresponding component of $\boldsymbol{x}_j$.

— The idea behind this construct is the following.

Since ϵ is very small, the generated neighbours of either design point are very close to the respective design point. Then predictions of the outputs at such neighbours of the i-th and j-th design points are s.t.

the estimate of the correlation computed using such output realisations predicted at the two design inputs, approximates the correlation between $\boldsymbol{Y}_i$ and $\boldsymbol{Y}_j$. So using the samples $\psi_c^{(i)}$ and $\psi_c^{(j)}$, we invoke the unbiased estimator of Pearson's correlation coefficient to estimate the correlation between $\boldsymbol{Y}_i$ and $\boldsymbol{Y}_j$.

— For an m-th ordered tensor-valued output, let $d_{\mathcal{Y}_{m+1}}(\boldsymbol{y}_c^{(i)}, \boldsymbol{y}_c^{(j)})$ be the distance between the $m+1$-th ordered tensors $\boldsymbol{y}_c^{(i)}$ and $\boldsymbol{y}_c^{(j)}$ that are respectively predicted at those neighbours of $\boldsymbol{x}_i$ and $\boldsymbol{x}_j$ – that are slightly different from $\boldsymbol{x}_i$ and $\boldsymbol{x}_j$ respectively – only in their c-th components. To clarify, these $m+1$-th ordered tensors are:

$$\boldsymbol{y}_c^{(i)} = (\boldsymbol{y}_c^{(i,1)}, \ldots, \boldsymbol{y}_c^{(i,N_\epsilon)})^T \text{ and } \boldsymbol{y}_c^{(j)} = (\boldsymbol{y}_c^{(j,1)}, \ldots, \boldsymbol{y}_c^{(j,N_\epsilon)})^T,$$

with

$$\boldsymbol{y}_c^{(i,k)} \in \psi_c^{(i)}, \quad \forall k = 1, \ldots, N_\epsilon,$$
$$\boldsymbol{y}_c^{(j,k)} \in \psi_c^{(j)}, \quad \forall k = 1, \ldots, N_\epsilon.$$

— Subsequently, this distance $d_{\mathcal{Y}_{(m+1)}}(\cdot, \cdot)$ is transformed into the magnitude of the estimated correlation between outputs generated at inputs that are different only in their c-th components:

$$\widehat{corr}_c(\boldsymbol{Y}_i, \boldsymbol{Y}_j) := \exp(-d_{\mathcal{Y}_{m+1}}(\boldsymbol{y}_c^{(i)}, \boldsymbol{y}_c^{(j)})).$$

— But this estimated correlation should concur with the correlation between $\boldsymbol{Y}_i$ and $\boldsymbol{Y}_j$ as parametrised by a non-stationary kernel computed at inputs $\boldsymbol{x}_i$ and $\boldsymbol{x}_j$, with the hyperparameter $\theta_c^{(i,j)} = q_c^{(i,j)} = 1/\ell_c^{(i,j)}$, so that we can solve this equation for $\theta_c^{(i,j)}$. We choose this kernel with the unknown length scale hyperparameters $\{\theta_c^{(i,j)}\}_{i<j;i,j=1}^n \equiv \{1/\ell_c^{(i,j)}\}_{i<j;i,j=1}^n$, to be SQE-looking. Thus,

$$\widehat{corr}_c(\boldsymbol{Y}_i, \boldsymbol{Y}_j) = \exp\left(-\frac{(x_c^{(i)} - x_c^{(j)})^2}{(\ell_c^{(i,j)})^2}\right),$$

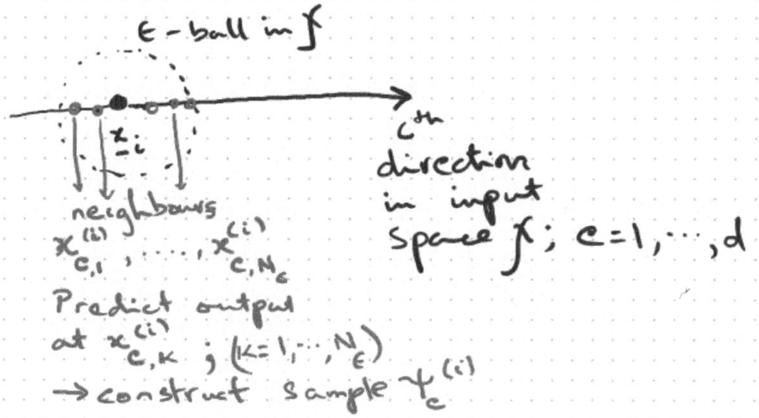

FIGURE 4.9
Cartoon to depict the construction of input values neighbouring a design (or test) point $\boldsymbol{x}_i$, s.t. each of the constructed N_ϵ neighbouring points differs from $\boldsymbol{x}_i$ only in the c-th component of the d-dimensional $\boldsymbol{x}_i$. The neighbouring points are constructed $\forall c = 1, \ldots, d$, for a small ϵ. The non-parametric kernel that underlines the 2-layered learning strategy, is used to predict the output at each constructed input, s.t. at $\boldsymbol{X} = \boldsymbol{x}_i^{(c,k)}$, the predicted output is $\boldsymbol{y}_i^{(c,k)}$. The (Pearson) correlation between $\boldsymbol{Y}_i$ (realised at $\boldsymbol{X} = \boldsymbol{x}_i$) and $\boldsymbol{Y}_j$ is estimated using the samples $\{\boldsymbol{y}_i^{(c,k)}\}_{k=1}^{N_\epsilon}$ and $\{\boldsymbol{y}_j^{(c,k)}\}_{k=1}^{N_\epsilon}$. This estimated correlation is then equated to the (SQE) kernel that bears the input-pair dependent hyperparameter $\ell_c^{(i,j)}$, where this kernel parametrises $corr(\boldsymbol{Y}_i, \boldsymbol{Y}_j)$, at inputs of $\boldsymbol{x}_i$ and $\boldsymbol{x}_j$, Such a construct then yields the value of $\ell_c^{(i,j)}$, $\forall i, j = 1, \ldots, n$, and $\forall c = 1, \ldots, d$.

which leads to the solution $\ell_c^{(i,j)} = -\dfrac{(x_c^{(i)} - x_c^{(j)})^2}{\log(\widehat{corr}_c(\boldsymbol{Y}_i, \boldsymbol{Y}_j))}$.
This is undertaken $\forall c = 1, \ldots, d;\ \forall i, j = 1, \ldots, n; i \neq j$.

Thus, the hyperparameters of the kernel – that parametrises the covariance function of $\mathcal{GP}_{non-stat}$ – are *computed* and not learnt, as we anticipated above. Thus, this kernel is clearly non-stationary. While we had stated earlier in Section 4.3.4 that the model of hyperparameter-dependence on the relevant *pair* of inputs is a bad model given the infeasibly huge learning burden that such a model entails, in this method, we notice that we *did not learn these input pair dependent hyperparameters* – we merely computed the hyperparameters.

In fact, although the above method development clarifies the computation of the hyperparameters that are dependent on pairs of design inputs, it can be extended to compute hyperparameters relevant to pairs of design and test inputs, as well as test input pairs. Such is possible since outputs at any two inputs can be predicted using our non-parametric kernel model in which we model $\theta_c = \eta_c(t)$, $\forall c = 1, \ldots, d \equiv h$, at the t-th iteration of the undertaken inference.

Thus, by this computation of input pair dependence of kernel hyperparameters, we have demonstrated that the non-parametric model of the kernel – with hyperparameters $\theta_c = \eta_c(t)$, $\forall c = 1, \ldots, d \equiv h$ – is equivalently non-stationary, with hyperparameters that vary with a pair of inputs. These inputs are s.t. the correlation between the outputs realised at these is sought via parametrisation with this kernel.

4.8.2 Construction of a non-stationary kernel in the simplest case

In the above approach in which a non-stationary kernel is generated from the non-parametric kernel discussed in Sections 4.5, we realise that the simplest case arises if the output variable is scalar-valued, and the input is also a scalar, i.e. $c = 1$. Then the estimated Pearson correlation coefficient $\widehat{corr}_c(\boldsymbol{Y}_i, \boldsymbol{Y}_j)$ is denoted $\widehat{corr}(Y_i, Y_j)$ nstead. This

$$\widehat{corr}(Y_i, Y_j) = \frac{\sum_{k=1}^{N_\epsilon} (y_c^{(i,k)} - \bar{y}_c^{(i)})(y_c^{(j,k)} - \bar{y}_c^{(j)})}{N_\epsilon - 1},$$

where $\bar{y}_c^{(i)}$ and $\bar{y}_c^{(j)}$ are the means of Y_i and Y_j in the samples $\psi_c^{(i)}$ and $\psi_c^{(j)}$ respectively. We then set this computed $\widehat{corr}(Y_i, Y_j)$ to a non-stationary kernel of an assigned parametric shape, parametrised with the vector of hyperparameters $\{\theta_c^{(i,j)}\}_{i<j;i,j=1}^{n}$, $c = 1, \ldots, d$. In this example, $d = 1$. Thus, for an SQE-looking kernel we set

$$\widehat{corr}(Y_i, Y_j) = \exp\left(-\frac{(x_i - x_j)^2}{2(\ell^{(i,j)})^2}\right),$$

where $1/(\ell_c^{(i,j)})^2 = (q_c^{(i,j)})^2 \equiv (\theta_c^{(i,j)})^2$. This equation then offers the value of $\ell^{(i,j)}$ $\forall i < j; i, j = 1, \ldots, n$, where $\ell^{(i,j)}$ is the hyperparameter to be used within the non-stationary kernel that is computed at the ij-th pair of inputs, to offer the correlation between the output variables generated at these inputs.

4.8.3 Why the non-parametric nature of the kernel is required for correct non-stationary construction

In Section 4.2 we began our discussions by suggesting a kernel with hyperparameters that are dependent on the *input pair* at which the inter-output correlation is sought. As we can also recall from above, such an input-pair-dependent kernel implies an infeasibly high computational burden, and we had avoided it by replacing sampling across inputs at one time-step, with sampling across time-steps at each input. *This was our non-parametric kernel.*

Thus, we can generate a non-stationary equivalent of this non-parametric kernel, using the latter kernel, where the hyperparameters of this non-parametric kernel have been learnt. If the model of this basal non-parametric kernel is wrong, the constructed non-stationary equivalence would not correctly parametrise the covariance function of $\mathcal{GP}_{non-stat}$, given the available data. *Could we not construct a non-stationary kernel – as discussed earlier in this section – had we modelled the sought function $\boldsymbol{f}(\cdot)$ with a parametric kernel, irrespective of how that parametric kernel was formulated?*

Of course we could have constructed *a* kernel with (length scale) hyperparameters that are dependent on the pair of inputs at which the constructed kernel is computed. But we question the relevance of any such constructed, input-pair-dependent kernel, given the data at hand. Only if the base kernel – that parametrises the covariance function of $\mathcal{GP}_{non-stat}$ – is correct given the data, will the input-pair-dependent kernel constructed as above, be correct. So the following are noted.

— Only if this base kernel – that parametrises the covariance function of $\mathcal{GP}_{non-stat}$ – correctly replicates the inhomogeneities of the data $\mathbf{D}$, will the predictions of output at the neighbours of a design input be accurate, within the construction of the input-pair-dependent kernel.

— Only then will the hyperparameters of the constructed non-stationary kernel be accurate in the data, to thereby make correct predictions of the output at a test input.

— Since the covariance structure of the data – that is marked by inhomogeneities – cannot be replicated by modelling the kernel as parametric, we need to construct the non-stationary kernel using the learnt non-parametric one only.

— If such constructed non-stationary kernel offers correct predictions at test inputs, all that this will inform on, is that the basal non-parametric kernel can be represented equivalently as a non-stationary one.

Two points emerge from the above.

- Firstly, accuracy of predictions at a set of chosen test inputs, using the constructed non-stationary kernel, indicates accuracy of the computed length scale hyperparameters of this constructed kernel, in the available data. This in turn will follow, only if the predicted mean of output values are correct at the inputs that lie within the chosen ϵ-neighbourhood of design points, and of test points. Such correctness of prediction is possible only if the non-parametric kernel that parametrises the covariance function of $\mathcal{GP}_{non-stat}$ is correct given the data at hand.

- The second point that arises from the first is as follows.

Since accuracy of predictions from the non-stationary kernel is predicated upon the correct learning of the hyperparameters of the preceding non-parametric kernel, there is no need to construct the non-stationary kernel, subsequent to the learning of the non-parametric kernel. In other words, we only need to learn the non-parametric kernel correctly, to perform accurate predictions of the output that is realised at a test input.

To reiterate the last sentence of the previous paragraph, *the constructed non-stationary kernel is useful only in showing that there exists a correct non-stationary representation of this originally learnt non-parametric kernel.*

4.8.4 Criticism of the non-stationary construction

When we scrutinise the protocol suggested for the construction of the non-stationary kernel, we realise that we have assumed that $\kappa_c(\boldsymbol{x}_i, \boldsymbol{x}_j; \ell_c^{(i,j)})$ is approximated by the correlation estimated using the predicted mean output values in the set $\{\boldsymbol{y}_c^{(i,1)}, \ldots, \boldsymbol{y}_c^{(i,N_\epsilon)}\}$ - predicted at N_ϵ inputs that lie along the c-th direction of input space at a distance of $< \epsilon$ from $\boldsymbol{x}_i$ – and the predictions at inputs that lie in the ϵ-neighbourhood of $\boldsymbol{x}_j$ in the same direction inside $\mathcal{X}$. Here i and j are indices of design or test points. But if the correlation structure of the data is s.t. the neighbourhood of the i-th input is marked by a high level of inhomogeneities in correlation, and sample functions in this neighbourhood are highly rough, then such an assumption might stand challenged.

Also, given the exponential in the SQE form that is used in kernel $\kappa_c(\cdot, \cdot; \cdot)$, a small fractional error in the estimated correlation $\widehat{corr}(\boldsymbol{Y}_i, \boldsymbol{Y}_j)$ – corresponding to a small deviation from the κ_c-parametrised correlation between $\boldsymbol{Y}_i$ and $\boldsymbol{Y}_j$ – will give rise to a high fractional change in the computed $\ell_c^{(i,j)}$. This is because, such fractional error in the length scale is the product of

$1/[\widehat{corr}(\boldsymbol{Y}_i, \boldsymbol{Y}_j)\ln(\widehat{corr}(\boldsymbol{Y}_i, \boldsymbol{Y}_j))]$ and the fractional error in the estimated correlation. Thus, if a highly non-linear variation in the output shows up around a chosen input, and the estimated correlation is anomalous (compared to the true correlation between a pair of outputs computed at/near the considered input), a large error can be induced in the constructed length scale relevant to that input. Indeed, such a possibility can be checked – and possibly mitigated – by using different ϵ and N_ϵ values, during an application.

4.8.5 Implementation to the Tetouan city data

We implement this non-stationary equivalent of the non-parametric kernel (that we developed above and discussed the implementation of, in Section 4.6), to make predictions at the test temperatures (i.e. the test inputs) that are relevant to the univariate Tetouan city data that we discussed in Chapter 3. The 1st exercise of this data discusses the thermal variation of the energy that was consumed in all zones of Tetouan city. We undertake this exercise here, using the non-stationary kernel, the input-dependent hyperparameters of which we compute, following the methodology discussed in Section 4.8.1. Then using the learnt (two) hyperparameters (or a length scale parameter in each of the two layers of the dual-layered learning strategy), we compute the $n(n-1)/2$ distinct length scale hyperparameters at the locations of the design input pairs, along with the $(n_{test}-1)n/2$ pairs of hyperparameters at the locations of the n design, and $(n_{test}-1)n_{test}/2$ hyperparameters at n_{test} test input locations. We present the length scale hyperparameters at each pair of design points, in a contour plot (in Figure 4.8).

4.8.6 Predictions

In the example application to learn the univariate function $f(\cdot)$ that is scalar-valued, the output predicted at a test input $X =$

$x^{(test)}$ is

$$f(x^{(test)}) = (e^{-(x_1 - x^{(test)})^2/\ell^2_{1,test}} \ldots e^{-(x_n - x^{(test)})^2/\ell^2_{n,test}}) \\ \boldsymbol{\Sigma}_{non-stat}(y_1, \ldots, y_n)^T,$$

where the training data $\mathbf{D} = \{(x_i, y_i)\}_{i=1}^n$, and the correlation matrix of $\mathcal{GP}_{non-stat}$ is $\boldsymbol{\Sigma}_{non-stat} = [\exp(-(x_i - x_j)^2/2\ell^2_{i,j})]$. In other words, the correlation function is parametrised by the non-stationary kernel with an SQE form. This kernel is manifestly non-stationary since the length scale hyperparameters are dependent on the (pair of) inputs at which the kernel is computed.

Then calling $\boldsymbol{\Sigma}_{non-stat}(y_1, \ldots, y_n)^T = (\alpha_1, \ldots, \alpha_n)^T$, we recall the definition of $\ell_{i,j}$ as

$$\ell_{i,j} = \frac{-(x_i - x_j)^2}{\ln(\widehat{corr}(Y_i, Y_j))},$$

so that output $f(x^{(test)})$ predicted at a test input $x^{(test)}$ reduces to

$$f(x^{(test)}) = \sum_{p=1}^{n} \widehat{corr}(Y_p, Y^{(star)})\alpha_p.$$

Closed-form predictions of the expected output value, at each of n_{test} test inputs are shown in Figure 4.7 – using the non-parametric model for the kernel and the non-stationary kernel. The true output values are depicted in black circles while the expected values are in green filled circles, that have the 2 times the predicted standard deviation of the output value superimposed on the predicted mean. In the left panel of this figure, results from using the non-parametric kernel are depicted, while the results of usage of the non-stationary model are shown in the middle panel. The right panel depicts results obtained with a blend of the non-parametric and non-stationary models. We will see this *blended model* in the next section.

4.9 Blending non-parametric & non-stationary models

In the last section we saw how a non-stationary equivalent of the non-parametric kernel could be realised. We could of course compute the $\ell_1^{(i,j)}, \ell_2^{(i,j)}, \ldots, \ell_d^{(i,j)}$ hyperparameters of the non-stationary kernel, $(\forall i < j; i,j = 1,\ldots,n)$, at each time-step during which we learn the hyperparameters $(\ell_1,\ldots,\ell_d;\delta_1,\ldots,\delta_d)$ of the non-parametric kernel and the stationary kernel of the inner layer of our 2-layered learning strategy. Thereafter, we can predict expectation and variance of the output at each test input, using the input-pair-dependent length scales computed with the current values of $\ell_1,\ldots,\ell_d;\delta_1,\ldots,\delta_d$. Additionally, we could predict the output values at all such test inputs using the current non-parametric kernel.

The possible lack of robustness in the computation of the input-pair-dependent hyperparameters of the non-stationary kernel could be a worry that we may wish to address. This is possible within a learning strategy in which difference between predictions from the non-parametric and the non-stationary kernels is monitored – and when such a difference, (or a function of it) stabilises, we perform high-accuracy predictions. We discuss such blending of the non-stationary and the non-parametric models next.

1. We learn hyperparameters of the non-parametric kernel using a 2-layered learning strategy, implemented within a Metropolis-within-Gibbs chain. Then in the N_p-th iteration, after this 2-blocked chain has converged, we stop undertaking this 2-blocked inference strategy.
2. The N_p+1-th iteration onward, we do not learn $\delta_1,\ldots,\delta_d$ any further. Thus, a single block is used in each iteration, subsequent to the N_p-th iteration.
3. In iteration t – where $t \geq N_p + 1$ – we update hyperparameters $\theta_1,\ldots,\theta_h$ of the generic non-parametric kernel, and construct the corresponding non-stationary kernel using this updated non-parametric kernel. Said updating of the non-parametric kernel is undertaken by propos-

ing $\theta_1, \ldots, \theta_h$ from their respective proposal densities, and computing the joint posterior of all these h parameters – at the proposed values of these parameters, and at their current values (that are current as of at the end of the $t-1$-th iteration) – using their likelihood in the data $\mathbf{D} = \{(\boldsymbol{x}_i, \boldsymbol{y}_i)\}_{i=1}^{N}$ and the chosen priors on these parameters.

4. We motivate the probability density of data $\mathbf{D}$ – conditional on $\theta_1, \ldots, \theta_h$ – to be s.t. it is maximal when the predictions from the non-stationary and non-parametric kernels are closest. Thus, the likelihood is motivated to be a decreasing function of the distance between $\boldsymbol{y}_k^{(non-param,t-1)}$ and $\boldsymbol{y}_k^{(non-stat,t-1)}$, $\forall k = 1, \ldots, N_{test}$, where the mean predicted output of $\boldsymbol{f}(\cdot)$ at test input $\boldsymbol{x}_k$, using the non-parametric kernel that is current at the beginning of the t-th iteration is $\boldsymbol{y}_k^{(non-param,t-1)}$. By "at the beginning of the t-th iteration", we imply before accept/reject on $\theta_1, \ldots, \theta_h$ has been undertaken in the t-th iteration. Again, $\boldsymbol{y}_k^{(non-stat,t-1)}$ is the mean output predicted at $\boldsymbol{x}_k$ using the non-stationary kernel that is current at the beginning of the t-th iteration.

5. Then we advance a model for the probability density of the data given values of the model parameters $(\theta_1, \ldots, \theta_h)$ at time-step $T = t$ as a decreasing function of the distance between $\boldsymbol{y}_k^{(non-param,t-1)}$ and $\boldsymbol{y}_k^{(non-stat,t-1)}$, over $k = 1, \ldots, N_{test}$, where this distance is scaled by the closed-form prediction of the variance $Var(\mathbf{Y}_k^{(non-param,t-1)})$ of the output variable $\boldsymbol{Y}_k$ made using the non-parametric kernel, at the beginning of the t-th iteration. In other words, likelihood of $\theta_1, \ldots, \theta_h$ in time-step t, in data $\mathbf{D}$, is this decreasing function. We choose this decreasing function to have the following form:

$$\mathcal{L}_t(\theta_1, \ldots, \theta_h; \mathbf{D}) =$$

$$\prod_{k=1}^{N_{test}} \exp\left(-\frac{(d_{\mathcal{Y}}(\boldsymbol{y}_k^{(non-param,t-1)} - \boldsymbol{y}_k^{(non-stat,t-1)}))^2}{2Var(\mathbf{Y}_k^{(non-param,t-1)})}\right),$$

where $d_{\mathcal{Y}}(\cdot,\cdot)$ is the distance function in the space $\mathcal{Y}$ of the output variable. The hyperparameters $\theta_1,\ldots,\theta_h$ and the training data $\mathbf{D}$ enter the RHS of this equation via the mean and variance predictions of the output variables that are realised at the considered test inputs.

6. Using the notation, $\boldsymbol{y}_k^{(non-param,t\star)}$, we denote the output predicted at the k-th test input using the values of the hyperparameters $\theta_1,\ldots,\theta_h$ that are proposed at the t-th iteration, which are used then in the non-parametric kernel. The resulting non-parametric kernel is then referred to as that "proposed" in the t-th iteration. The non-stationary kernel that is constructed using this "proposed" non-parametric kernel in the t-th iteration, is then used to make the predictions at the N_{test} inputs; such a prediction made at test input $\boldsymbol{x}_k$ is denoted $\boldsymbol{y}_k^{(non-stat,t\star)}$. Likelihood of $\theta_1,\ldots,\theta_h$ is computed at the proposed values of these hyperparameters at time-step $T=t$. As we saw above, the current values of the hyperparameters lead to the variance and the mean outputs $\boldsymbol{y}_1^{(non-param,t-1)},\ldots,\boldsymbol{y}_{N_{test}}^{(non-param,t-1)}$ – predicted at the N_{test} test inputs using the non-parametric kernel that is current at the beginning of the t-th iteration. Similarly, the predictions made using the current non-stationary kernel, are also recorded. Then the likelihood of the model parameters is also computed at their proposed, as well as their current values. Using the likelihood computed at the proposed and current model parameters in data $\mathbf{D}$, the joint posterior $\pi(\theta_1,\ldots,\theta_h|\mathbf{D})$ is thereby computed at the current and proposed values of these parameters given the data $\mathbf{D}$, where we choose priors on $\theta_1,\ldots,\theta_h$.

7. We choose generic priors $\pi_0(\theta_1,\theta_h)$ on the hyperparameters – which we retain as weak if we do not have information on these parameters, as would be the case in a new application. Then using the product of the priors and the likelihood, we compute the joint posterior of the hyperparameters given the training set, at the current and proposed values of these hyperparameters –

as $\pi(\theta_1^{(t-1)}, \ldots, \theta_h^{(t-1)}|\mathbf{D})$ and $\pi(\theta_1^{(t\star)}, \ldots, \theta_h^{(t\star)}|\mathbf{D})$ respectively.

8. We undertake a Random Walk Metropolis Hastings in the $N_p + 1$-th iteration onward – unless there is a constraint on any of the hyperparameters, such as one of positivity etc. In such a case, we do not propose the relevant hyperparameter(s) from a Normal proposal density, but choose the proposal density as relevant, (such as truncated Normal, in the example case of positivity of the hyperparameter). In that case the chain implements vanilla MCMC. We insert the joint posterior computed at the proposed and current values of the hyperparameters, in the acceptance ratio relevant to the MCMC algorithm we use, and update the hyperparameters accordingly.
9. Thus the learning that we undertake the $N_p + 1$-th iteration onward, blends the outputs that emanate from a non-parametric kernel and the non-stationary kernel that is constructed using that non-parametric kernel. We refer to this model as the *blended model.*
10. Once the hyperparameters are updated within the implementation of the *blended model,* in the t-th iteration $(t > N_p)$, we record $\boldsymbol{y}_k^{(non-param,t)}$ $\forall k = 1, \ldots, N_{test}$.

4.9.1 Implementation of the *blended model* to Tetouan city data

We illustrate the *blended model* based learning strategy to perform the learning of the thermal variation in the energy consumed over an identified time interval, in Tetouan city. The training data for this exercise is the same as that clarified in Section 3.7 of Chapter 3, and used for illustration of the non-parametric and the non-stationary kernels. Predictions made with the *blended model* at the 48 test temperature values – at which predictions of energy consumption were made using the non-parametric, as well as the non-stationary kernels – are displayed in the right panel of Figure 4.7. The predictions made with the non-parametric kernel are displayed in the left panel, while the results from the non-stationary kernel are in the middle panel. The true values of the energy consumption are in green and these are superimposed on the uncertainty-included predictions shown in black. The 95% HPDs are the displayed uncertainties. Results from a preliminary run undertaken with the *blended model* are displayed in Figure 4.10; traces of prediction of the mean output value, made at 15 different test inputs, are displayed in this figure, along with the trace of the log likelihood (in the top right panel).

4.10 Summarising the kernel explorations

In the last three sections of this chapter, we discussed three kernels and applied these to the learning of a real-world univariate and scalar-valued function, to then make prediction of the output of this function, at a chosen input. We realise that the non-stationary version of the kernel presented in Section 4.8.1 is one that represents the information included within the non-parametric kernel. In this sense, the only distinct contribution of the non-stationary kernel is to illustrate that the non-parametric kernel is equivalently non-stationary. Again, if the numerical instability inherent in the construction of the non-stationary kernel is to be addressed, then a blending of the non-parametric and non-stationary kernels is also possible – as suggested in the "blended" kernel model that is discussed Section 4.9

Ultimately, the non-stationary and blended extensions of the non-parametric kernel reinforce the dependence of the non-parametric kernel, to a translation in the space of the input variable, thereby rendering the non-parametric kernel capable of modelling functions given training data that bears inhomogeneities in its correlation structure. This non-parametric kernel is the tool that underlines the 2-layered learning strategy – the sufficiency of which we have justified (as follows from Theorem 4.5, in the gray box following proof of this theorem). Thus, it is only the non-parametric kernel-based learning that we will undertake in

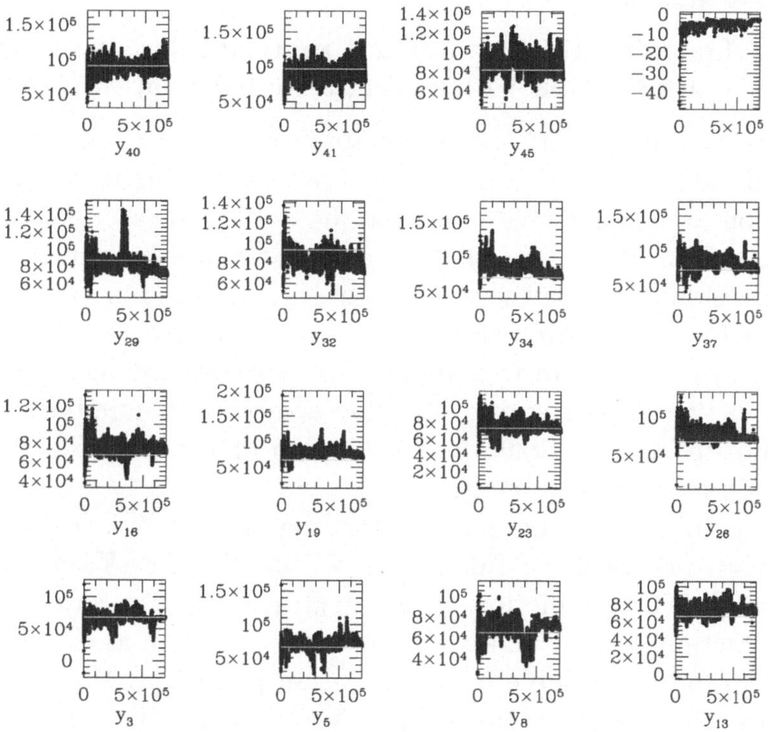

FIGURE 4.10
Plots of traces of mean of the output predicted at iterations of a chain conducted to learn kernel hyperparameter ℓ, using the blended kernel model, at 15 different test inputs. True value of the output is overplotted in a yellow (or white in the monochromatic version) line. The top right panel displays the trace of logarithm of the likelihood.

applications, within the 2-staged learning strategy. We examine a few other applications below.

4.11 Learning the sub-surface material density function – application to learning a multivariate function

In this section we will consider an application on learning a scalar-valued, multivariate function that represents the material density function at a point that is under the surface of a cuboidal slab of a material. Since (such sub-surface) location is specified with a 3-dimensional vector, the sought material density function at sub-surface locations within the material sample takes a 3-dimensional vector as its input. In this application, we start by declaring that we already possess training data that comprises N number of pairs of: sub-surface locations, and the value of the material density function at each such location.

A point inside the bulk of the material slab is referred to as $\boldsymbol{x} = (x_1, x_2, x_3)^T$, i.e. $\boldsymbol{X} \in \mathcal{X} \subseteq \mathbb{R}^3$. At $\boldsymbol{X} = \boldsymbol{x}$, the material density function is $\rho = f(x_1, x_2, x_3) \equiv f(\boldsymbol{x})$, where the function $f(\cdot)$ is unknown, s.t. it is treated as a function-valued random variable in our Bayesian paradigm. We will standardise the data on the output variable, so that $f : \mathcal{X} \subset \mathbb{R}^3 \longrightarrow \mathbb{R}$. We will model $f(\cdot)$ as a sample function from a scalar-variate GP. In our 2-layered learning strategy, we model $f(\cdot) \sim \mathcal{GP}_{non-stat}(\mu(\cdot), Cov(\cdot, \cdot; \theta_1, \ldots, \theta_h))$, where $\theta_c = \eta_c(t)$, at time-step $T = t$, with $\eta_c(\cdot) \sim \mathcal{GP}_c(\mu_c(\cdot), \kappa_c(\cdot, \cdot)), \forall c = 1, \ldots, h$. As we will soon see, in this application, $h = 3$. Here $\mathcal{GP}_{non-stat}$ is a non-stationary, scalar-variate GP and each of $\mathcal{GP}_1, \ldots, \mathcal{GP}_h$ is a stationary, scalar-variate GP. Also, the available training dataset $\mathbf{D}$ offers a value of ρ at a given design input, s.t. $\mathbf{D} = \{(\boldsymbol{x}_i, \rho_i)\}_{i=1}^N$.

The value of $f(\boldsymbol{x}_i)$ was obtained by performing a Bayesian learning of the material density function at the chosen input $\boldsymbol{x}_i = (x_1^{(i)}, x_2^{(i)}, x_3^{(i)})^T$; this followed the experiment suggested in [27] in which the same part of a material sample was imaged multiple times by a Scanning Electron Microscope (SEM), using a set of N_ε

electron beams with increasingly higher energies, s.t. information is collected in the form of N_ε distinct images, from further inside the material sub-sample volume than the surface of the material sample.

To obtain image data in a specified pixel of these N_ε distinct images, all the electron beams (of varying energies) are made incident on a given superficial point. The maximal volume – within the material bulk – that an incident beam can undergo interactions within, is referred to as the "interaction volume" or IV. To clarify, an incident electron beam loses its energy via the atomistic interactions it undergoes with the material, s.t. the volume of the IV increases with energy that is inherent in the incident beam. Here, the j-th electron beam has energy ε_j, $j = 1, \ldots, N_\varepsilon$. Then inter-comparison of the images obtained with electron beams at energy ε_j and $\varepsilon_j + 1$, allows for information on the sub-surface volume that is confined within the j-th "interaction volume" (IV) and the $j + 1$-th IV, for $j = 1, \ldots, N_\varepsilon - 1$.

As described in the last paragraph, an IV is the volume within the bulk of the material that interacts with the electron beam of energy ε_j, which upon incidence on the surface at point $(x_1^{(i)}, x_2^{(i)}, 0)^T$, interacts with the material inside the IV. [27] modelled an IV as a hemisphere with centre $(x_1^{(i)}, x_2^{(i)}, 0)^T$ and radius $r_i^{(j)}$ that is an identified function of ε_j. Indeed, if the electron beam incident at the same point bore higher energy, it would interact with a larger volume inside the material bulk.

The inter-comparison of the image data produced from the interactions of electron beams of energies in $\{\varepsilon_1, \ldots, \varepsilon_{N_\varepsilon}\}$ incident at points $(x_1^{(1)}, x_2^{(1)}, 0)^T, \ldots, (x_1^{(M)}, x_2^{(M)}, 0)^T$, allowed for the learning of the values of the material density at points

$$\{(x_1^{(1)}, x_2^{(1)}, x_3^{(i)})^T\}_{i=1}^{N_z}, \ldots, \{(x_1^{(M)}, x_2^{(M)}, x_3^{(i)})^T\}_{i=1}^{N_z},$$

where $MN_z = N$. Then at the design input $(x_1^{(j)}, x_2^{(j)}, x_3^{(k)})^T$, which is the i-th design input, (for $j = 1, \ldots, M$; $k = 1, \ldots, N_z$; $i = 1, \ldots, MN_z = N$), the material density was learnt to have the value ρ_i.

1. Thus, training set $\mathbf{D} = \{(\boldsymbol{x}_i, \rho_i)\}_{i=1}^{N}$.

2. As stated above, we will employ this training set, to learn the function $f(\cdot)$ within our 2-layered learning strategy.
3. We will standardise the outputs in the training set, and use a zero-mean GP for $\mathcal{GP}_{non-stat}$.
4. Then the covariance function of this non-stationary GP is also rendered its correlation function, which is parametised using the kernel $K(\cdot, \cdot; \theta_1, \ldots, \theta_h)$.
5. We choose to model this kernel non-parametrically, by assigining an SQE-looking form to this kernel, while modelling each relevant hyperparameter as a function of time-step or iteration index T. Then this SQE-looking kernel will bear three hyperparameters ℓ_1, ℓ_2, ℓ_3, which at iteration index (or time-step) $T = t$, enter the multivariate Normal likelihood as:

$$\mathcal{L}(\ell_1, \ell_2, \ell_3; \mathbf{D}) =$$

$$(2\pi)^{-n/2}|\mathbf{\Sigma}_t|^{-1/2} \exp\left(-\left((\boldsymbol{\rho} - \boldsymbol{\mu})^T \mathbf{\Sigma}_t^{-1} (\boldsymbol{\rho} - \boldsymbol{\mu})\right)\right),$$

with $\boldsymbol{\mu} = \mathbf{0}$; $\boldsymbol{\rho} = (\rho_1, \ldots, \rho_N)^T$; $\mathbf{\Sigma} = [\exp(-(\boldsymbol{x}_i - \boldsymbol{x}_j)^T \boldsymbol{Q}(\boldsymbol{x}_i - \boldsymbol{x}_j))]$ with the 3×3-dimensional diagonal matrix $\boldsymbol{Q}$ defined as one with diagonal elements $1/\ell_1(t)^2$, $1/\ell_2(t)^2$, $1/\ell_3(t)^2$, given the 3-dimensional input variable. Here

$$\ell_c(t) \equiv \eta_c(t), \text{ with } \eta_c(\cdot) \sim \mathcal{GP}_c(\mu_c(\cdot), \kappa_c(\cdot, \cdot; \delta_c),$$

for $c = 1, 2, 3$.
6. The training data for the learning of $\eta_c(\cdot)$ is the lookback data $\mathbf{D}_c^{(t)} = \{(i, \ell_c^{(i)}\}_{i=t-N_{lookback}}^{t-1}$. The output values in this training data are standardised at the beginning of the 2nd block of the t-th iteration, and the correlation matrix $\kappa_c(\cdot, \cdot; \delta_c)$ is parametrised by an SQE kernel with the (unknown) constant hyperparameter δ_c. Here $c = 1, 2, 3$.
7. Likelihood of δ_c, in lookback data $\mathbf{D}_c^{(t)}$ is again multivariate Normal, with a zero-mean vector and an $N_{lookback} \times N_{lookback}$-dimensional covariance matrix $\mathbf{\Delta}_c$

that is parametrised using a stationary kernel with hyperparameters δ_c.

8. Then ℓ_1, ℓ_2, ℓ_3 are updated using $\mathbf{D}$ in the first block of the t-th iteration, while the hyperparameters $\delta_1, \delta_2, \delta_3$ are updated in the 2nd block, using the respective lookback data that this relevant to this iteration. Then, at the updated GPs of the inner layer – that generate the functions $\eta_1(\cdot), \eta_2(\cdot), \eta_3(\cdot)$ – the mean output from each such function – namely ℓ_1, ℓ_2 and ℓ_3 – are predicted. These predicted values are then considered the current values of ℓ_1, ℓ_2 and ℓ_3 at the end of the t-th iteration, and each of these predicted values is included as the latest element of the respective lookback data that would be relevant at the 2nd block of the $t+1$-th iteration.

9. Thus, in the t-th iteration, at a sub-surface location $\boldsymbol{x}^{(test)}$, the mean material density function is

$$f(\boldsymbol{x}^{(test)}) =$$

$$\sum_{p=1}^{N} e^{(-(\boldsymbol{x}_p - \boldsymbol{x}^{(test)})^T \boldsymbol{Q}^{(test)}(\boldsymbol{x}_p - \boldsymbol{x}^{(test)}))} \boldsymbol{\Sigma}^{(test)}(\rho_1, \ldots, \rho_N)^T,$$

where $\boldsymbol{\Sigma}^{(t)} = [e^{(-(\boldsymbol{x}_i - \boldsymbol{x}_j)^T \boldsymbol{Q}^{(t)}(\boldsymbol{x}_i - \boldsymbol{x}_j)}]$;

$$\boldsymbol{Q}^{(t)} = [\delta(i,j)/(\ell_i^{(t)})^2],$$

with $\delta(i,j) = 0\ \forall i \neq j; i,j = 1,2,3$ and $\delta(i,j) = 1\ \forall i = j; i,j = 1,2,3$. Also,

$$\ell_c^{(t)} = \quad (e^{-(t-1-t^{(test)})^2/2\delta_c^2} \ldots e^{-(t-N_{lookback}-t^{(test)})^2/2\delta_c^2})\boldsymbol{\Delta}_c \\ (\ell_c^{(t-1)}, \ldots, \ell_c^{(t-N_{lookback})})^T,$$

with the test time point at which prediction is sought being $t^{(test)} = t$ and $\boldsymbol{\Delta}_c = [\exp(-\delta(i,j)/2\delta_c^2]$, for $c = 1,2,3$.

10. We record the mean predicted material density function at each considered test input, in each iteration of the

MCMC chain used to undertake inference of the model parameters. The sample of values of the predicted mean material density function $f(\boldsymbol{x}^{(test)})$, recorded over all iterations at the test input $\boldsymbol{x}^{(test)}$, is employed to compute the relative frequency distribution of this output variable, and this is employed to learn 95% HPD of this variable, along with the sample mean.

The image data for this application was recorded with an SEM, at 11 different energies (10 keV, 11 keV, ..., 20 keV), of a material nano-brick (composed of Nickel and Silver) that was grown in the laboratory. Some of these images are shown in Figure 4.2 of [27]. Results of learning the length scale hyperparameters ℓ_1, ℓ_2 and ℓ_3 are presented in Figure 4.3 of [27] which distinguishes said results from attempts made at learning the length scales under the assumption that a stationary GP can model the sub-surface material density function. Predictions of mean of this function made at locations on $X_1 = x$ planes, for various values of x, are presented in Figures 4.5 onward in [27].

Thus, this application demonstrates the capacity for the dual-staged learning strategy to successfully learn a multivariate function that manifests an inhomogeneous correlation structure, to thereafter undertake accurate and fast predictions. The method enables a completely non-destructive way for learning the highly nonlinear material density function at sub-surface locations, using data on designed sub-surface locations that span a small sub-volume of the bulk of the considered material sample. Such data is generated using multiple and sequential inversions of the image data that results from a novel imaging - done with Scanning Electron Microscope – of such a small sub-volume of the material bulk. Then possessing such data from a small number of non-overlapping (small) sub-volumes, we can now reliably predict the material density at intermediate sub-surface locations. This is in spite of the very highly inhomogeneous correlation structure of the density function. Such predictions can then very usefully offer the material density function for the whole material sample.

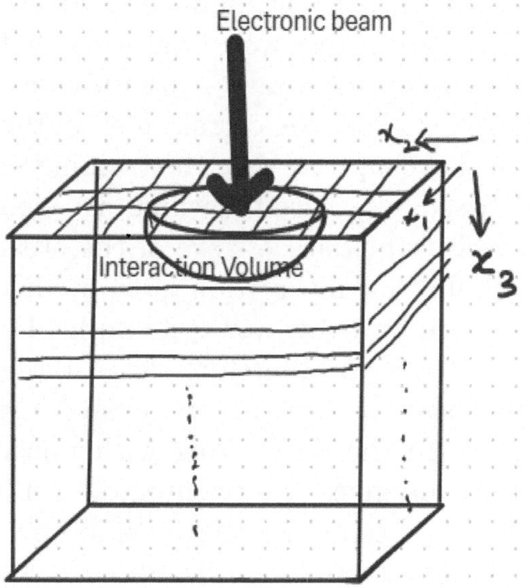

FIGURE 4.11
Cartoon to depict the 3-dimensional grid that the material sample is placed in. The gridding along the X_1 and X_2 directions are uniform, but the grids along X_3 get narrower with increasing x_3. The Interaction Volume (IV) is modelled as a hemisphere centred at the point of impingement of the electron beam on the$X_3 = 0$ surface. Radius of this IV increases with energy of the electrons in the incident beam. The beam penetrates the surface of the material slab and interacts with the sub-surface material within the IV. The outcome of said interaction is recorded as a pixel of the resulting image.

4.12 Summary

In this chapter we have come across some important considerations that are highly relevant to the exercise of mechanistic, Stochastic Process-modelled learning of nonlinear and inhomogeneously-correlated functions, given real-world datasets that bear inhomogeneities in their correlation. A new - and fully data-driven – definition of "inhomogeneities in the correlation" of the available data was introduced in this chapter. Also, the training sets that are relevant to the discussed supervised learning are small to moderate in size – to be precise, the datasets are of the order of magnitude of 10^1 to 10^3. The ulterior motivation is to predict the output of such a learnt function, at a new, (or test) input. Indeed, such prediction needs to be fast – and therefore closed-form – and needs to be reported with objective and comprehensive uncertainties that should fold in noise in the data along with uncertainties in the learning of the function given the training data. If the learning of the function – that represents the relationship between the output and input of this function – is done without correctly capturing the correlation structure of the available training set, the predictions will be incorrect. Therefore, it is crucially important to learn the empirical correlation structure correctly, and transfer this information towards the learning of the sought function – or rather of the correlation structure of the Stochastic Process that generates the sought function modelled as a sample function of this process. Such a process is motivated to a Gaussian Process (or GP) since a GP imposes the fewest constraints on the sought function – in fat, the sole constraint is that the joint distribution of any finite set of realisations of this function, is a (high-dimensional equivalent) of the multivariate Normal distribution.

This chapter introduces a parameter that quantifies the inhomogeneities of the correlation of a given training set, using nothing other than the data on the outputs that are realised at respective design points, where the available training set comprises pairs of a design point and the corresponding realisation of the output. When a given training set is found to bear inhomogeneities in its correlation, it is understood that the stochastic process – namely a GP – that generates the sought function, must also be parametrised by a non-stationary covariance function. While a parametric non-stationary form could be used for the covariance kernel that parametrises the covariance function, sensitivity of predictions to the chosen form of the non-stationary covariance function, motivates the alternative non-stationary approach. Indeed, we then opt for a fully non-parametric framework – not only is the sought function not ascribed any parametric form, but is instead learnt as a sample function of a GP. So, the kernel used in the learning is effectively not ascribed a parametric form, but each of its parameters is instead learnt as the output of a GP-generated (unknown or) random function. The simultaneous demands of non-stationarity and non-parametric-ness on the kernel are addressed in the most comprehensively generalised model, if a distinct kernel parameter is set as the output of a freshly-drawn sample function of the GP (that generates the sought function).

However, implementation of such a comprehensive model is not possible. Instead, we appreciate that assuming the system to be ergodic, multiple random draws from a GP, is on the average, equivalent to a single draw from each (randomly selected) GP within a set of multiple GPs – as the set of GPs attained at distinct iterations of the equilibrium stage of the inference exercise that is undertaken to perform the considered learning. Then we set each distinct kernel parameter as a random function of the parameter T, where T is s.t. at a distinct value of T, a distinct sample function is drawn from the GP that is invoked to model the sought function. T is then motivated to be an iteration index of the inference undertaken in the learning. To be precise, T is an index of in iteration during the equilibrium stage of this inference. Then each of the h kernel parameters is a distinct random function that takes T as its input, where each such random function is modelled as a sample function of a distinct GP. We prove each such GP to be stationary, and each such GP is also scalar-variate since the output of the random function is a scalar-valued kernel parameter. The learning strategy is then fully non-parametric and entails the nesting of h distinct stationary scalar-variate GPs within a non-stationary (and generally) tensor-variate GP. Modelling each kernel hyperparameter as a GP-driven random function of the integer-valued variable T

1. enhances parsimoniousness of the model;
2. contributes towards streamlining and ease of computation of the model;
3. provides sound justification for the usage of only 2 layers of GPs in the learning strategy, with GPs of the inner layer requiring the learning of only constant hyperparameters of their respective covariance functions;
4. and most importantly, encourages accuracy of predictions via the detailed learning of the covariance functions of the inner layer GPs – even in the face of training sets that bear inhomogeneities in their correlation, leading to a highly inhomogeneously-correlated inter-variable function that we seek to begin with.

The explicitly non-stationary nature of this fully non-parametric kernel is demonstrated using a construction, while we have also discussed another model that is a composite of this non-stationary construct and the non-parametric kernel.

Empirical illustrations have been discussed to demonstrate the capacity in the model to learn a scalar-valued univariate, and another scalar-valued multivariate function, where such learnt functions are highly nonlinear and inhomogeneously-correlated, given the available real-world datasets that bear inhomogeneities in their correlation. We will come across another multivariate functional learning – and prediction – exercise in the 6th chapter, where we will come across a very useful prior construct to avail of even better predictions. However, the learning of a high-dimensional function has not been discussed yet – this will be undertaken in the following chapter. In fact, the next chapter is dedicated to a real-world application that discusses the learning of a matrix-valued function using a three-dimensional, or cuboidally-shaped training set. Interestingly, such learning will be undertaken using the dual-staged learning strategy, to undertake prediction of the input, at which a newly-recorded output is realised. Indeed, such an exercise is not achievable via closed-form prediction any more. In fact, the sought input will be learnt, rather than predicted.

5

Application of dual-layered learning strategy to learning a high-dimensional function – joint work with Dr. Kangrui Wang

5.1 Introduction

We have been introduced to the dual-layered learning strategy in the last chapter, where we pursued learning of the function that links an input $\boldsymbol{X}$ to an output $\boldsymbol{Y}$. In generalised considerations, the output variable $\boldsymbol{Y}$ is a tensor while the input $\boldsymbol{X}$ is a vector. In the dual-layered strategy, the inner layer comprises multiple scalar-variate stationary GPs that are all nested within a generally tensor-variate non-stationary GP that comprises the "outer" layer. Each hyperparameter of the kernel that parametrises the covariance function of the outer non-stationary GP is then modelled as an unknown function of a sample function of this GP, such that (s.t.) at every input pair that the kernel is computed at, a new sample function is drawn from the outer GP. This maintains non-stationarity of the outer GP. The aforesaid "unknown function" is in turn modelled as a sample function from one of the GPs in the inner layer.

However, difficulties were envisaged in the implementation of the model that sets a kernel hyperparameter as an unknown function of the sample function of the outer GP. So we motivated an alternative model for any hyperparameter: we invoked ergodicity of the system to suggest that drawing multiple samples from the outer GP at one instance, (at each of the relevant input pairs), is equivalent to drawing a sample from each of slightly varying non-stationary GPs, over multiple instances. Here an "instance"

DOI: 10.1201/9781003296331-5

refers to a time-step or iteration index of the equilibrium stage of an inferential scheme within which, the GP covariance function is being updated. Then in the inference we discuss in this book - namely MCMC – an "instance" is a post-burnin time-step. Then, in the model we use, each kernel hyperparameter is an unknown function of the time-step, where this unknown function is modelled as a sample function of an inner layer GP.

We discussed an application of this learning strategy to learn a scalar-valued univariate function given a real-world dataset (a subset of the Teoutan city data). We also discussed an application in Materials Science – to learn a scalar-valued multivariate function. In this chapter, we discuss an application in Astronomy, within which we learn a matrix-valued function that takes a vector as its input.

5.2 Learning a high-dimensional function

One inevitable complication of the pursuit of a high-dimensional (i.e. tensor-valued in general) function is that we need to learn multiple correlation matrices of the likelihood (that is tensor Normal in general) – not all of which can be kernel parametrised. While the inter-realisation correlation matrix is kernel parametrisable, the other correlation matrices are generally not. As stated in the previous chapter, the inter-realisation correlation matrix is s.t. its ij-th element is the correlation between the outputs $\boldsymbol{Y}_i$ and $\boldsymbol{Y}_j$ that are realised as $\boldsymbol{X} = \boldsymbol{x}_i$ and $\boldsymbol{X} = \boldsymbol{x}_j$ respectively. Then this ij-th element of this correlation matrix is $corr(\boldsymbol{Y}_i, \boldsymbol{Y}_j)$ which can be parametrised as a decreasing (kernel) function of the difference between $\boldsymbol{x}_i$ and $\boldsymbol{x}_j$, while affected by certain hyperparameters of this kernel.

Basically, the more different the inputs are from each other, lower the correlation is likely to be, where such correlation refers to that between the outputs that are realised at these inputs. However, the correlation between a particular component of the tensor-valued output $\boldsymbol{Y}$, and another component of $\boldsymbol{Y}$, cannot generally be parametrised by a function of the difference between a pair of values of any other variable, s.t. the difference between the values of this "other variable", corresponds to the correlation between said components. Indeed, such an "other variable" is generally absent. Certain applications may offer information on such a variable, but that would be an exception. The indices that mark the ordering of values of a component of $\boldsymbol{Y}$, do not always hold information that is meaningful enough to be used as the input that a kernel can be proposed with. Then, correlation matrices informing on the correlation between pairs of components of $\boldsymbol{Y}$, are not the inter-realisation correlation matrix, and we can undertake either of the two below:

- treat each of the elements of the upper (or lower) triangle of off-diagonal elements of this correlation matrix as an unknown and learn it directly within the inferential exercise undertaken, e.g. MCMC-based inference.
- estimate each element of this matrix in the sample of observations on the relevant variables.

The latter is of course not the best of options as the computed correlation matrix will then be specific to the sample at hand. At the same time, the larger the sample size, the unbiased estimation of the correlation is less worse an approximation of the true correlation matrix.

Again, learning the distinct elements of the sought correlation matrix as unknown parameters (on which inference is made) is an excellent idea. However, we can undertake this task only as long as the size of the sought correlation matrix $\lesssim 6 \times 6$, which corresponds to the learning of 15 parameters. This number is only to be added to, since we will still need to learn hyperparameters of the kernel that parametrises the inter-realisation correlation matrix. It is difficult to make inference on $\geq$30 parameters with an inferential tool such as MCMC.

In the application that we will discuss in this chapter, we will come across the need to learn/estimate correlation matrices via:

— kernel parametrisation;

— directly from MCMC;
— unbiased estimation.

5.2.1 Learning input at which new output is recorded

An interesting aspect of the application that we discuss here, is that while we are learning the functional relation between a vector-valued input and a matrix-valued output, we will learn the input at which a value of the realised output is as recorded. Thus, we are not going to indulge in closed-form predictions of the learnt function computed at given test inputs, but perform the learning of the inputs at which relevant outputs are observed.

5.3 Application

Let us now discuss the dual-staged learning method discussed in the last chapter, in the context of an application made on astronomical data.

5.3.1 Background

Galactic dynamical theory tells us that if a set of stars is evolved from some primordial time, in the disk of the Milky Way galaxy, then the velocity vectors of the stars that land in an identified region of the Milky Way disk, at the current time, are affected by the features of the Milky Way. Thus, if the velocity vectors of these stars are collated to form the velocity matrix $\boldsymbol{V}$, then $\boldsymbol{V}$ is affected by the feature parameter vector $\boldsymbol{S}$ of the Galaxy. So we model that $\boldsymbol{V}$ and $\boldsymbol{S}$ exist in a relationship, and express this by stating $\boldsymbol{V} = \boldsymbol{\xi}(\boldsymbol{s})$, where $\boldsymbol{\xi}(\cdot)$ is an unknown random function.

Focusing on the immediate neighbourhood of the Sun in the Milky Way disk, the velocity matrix of a sample of our stellar neighbours was observed by the *Hipparcos* satellite, though we have no measurement on the value of $\boldsymbol{S}$ at which this observed value of $\boldsymbol{V}$ was recorded.

Astronomical simulations [25] can generate the value of the matrix $\boldsymbol{V}$, at a chosen value of $\boldsymbol{S}$. However, astronomical simulations/models cannot predict the value of $\boldsymbol{S}$ at which a value of $\boldsymbol{V}$ is realised, s.t. reliable estimation/astronomical-modelling of the feature parameters of the Milky Way is not possible, given the observed velocity matrix that comprises velocity information of some of our stellar neighbours. Then to undertake the prediction of this value of $\boldsymbol{S}$, we aim to first learn $\boldsymbol{\xi}(\cdot)$ – by modelling it as a random realisation from a correspondingly high-dimensional GP – and then undertake the inverse prediction (or rather learning) of the value of $\boldsymbol{S}$ at which the observed value of $\boldsymbol{V}$ (test data) is realised.

The training data used to learn $\boldsymbol{\xi}(\cdot)$ is provided by astronomical simulations that generate the value $\boldsymbol{v}_i$ at the i-th design point $\boldsymbol{s}_i$; $i = 1, \ldots, n$. Since $\boldsymbol{V}$ is a matrix-valued r.v., collating n realisations of it results in the data $\boldsymbol{D_V}$ that is cuboidally-shaped. i.e. is a 3rd-ordered tensor. Having modelled $\boldsymbol{\xi}(\cdot)$ using a GP, likelihood of the model parameters given the data is then a 3rd-order tensor Normal density, parametrised by a mean tensor and 3 covariance matrices.

In these astronomical simulations, the sought Milky Way feature parameters include the angular speed of the bar – which is an elongated (triaxially-shaped) structure built of stars, that rotates, centred at the galactic centre, affecting the dynamics of the Milky Way. In fact, the spiral pattern of the Milky Way is considered rotating at a fraction of the bar's angular speed, where this fraction is fixed (to about one-third) in the astronomical base model used to undertake these simulations. The second sought feature parameter is the angular distance of the long-axis of the bar, to the line that joins the Sun to the Milky Way centre, as this angle also affects Milky Way dynamics. All other Milky Way parameters are input as known constants in the base astronomical model that is used to undertake these simulations.

Now, galactic astronomy allows for the existence of a known relation between the bar angular speed at the distance of the Sun from the centre of the Milky Way (discussed in Section 5.5.4). So, the sought Milky Way feature parameters are:

1. bar rotational frequency $\equiv$ radial location S_1 of the Sun from the Milky Way centre;
2. the angular separation S_2 of the Sun from an identified axis in the Milky Way disk, namely the bar long-axis.

Thus the location vector $\boldsymbol{S} = (S_1, S_2)^T$ of the Sun, in the 2-dimensional Milky Way disc, is the feature parameter vector in our work.

On galactic length scales, the solar location vector – with respect to the centre of the Galaxy – is equivalent to such location of the observer on Earth, who observes the test data comprising velocities of stellar neighbours of the Sun. So prediction of the Milky Way feature parameter vector is equivalent to the prediction of the value of the observer's location vector variable, where this observer observes the matrix of velocities of some of the Sun's stellar neighbours.

Data on values of velocity matrix variable ($\boldsymbol{V}$), generated at the n different values of the solar (or observer) location vector $\boldsymbol{S} = (S_1, S_2)^T$ in the astronomical simulations, are s.t. there may exist *inhomogeneities in correlation* structure of data on the generated $\boldsymbol{V}$ values, as displayed in Figure 8 of [25].

5.3.2 Details

In this application, the aim is to learn location vector $\boldsymbol{S}$ of the Sun in the Milky Way modelled as a 2-dimensional disk. The training data $\mathbf{D}$ is cuboidally-shaped, and is of dimensionalities $m_1 \times m_2 \times m_3$, where

1. dimension of a stellar velocity vector is $m_1 = 2$;
2. the number of stars for which velocity vectors are generated at each design point is $m_2 = 50$;
3. the number of design points is $m_3 \equiv n = 216$.

So the 3-rd ordered tensor $\boldsymbol{D_V}$ comprises $n = 216$ matrices of dimension 50×2 each, where i-th value $\boldsymbol{v}_i$ of the matrix-valued observable $\boldsymbol{V}^{(50\times2)}$ is realised at the i-th value of system parameter vector $\boldsymbol{S}$, s.t.

$$\mathbf{D} = \{(\boldsymbol{s}_i, \boldsymbol{v}_i)\}_{i=1}^{n}.$$

The 3rd-ordered tensor

$$\boldsymbol{D}_{\boldsymbol{V}}^{(m_1\times m_2\times n)} = (\boldsymbol{v}_1, \vdots, \ldots, \vdots \boldsymbol{v}_n).$$

Numerical simulations are conducted with $n = 216$ different astronomical models of the Galaxy, with each such model of the Galaxy distinguished by a value of the Milky Way feature parameter vector $\boldsymbol{S} \in \mathbb{R}^d$, d=2 [25] – or equivalently, the location vector of the Sun, which is also approximately the location of the observer who observes the stellar velocity matrix. Thus, at the i-th design observer location $\boldsymbol{s}_i$, $i = 1, \ldots, 216$, $\boldsymbol{V} = \boldsymbol{v}_i$. Since $\boldsymbol{V}$ is affected by $\boldsymbol{S}$, we write $\boldsymbol{V} = \boldsymbol{\xi}(\boldsymbol{s})$, and aim to learn the high-dimensional function $\boldsymbol{\xi}(\cdot)$, with the aim of predicting value of either $\boldsymbol{V}$ or $\boldsymbol{S}$, at a given value of the other.

In particular, there exists the test data $\boldsymbol{v}^{(test)}$ that comprises the 2-dimensional velocity vectors of the 50 identified, stellar neighbours of the Sun, as measured by the *Hipparcos* satellite [25]. It is the same 50 stars for which velocity vectors are simulated at each design point. *However, we do not know the real Milky Way feature parameter vector* $\boldsymbol{s}^{(test)}$ *at which* $\boldsymbol{V} = \boldsymbol{v}^{(test)}$ *is realised.*

Since we are observing velocities of stars around the Sun, the observed velocities will be impacted by certain Galactic features. These features include location $\boldsymbol{S}$ of the Sun. Thus, the observed matrix $\boldsymbol{v}^{(test)}$, can be regarded as resulting from the Galactic features (including the sought solar location) attaining certain values.

So, fixing all Galactic features other than the location $\boldsymbol{S}$ of the Sun in the simulations that generate the training data, the matrix $\boldsymbol{V}$ of stellar velocities is related to $\boldsymbol{S}$, i.e. $\boldsymbol{V} = \boldsymbol{\xi}(\boldsymbol{s})$. The input variable $\boldsymbol{S}$ is then also the location from which an observer on Earth (or equivalently the Sun, on Galactic length scales), observes the 2-dimensional velocity vectors of m_2 (=50) of our stellar neighbours.

[25] generated the training data by first placing a regular 2-dimensional, galactocentric polar grid, over a chosen annulus in a 2-dimensional astronomical model of the MW disk. This polar grid is uniformly partitioned into $n = 216$ grid cells, s.t. the spatial vector $\boldsymbol{s}_i$ joining the origin (i.e. the centre of the Galaxy), to the centroid of the i-th grid cell, is the i-th design point that represents the solar location vector – which is the galactic feature parameter vector – in the i-th model of the Galaxy, $\forall i = 1, \ldots, 216$. So in the i-th model of the Milky Way, the observer – who is at the Sun on Galactic length scales – is placed at the centroid $\boldsymbol{s}_i$ of the i-th grid cell.

The observer in the i-th grid cell measures velocities of m_{2i} stars that land in that grid cell, at the end of a simulated evolution of a sample of stars that are evolved in this i-th model of the MW disk, under the influence of the feature parameters $\boldsymbol{s}_i$. The m_{2i} number of stars are indexed by their locations with respect to the observer inside the grid cell, and a random sample of $m_2 = 50$ stars from this collection of m_{2i} stars is drawn, where $m_{2i} \geq m_2 = 50$, $\forall\, i = 1, \ldots, n = 216$. Thus, the observer in the i-th model, records a matrix (or sheet) of 2-dimensional velocity vectors of m_2 stars, given the design vector $\boldsymbol{s}_i$.

The test data measured by the *Hipparcos* satellite is then the 217-th such sheet, except we are not aware of the value of $\boldsymbol{S}$ that this sheet is realised at.

In [29], the matrix of velocities was vectorised, so that the observable was then a vector. In this application, the observable $\boldsymbol{V}$ is a matrix. The process of vectorisation, causes [29] to face loss of correlation information since vectorisation of the stellar velocity matrix may imply the flattening of the inter-component correlation structure of the velocity data of the stars, which may be a misrepresentation. However, the high-dimensional learning strategy that we discuss above, allows for clear quantification of inter-component covariances. More importantly, such a strategy provides a clear template for implementing learning given high-dimensional data that comprise measurements of a tensor-valued observable. As mentioned above, the empirical estimate of the mean tensor is obtained, and used as the mean of the Tensor Normal density that represents the likelihood.

To learn $\boldsymbol{\xi}(\cdot)$, we model it as a realisation from a high-dimensional GP, s.t. joint of n values of $\boldsymbol{\xi}(\cdot)$ – computed at $\boldsymbol{s}_1, \ldots, \boldsymbol{s}_n$ – is 3rd-order Tensor Normal, with 3 covariance matrices that inform on:

— amongst-observer-location, or *inter-realisation* covariance ($\boldsymbol{\Sigma}_3^{(216\times 216)}$);

— amongst-stars-at-different-relative-position-w.r.t.-observer, or *inter-star* covariance ($\boldsymbol{\Sigma}_2^{(50\times 50)}$);

— amongst-velocity-component, or *inter-component* covariance ($\boldsymbol{\Sigma}_1^{(2\times 2)}$).

Upon standardisation of the data on $\boldsymbol{V}$ – by the mean and standard deviation of the sample o observations that are collected in this data on each component of $\boldsymbol{V}$, at each design point – the above covariance matrices reduce to the respective correlation matrices.

The cartoons in Figure 5.1 and Figure 5.2, depict the modelling of the Milky Way disc as 2-dimensional, with the proposed solar neighbourhood placed on a grid cell of a uniform polar grid. The radial range of this polar grid is [1.7, 2.3] in model units, i.e. the random variable S_1 that represents the radial location of the Sun - which informs on the bar rotational frequency – is in [1.7, 2.3], where the radial partition is 0.025 in width. The angle between the Sun-galactic centre line and the long axis of the bar is denoted S_2, which takes values in [0, 90°] in this grid, and each angular bin is 10° wide. The schematic diagram for the three correlation matrices $\boldsymbol{\Sigma}_1, \boldsymbol{\Sigma}_2, \boldsymbol{\Sigma}_3$ is also depicted.

5.3.2.1 Estimating the inter-star covariance

Given the size of the matrix $\boldsymbol{\Sigma}_2$, its elements are not learnt directly using MCMC. Indeed, for the standardised data, direct learning

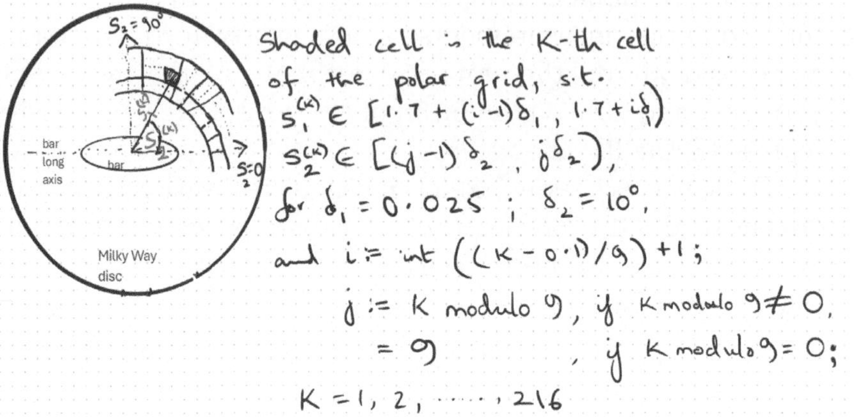

FIGURE 5.1
Cartoon to depict the modelling of the Milky Way disk as 2-dimensional. A uniform polar grid is set up in S_1 and S_2, where S_1 is the radial location (in model units) of a grid cell in this polar grid, while S_2 is the angular distance between the line joining the galactic centre to this grid cell and the long axis of the bar. Here,any grid cell is centred around a possible candidate location for the Sun in the galactic disk. Thus, $\boldsymbol{S} = (S_1, S_2)^T$ is the solar location variable and a grid cell located at $\boldsymbol{S} = \boldsymbol{s}_k$ represents the k-th possible candidate for the solar neighbourhood. We randomly sample 50 stars from all the stars that land in a grid cell – i.e. a candidate solar neighbourhood – at the end of an astronomical simulation that emulates the distribution of stars in the Milky Way disc, under the effect of Milky Way features. Then the observer in the grid cell located at $\boldsymbol{S} = \boldsymbol{s}_k$ matrix-valued random variable (r.v.) $\boldsymbol{V}$. Since the solar location variable also informs on the Milky Way feature parameters, we suggest $\boldsymbol{V}$ is associated with $\boldsymbol{S}$, and learn this functional relationship using the training data $\{(\boldsymbol{s}_k, \boldsymbol{v}_k)\}_{k=1}^n$, where there are $n = 216$ grid cells arranged in a 24-layered radial sequence of 9 grid cells in each layer, i.e. the i-th set of 9 grid cells are distinguished in values of S_2 of their respective centroids with a given S_1 value, for $i = 1, \ldots, 24$. Once the relation between $\boldsymbol{S}$ and $\boldsymbol{V}$ is learnt, we learn the value $\boldsymbol{S}$ of the Milky Way feature parameters (or equivalently the solar location), at which the measured – by the *Hipparcos* satellite – matrix $\boldsymbol{v}^{(test)}$ of 50 stellar neighbours of the Sun is realised.

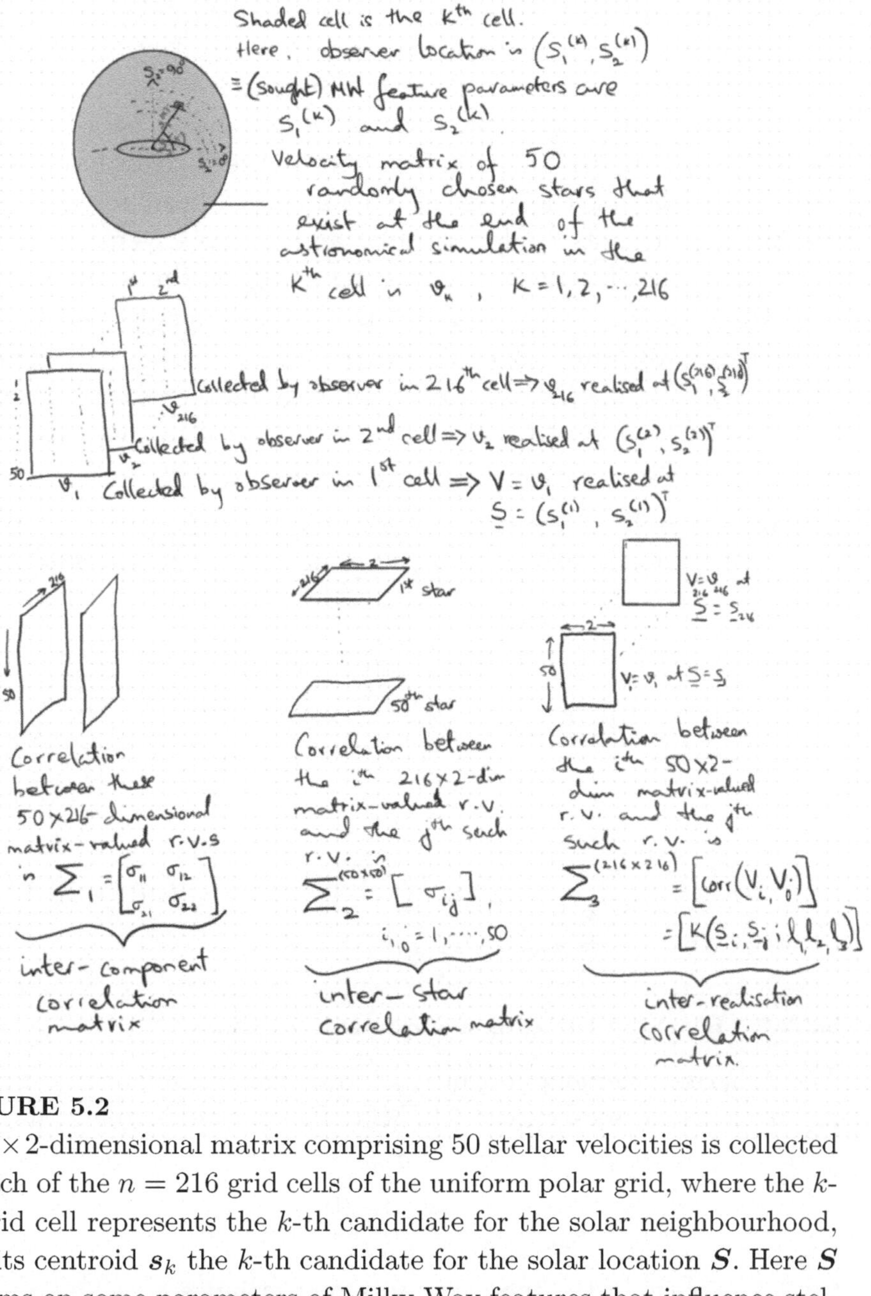

FIGURE 5.2

A 50×2-dimensional matrix comprising 50 stellar velocities is collected at each of the $n = 216$ grid cells of the uniform polar grid, where the k-th grid cell represents the k-th candidate for the solar neighbourhood, and its centroid $\boldsymbol{s}_k$ the k-th candidate for the solar location $\boldsymbol{S}$. Here $\boldsymbol{S}$ informs on some parameters of Milky Way features that influence stellar velocities that are simulated to evolve under their influence. Then the (standardised) correlation structure of the 3-dimensional training set is marked by 3 correlation matrices that are: the inter-component ($\boldsymbol{\Sigma}_1^{(2\times 2)}$); inter-star ($\boldsymbol{\Sigma}_2^{(50\times 50)}$); and inter-realisation correlation matrix ($\boldsymbol{\Sigma}_3^{(216\times 216)}$) that is kernel parametrised.

of the 50×49/2 distinct elements of $\boldsymbol{\Sigma}_2$, using MCMC, is ruled out, given that this is a large number. Also, there is no input space variable that can be identified, at which the ij-th element of $\boldsymbol{\Sigma}_2$ can be considered to be realised; $i, j = 1, \ldots, 50$, where this ij-th element gives the covariance amongst the i-th and j-th, 216×2-dimensional matrices within the 3-rd ordered tensor $\boldsymbol{D}_{\boldsymbol{V}}$. For example, the 41-st star could have been referred to as the 3rd star in this stellar sample, and the vice versa, i.e. there is no meaningful ordering in the labelling of the sampled stars with these indices. So we cannot use these labels as values of an input space variable, in terms of which, covariance between the i-th and j-th 216×2-dimensional velocity matrices can be kernel-parametrised. In light of the material discussed in this paragraph, we will undertake empirical estimation of elements of $\boldsymbol{\Sigma}_2$.

Definition 5.1 *Covariance between the 216×2-dimensional stellar velocity matrix $\boldsymbol{W}_i := [v_{pq}^{(i)}]$ of the sampled star labelled by index i, and the matrix $\boldsymbol{W}_j := [v_{pq}^{(j)}]$ of the star labelled as j, ($p = 1, \ldots, 216$; $q = 1, 2$), is estimated as $\widehat{\sigma_{ij}^{(2)}}$, where:*

$$\widehat{\sigma_{ij}^{(2)}} = \frac{1}{2} \times \sum_{q=1}^{2} \left[\frac{1}{215} \times \left(\sum_{p=1}^{216} (v_{pq}^{(i)} - \bar{v}_q^{(i)}) \times (v_{pq}^{(j)} - \bar{v}_q^{(j)}) \right) \right],$$

where $\bar{v}_q^{(i)} = \dfrac{\left(\sum_{p=1}^{216} v_{pq}^{(i)}\right)}{216}$ is the sample mean of the q-th column of the matrix $\boldsymbol{V}_i = [v_{pq}^{(i)}]$, and similarly for $\bar{v}_q^{(i)}$.

5.3.2.2 Learning the inter-component covariance

The 3 distinct elements of the 2×2-dimensional covariance matrix $\boldsymbol{\Sigma}_1$ are learnt directly from MCMC. These include the 2 diagonal elements $\sigma_{11}^{(1)}$, $\sigma_{22}^{(1)}$ and $\rho := \dfrac{\sigma_{12}^{(1)}}{\sqrt{\sigma_{11}^{(1)}\sigma_{22}^{(1)}}}$

5.3.2.3 Learning the inter-realisation covariance

If we choose to model $\boldsymbol{\xi}(\cdot)$ as a sample function of a *stationary* GP, $\boldsymbol{\Sigma}_3$ is kernel-parametrised, using the SQE kernel such that the jp-th element of $\mathbf{\Sigma}_3$ is $\sigma_{jp} = \exp\left(-(\boldsymbol{s}_j - \boldsymbol{s}_p)^T \boldsymbol{Q}^{-1}(\boldsymbol{s}_j - \boldsymbol{s}_p)\right)$ $j, p = 1, \ldots, 216$. Since $\boldsymbol{S}$ is a 2-dimensional vector, $\boldsymbol{Q}$ is a 2×2 square diagonal matrix, elements $1/\ell_1^2, 1/\ell_2^2$ of which, informs on the correlation length scale hyperparameters ℓ_1, ℓ_2. Then in this stationary model, we learn the (modelled as constant) ℓ_1, ℓ_2, along with $\sigma_{11}^{(1)}$, $\sigma_{22}^{(1)}$ and ρ.

However, *under the fully non-parametric model* – that invokes the dual-layered learning strategy -

- $\boldsymbol{\xi}(\cdot)$ is modelled using the matrix-variate non-stationary $\mathcal{GP}_{non-stat}$ and
- ℓ_c is modelled as $\ell_c = \eta_c(t)$, where at iteration index $T = t$. $\eta_c(t)$ is sampled from a zero-mean, scalar variate process $\mathcal{GP}_c$, $c = 1, 2$, $\forall t = n_0 + 1, \ldots, N_{iter}$, for a chosen iteration index n_0.
- In this application we do not standardise the lookback data at the beginning of the 2nd block of any iteration, (to unstandardise it at the end of this block). Instead, we use the unstandardised lookback data in the 2nd block of the k-th iteration, and learn the amplitude a_c and correlation length scale δ_c, for $c = 1, 2$. Such parameters are learnt in the 2nd block of the k-th iteration, while in the first block of this iteration, we learn the parameters $\sigma_{11}^{(1)}$, $\sigma_{22}^{(1)}$ and ρ of $\mathbf{\Sigma}_1$.
- Here $k = n_0 + 1, \ldots, N_{iter}$, i.e. the inference proceeds over N_{iter} number of iterations and the dual-staged learning is switched on after the n_0-th iteration, where $n_0 > N_{lookback}$ with the size of the lookback data used in the k-th iteration denoted $N_{lookback}$.
- ℓ_1, ℓ_2 are also updated in this 1st block of the k-th iteration, and the hyperparameters $a_1, \delta_1, a_2, \delta_2$ of the kernels that parametrise the covariance functions of the inner layer processes ($\mathcal{GP}_1$ and $\mathcal{GP}_2$) ar updated in the 2nd block of this iteration. Using these updated inner layer GPs, the expected ℓ_1, ℓ_2 are predicted at the end of the 2nd block of this iteration. Then at the beginning of the 1st block of the $k+1$-th iteration, these predicted means are treated as the current length scale hyperparameters of the kernel that parametrises $\mathcal{GP}_{non-stat}$, and the freshly updated value of ℓ_c

is added to the lookback data $\mathbf{D}^{(c)}_{lookback,k+1}$ that is relevant for the c-th direction of input space, at the $k+1$-th iteration; $c=1,2$. Here $k=n_0+1,\ldots,N_{iter}$.

5.3.2.4 Posterior of unknowns given stationary GP

When the stationary GP is used to model $\boldsymbol{\xi}(\cdot)$, likelihood of the covariance matrices of the tensor-variate $\mathcal{GP}_{non-stat}$, in the training data, is:

$$\mathcal{L}(\ell_1,\ell_2,\sigma^{(1)}_{11},\sigma^{(1)}_{22},\rho;\mathbf{D}) = (2\pi)^{-m/2}(\prod_{i=1}^{3}|\mathbf{\Sigma}_i|^{-m/2m_i}) \\ \times \exp(-\|(\boldsymbol{D_V}-\hat{\boldsymbol{M}})\times_1 \boldsymbol{A}_1^{-1}\times_2 \hat{\boldsymbol{A}}_2^{-1}\times_3 \boldsymbol{A}_3^{-1}\|^2/2). \tag{5.1}$$

where $\mathbf{\Sigma}_p=\boldsymbol{A}_p\boldsymbol{A}_p^T$, $p=1,2,3$ and $\hat{\boldsymbol{M}}$ is the empirical estimate of the mean tensor and $\hat{\mathbf{\Sigma}}_2$ is the empirical estimate of the covariance matrix $\mathbf{\Sigma}_2$ such that $\hat{\mathbf{\Sigma}}_2=\hat{\boldsymbol{A}}_2\hat{\boldsymbol{A}}_2^T$. Here $m_3=216$, $m_2=50$, $m_1=2$, and $m=m_1m_2m_3$. $\mathbf{\Sigma}_3$ is kernel parametrised, where the kernel is a function of pairs of values of the input variable $\boldsymbol{S}$ – this explains the dependence of the RHS of this equation on the whole of $\mathbf{D}$, with the data tensor $\boldsymbol{D_V}$ comprising velocity matrix values realised at corresponding design locations.

This allows us to write the joint posterior probability density of the unknown parameters given training data $\mathbf{D}$. To write this posterior, we impose non-informative priors $\pi_0(\cdot)$ on each unknown. While other applications may be able to avail of informative priors, in this application there is little prior information on the parameters that we wish to learn. This prompts the usage of Gaussian priors with wide, experimentally chosen variance, and mean that is the arbitrarily chosen seed value of $\ell_\cdot$. Also, we use Jeffry's priors on $\mathbf{\Sigma}_1$. The posterior probability density of the unknown GP parameters, given the training data is then given in this implementation of the stationary GP, as:

$$\pi(\ell_1,\ell_2,\sigma^{(1)}_{11},\sigma^{(1)}_{22},\rho|\mathbf{D}) \propto \mathcal{L}(\mathbf{\Sigma}_1,\mathbf{\Sigma}_3;\mathbf{D})\times\pi_0(\ell_1)\pi_0(\ell_2)\pi_0(\mathbf{\Sigma}_1). \tag{5.2}$$

The results of this learning and estimation of the covariance and mean of the GP invoked in the modelling, are discussed below in Section 5.5.

5.3.2.5 Posterior of unknowns in the dual-staged learning strategy

Definition 5.2 *Under the non-parametric model, inference is undertaken with Metropolis-within-Gibbs, using two distinct blocks.*

In the first block, we make inference on the parameters $\sigma_{11}^{(1)}, \sigma_{22}^{(1)}, \rho, \ell_1, \ell_2$*, given the training data* $\mathbf{D}$ *that contains the elements of the velocity tensor* $\boldsymbol{D}_{\mathbf{V}}$*. Then in the t-th iteration, for* $t > n_0$*, the joint posterior probability density of these parameters is*

$$\pi(\ell_1, \ell_2, \sigma_{11}^{(1)}, \sigma_{22}^{(1)}, \rho|\mathbf{D}) \propto (2\pi)^{-m/2}\left(\prod_{i=1}^{3}|\boldsymbol{\Sigma}_i|^{-m/2m_i}\right)$$
$$\times \exp(-\|(\boldsymbol{D}_{\boldsymbol{V}} - \hat{\boldsymbol{M}}) \times_1 \boldsymbol{A}_1^{-1} \times_2 \hat{\boldsymbol{A}}_2^{-1} \times_3 \boldsymbol{A}_3^{-1}\|^2/2)\times$$
$$\pi_0(\ell_1)\pi_0(\ell_2)\pi_0(\boldsymbol{\Sigma}_1), \tag{5.3}$$

In the 2nd block of this t-th iteration, the parameters that we make inference upon are the hyperparameters $(\delta_1, a_1, \delta_2, a_2)$ *of the kernels that parametrise the covariance function of the inner layer GPs, namely,* $\mathcal{GP}_1$*, and* $\mathcal{GP}_2$*. The joint posterior of* δ_c, a_c*, given the* $N_{lookback}$*-sized lookback data* $\mathbf{D}_{lookback,t}^{(c)} = \{\ell_c^{(t-N_{lookback})}, \ell_c^{(t-N_{lookback}+1)}, \ldots, \ell_c^{(t-N_{lookback}+N_{lookback}-1)}\}$ *is then formulated for* $c = 1, 2$*. The joint posterior in this t-th iteration is given as*

$$\pi(\delta_c, a_c|\mathbf{D}_{lookback,t}^{(c)}) \propto$$
$$\frac{1}{\sqrt{\det(2\pi\boldsymbol{\Psi}_{a_c,\delta_c})}}\exp\left[-\frac{1}{2}(\boldsymbol{\ell}_c^{(N_{lookback})})^T\,(\boldsymbol{\Psi}_{a_c,\delta_c})^{-1}\,(\boldsymbol{\ell}_c^{(N_{lookback})})\right]$$
$$\times\pi_0(a_c)\pi_0(\delta_c),$$

where $\boldsymbol{\ell}_c^{(N_{lookback})} := (\ell_c^{(t-N_{lookback})}, \ldots, \ell_c^{(t-1)})^T$*, and ij-th element of the covariance matrix* $\boldsymbol{\Psi}_{a_c,\delta_c}$ *is* $\left[a_c \exp\left[-\frac{(t_i - t_j)^2}{2(\delta_c)^2}\right]\right]$, $i, j = 1, \ldots, N_{lookback}$.

Once δ_c, a_c *are updated in the 2nd block of this t-th iteration, mean and variance of* ℓ_c *is predicted, and this newly predicted mean* ℓ_c *is added to the lookback data* $\mathbf{D}_{lookback,t+1}^{(c)}$*, while* $\ell_c^{(t-N_{lookback})}$ *is dropped from* $\mathbf{D}_{lookback,t+1}^{(c)}$*.*

N.B. In the definition of $\boldsymbol{\Psi}_{a_c,\delta_c}$ the dependence on the current time-step t is effectively suppressed, given that this dependence comes in the form $t - t_i - (t - t_j)$.

Posterior samples are generated using Metropolis-within-Gibbs, to identify the marginal posterior probability distribution of each unknown. The marginal then allows for the computation of the 95% HPD.

Also $n_0 \in \mathbb{Z}$ is s.t. the dual-staged learning is relevant only for $t > n_0$, with the learning performed with a stationary GP at all iterations before the $n_0 + 1$-th iteration.

5.4 Inverse prediction, aka learning the input given a test output: 2 ways

We aim to predict the location vector $\boldsymbol{s}^{(test)}$ of the Sun in the Milky Way disk, at which real (test) data $\boldsymbol{v}^{(test)}$ on the 2-dimensional velocity vectors of 50 identified stellar neighbours of the Sun, are measured by the *Hipparcos* satellite. We undertake this, subsequent to learning the relation $\boldsymbol{\xi}(\cdot)$ between solar location variable $\boldsymbol{S}$ and matrix-valued stellar velocity variable $\boldsymbol{V}$, using astronomically-simulated training data.

5.4.1 First way of learning $\boldsymbol{s}^{(test)}$

Definition 5.3 *The tensor that includes both test and training data has dimensions of* $217 \times 50 \times 2$*. We call this augmented data* $\boldsymbol{D}^* = \{\boldsymbol{v}_1, \ldots, \boldsymbol{v}_{50}, \boldsymbol{v}^{(test)}\}$*, to distinguish it from the tensor* $\boldsymbol{D}_{\boldsymbol{V}}$ *that comprises the training values of* $\boldsymbol{V}$ *realised at each of the 216 design values of* $\boldsymbol{S}$*.* $\boldsymbol{v}_i$ *is realised at design point* $\boldsymbol{s}_i$*, but* $\boldsymbol{s}^{(test)}$ *– at which* $\boldsymbol{v}^{(test)}$ *is realised – is unknown.*

This 217-th sheet of (test) data is realised at the unknown value $\boldsymbol{s}^{(test)}$ of $\boldsymbol{S}$, and upon its inclusion, the updated inter-realisation covariance matrix is renamed $\boldsymbol{\Sigma}_3^*$, which is now rendered 217×217-dimensional. Then $\boldsymbol{\Sigma}_3^*$ includes information about $\boldsymbol{s}^{(test)}$ via the kernel-parametrised covariance matrix $\boldsymbol{\Sigma}_3$. The effect of inclusion of the test data on the other covariance matrices is weaker. We refer to them as (empirically estimated) inter-star covariance matrix $\hat{\boldsymbol{\Sigma}}_2^*$ and (the directly learnt) inter-component covariance matrix $\boldsymbol{\Sigma}_1^*$. The updated (empirically estimated) mean tensor is $\hat{\boldsymbol{M}}^*$.

The likelihood in the augmented data is:

$$\mathcal{L}(\boldsymbol{s}^{(test)}, \boldsymbol{\Sigma}_1^*, \boldsymbol{\Sigma}_3^*; \boldsymbol{D}^*) = (2\pi)^{-m/2}\left(\prod_{i=1}^{3}|\boldsymbol{\Sigma}_i^*|^{-m/2m_i}\right)\times \\ \exp\left[-\|(\boldsymbol{D}^* - \hat{\boldsymbol{M}}^*)\times_1(\boldsymbol{A}_1^*)^{-1}\times_2(\hat{\boldsymbol{A}}_2^*)^{-1}\times_3(\boldsymbol{A}_3^*)^{-1}\|^2/2\right] \tag{5.4}$$

where $\hat{\boldsymbol{A}}_2^*$ is the square root of $\hat{\boldsymbol{\Sigma}}_2^*$. Here $m_3 = 217$, $m_2 = 50$, $m_1 = 2$, and $m = m_1 m_2 m_3$. Here $\boldsymbol{A}_3^*$ is the square root of $\boldsymbol{\Sigma}_3^*$ and depends on $\boldsymbol{s}^{(test)}$.

Then following Equation 5.4, joint posterior *pdf* of the unknowns given the augmented data is:

$$\pi(s_1^{(test)}, s_2^{(test)}, \boldsymbol{\Sigma}_1^*, \boldsymbol{\Sigma}_3^*|\boldsymbol{D}^*) \propto \mathcal{L}(s_1^{(test)}, s_2^{(test)}, \boldsymbol{\Sigma}_1^*, \boldsymbol{\Sigma}_3^*; \boldsymbol{D}^*)\times \\ \pi_0(s_1^{(test)})\pi_0(s_2^{(test)})\pi_0(\ell_2^{(*)})\pi_0(\ell_1^{(*)})\pi_0(\boldsymbol{\Sigma}_1^*). \tag{5.5}$$

In this application, we use the prior probability density $\pi_0(s_p^{(test)}) = \mathcal{U}(l_p, u_p)$, $p = 1, 2$, where l_p and u_p are chosen depending on the spatial boundaries of the fixed area of the Milky Way disk that was used in the astronomical simulations by [25]. Recalling that the observer is located in a quadrant of a two-dimensional polar grid, [25] set the lower boundary on the angular position of the observer to 0, and upper to $\pi/2$ radians, where the observer's angular coordinate is the angle included between the observer-Galactic centre line and the bar long axis in the Milky Way disk.

The observer's radial location is maintained within the interval [1.7, 2.3] in model units, where model units for length are related to galactic unit for length, as is discussed in Section 5.5.4.

5.4.2 Second way of learning $\boldsymbol{s}^{(test)}$

In the second method for prediction – or rather learning – of the unknown input value $\boldsymbol{s}^{(test)}$ – we infer $\boldsymbol{s}^{(test)}$ by sampling from the posterior of this variable, given the test data and the modal values of the parameters $q_1, q_2, \sigma_{11}^{(1)}, \rho, \sigma_{22}^{(1)}$ that were learnt using the training data. Let modal value of $\boldsymbol{\Sigma}_3$, learnt using $\mathbf{D}$ be $\boldsymbol{\Sigma}_3^{(M)} = [(\sigma_3^{(M)})_{jp}]_{j=1;p=1}^{216,216}$, Similarly, the modal value $\boldsymbol{\Sigma}_1^{(M)}$ that was learnt using the training data, is used. The posterior of $\boldsymbol{s}^{(test)}$, at learnt (modal) values is then

$$\begin{aligned} &\pi(s_1^{(test)}, s_2^{(test)} | \boldsymbol{D}^*, \boldsymbol{\Sigma}_1^{(M)}, \boldsymbol{\Sigma}_3^{(M)}) \propto \\ &\mathcal{L}(s_1^{(test)}, s_2^{(test)}; \boldsymbol{\Sigma}_1^{(M)}, \boldsymbol{D}^*, \boldsymbol{\Sigma}_3^{(M)}) \times \pi_0(s_1^{(test)})\pi_0(s_2^{(test)}) \end{aligned} \tag{5.6}$$

where $\mathcal{L}(s_1^{(test)}, s_2^{(test)}; \boldsymbol{D}^*, \boldsymbol{\Sigma}_1^{(M)}, \boldsymbol{\Sigma}_3^{(M)})$ is as given in Equation 5.1, with $\boldsymbol{\Sigma}_3$ replaced by $\boldsymbol{\Sigma}_3^{(M)}$, and $\boldsymbol{\Sigma}_1$ replaced by its modal value $\boldsymbol{\Sigma}_1^{(M)}$. The priors on $s_1^{(test)}$ and $s_2^{(test)}$ are as discussed above. For all real-valued parameters, we use Normal proposal densities that have experimentally chosen variances, while non-negative parameters are proposed from truncated Normal densities, left truncated at 0.

5.5 Results

In this section, we present results of the learning methodology described above, implemented within the considered application. Results of learning the unknown parameters of the 3rd-order tensor Normal likelihood, given training as well as training+test data, are discussed here.

While Figure 5.5 and Figure 5.3 depict results obtained from the stationary GP model, Figure 5.4 includes results of the learning of all relevant unknown parameters, using the non-parametric, dual-layered model. Such results pertain to the learning of the amplitude parameter a_c, and the smoothing parameter $d_c := 1/\delta_c$, $c = 1, 2$. Also, our modelling under the dual-layered paradigm

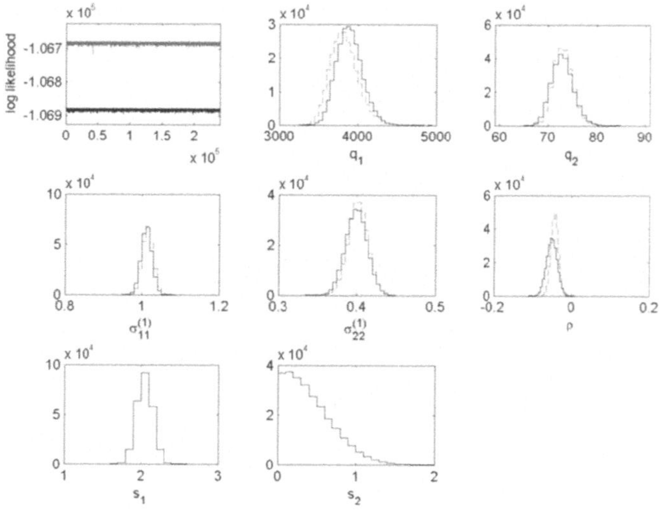

FIGURE 5.3

Results from chain run with training data $\mathbf{D}$ with the stationary GP model, are shown in red (or grey in the monochromaic version), while results from chain undertaken with augmented data, $\boldsymbol{D}^{\star}$, in this model, are depicted in black. Traces of the logarithm of the likelihood are displayed from the two chains in the top left panel. Reciprocals of length scale parameters are the shown in the top middle and right panels; here $q_c = \ell_c^{-1}$, $c = 1, 2$. Histograms representing the marginal posterior probability density of the learnt diagonal elements $\sigma_{11}^{(1)}$ and $\sigma_{22}^{(1)}$, of the covariance matrix $\boldsymbol{\Sigma}_1$, (given the relevant data set) are shown in the middle row, in the left and middle panels. Histograms representing the marginals of parameter $\rho = \sigma_{12}/\sqrt{\sigma_{11}^{(1)}\sigma_{22}^{(1)}}$, given the training and the augmented datasets, are displayed in the right panel of the middle row. Prediction – or rather learning – of the values of the input parameter $\boldsymbol{S} = (S_1, S_2)^T$ is possible only in the run performed with both training and test data. Marginals of S_1 and S_2 – approximated as histograms – learnt by MCMC-based sampling from the joint of all unknown parameters given $\boldsymbol{D}^{\star}$, are shown in the lower panels.

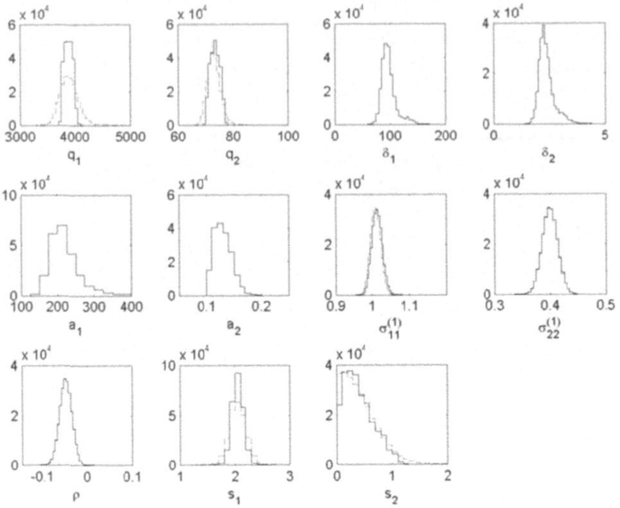

FIGURE 5.4
Results from the chain run with test+training data (i.e. augmented data) $\boldsymbol{D}^{\star}$ within the dual-layered (i.e. non-stationary) model, are shown in black, as distinguished from the results of learning given the same data, using the stationary GP model, that are depicted in red (or grey in the black and white version) broken lines. Here the used value of $N_{lookback}$ is 200 iterations. Histograms approximating the marginal posterior probability densities of each sought unknown is depicted. Here, the sought hyperparameter values a_c and δ_c are relevant only to the non-parametric model ($c = 1, 2$). Also, we have undertaken sampling from the joint posterior of all parameters to learn values of all unknowns, including the input parameter values $s_1^{(test)}$ and $s_2^{(test)}$ at which the test data are realised. Histograms approximating marginal posterior of the learnt parameters are presented.

relies on a lookback-time $N_{lookback}$ which gives the number of iterations over which we gather generated ℓ_c values – dependence on $N_{lookback}$ is depicted in Figure 5.7.

Figure 5.5 displays traces of the parameters learnt using the stationary GP model. Figure 5.6 depicts histogram representation of the marginal posterior probability density of the sought parameters, given training data, as obtained using the stationary GP model. These results are compared to the corresponding result obtained from the dual-layered non-parametric model. In this dual-layered model, the inter-realisation covariance matrix $\boldsymbol{\Sigma}_3$ (that bears information about the covariance structure between pairs of data sheets realised at pairs of values of the input variable $\boldsymbol{S} = (S_1, S_2)^T$), is parametrised using a kernel, each length-scale hyperparameter of which, is itself modelled as a dynamically-varying function that is considered sampled from a GP.

95% HPD credible regions computed on each learnt parameter given the stationary GP model, are displayed in Table 5.1. Again, a similar set of results from the chains run with the dual-layered model are displayed in Table 5.2. The results on prediction of $\boldsymbol{s}^{(test)}$ are also presented in Table 5.1 and Table 5.2.

5.5.1 Effect of inhomogeneities in correlation structure of data, manifest in our results

One difference between the learning of parameters from the dual-layered, as distinguished from the stationary GP model is the quality of the inference, in the sense that the uncertainty of parameters (i.e. the 95% HPDs) learnt using the dual-layered models, is less than that learnt using the stationary GP models. This difference in the learnt HPDs is most marked for the learning of values of ℓ_1 and S_1, and S_2 to a lesser extent.

To analyse possible inhomogeneities in the correlation structure of the training data used in the application, we refer to Figure 8 of [25], that is available at `https://www.aanda.org/articles/aa/pdf/2007/19/aa6677-06.pdf`. This figure informs on the compatibility of the value $\boldsymbol{v}$ of the stellar velocity matrix variable $\boldsymbol{V}$ – realised in these astronomical simulations at a given $\boldsymbol{s}$ – to the test velocity matrix $\boldsymbol{v}^{(test)}$ (recorded by the *Hipparcos* satellite). The parametrisation of this compatibility is in terms of a "compatibility parameter" that is plotted as a function of the two components S_1 and S_2 of the vector $\boldsymbol{S}$, as a contour plot in the space $\mathcal{D}$, where $\boldsymbol{S} \in \mathcal{D} \subset \mathbb{R}^2$. Here, the two components of $\boldsymbol{S}$ are represented in polar coordinates, with S_1 the radial and S_2 the angular component. From this figure, we appreciate the lack

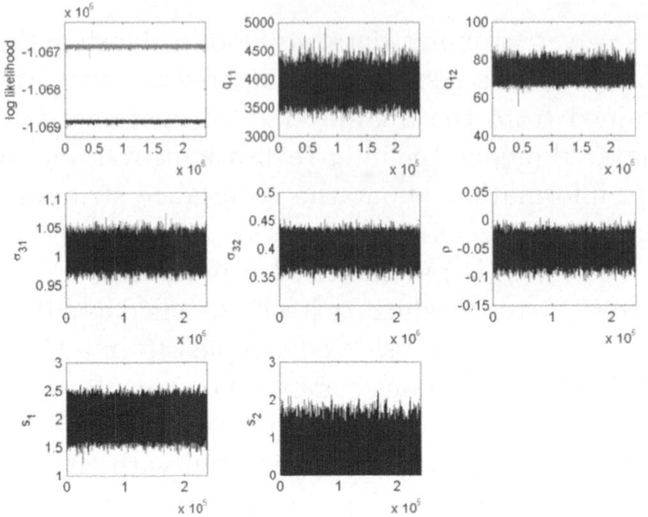

FIGURE 5.5
Results from the chain undertaken using data $\boldsymbol{D_V}$ – that helps populate the training data – within the stationary GP model, are shown in red (or grey) while results from the run undertaken with training+test (i.e. augmented) data, $\boldsymbol{D}^{\star}$, in this stationary GP model, are depicted in black. Traces of logarithm of the likelihood are displayed from the two runs in the top left panel. Reciprocal of the length scale parameters are shown in the top middle and right panels; here $q_c = \ell_c^{-1}$, $c = 1, 2$. Traces of the learnt diagonal elements $\sigma_{11}^{(1)}$ and $\sigma_{22}^{(1)}$, of the covariance matrix $\boldsymbol{\Sigma}_1$, are shown in the middle row, left and middle panels. Trace of the correlation $\rho = \sigma_{12}/\sqrt{\sigma_{11}^{(1)}\sigma_{22}^{(1)}}$ is displayed in the right panel of the middle row. Prediction of the values of the input parameter $\boldsymbol{S} = (S_1, S_2)^T$ is possible only in the run performed with both training and test data. Traces of S_1 and S_2 values learnt via MCMC-based sampling from the joint of all unknown parameters given $\boldsymbol{D}^{\star}$, are shown in the lower panel.

.

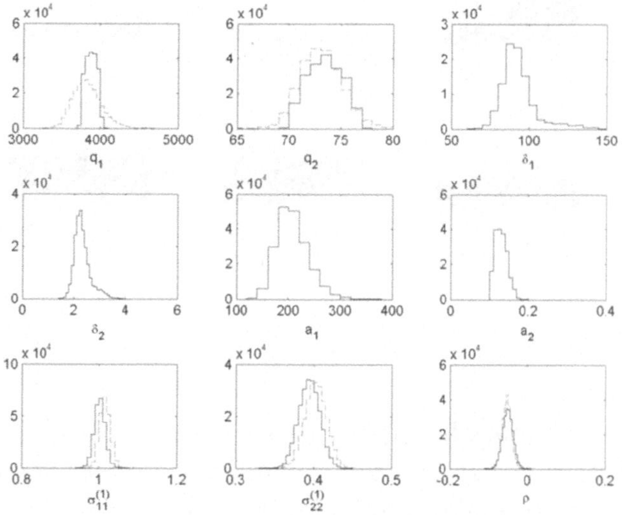

FIGURE 5.6
Marginal posterior probability densities of parameters learnt given training data **D**, are depicted as histograms. Histograms obtained from the run done within the dual-layered model, are shown in black, as distinguished from the results depicted in red (or grey in the monochromaic version), that pertain to results obtained using the same data, but within the stationary GP model. Given the data used here, prediction – or rather learning – of $s_1^{(test)}$ and $s_2^{(test)}$ is not of relevance here.

.

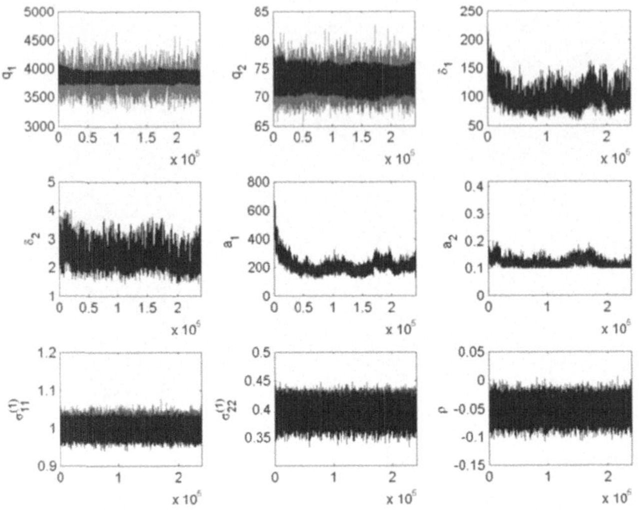

FIGURE 5.7
Traces of parameters learnt using the training data $\mathbf{D}$, in the run performed with the stationary GP model, are compared to traces of the corresponding parameter obtained in the run performed with the non-parametric dual-layered model. Implementation of the latter model uses $N_{lookback} = 200$. Traces of parameters learnt within the stationary GP model are in red, while the traces obtained using the non-parametric model are shown in black.

of homogeneity in the correlation across pairs of S_1 values, at a given value of S_2 (i.e. at fixed angular bins). In fact, this distribution is visually more discontinuous than the distribution across S_2, at given values of S_1, i.e. at fixed radial bins (each of which is represented by the space between two bounding arcs).

Velocity matrices that are astronomically simulated at different $\boldsymbol{S}$ values, are differently compatible with a given reference velocity matrix, (such as $\boldsymbol{v}^{(test)}$). Distribution of the velocity matrix variable $\boldsymbol{V}$, is discontinuous across values of $\boldsymbol{S}$, and in fact, less smoothly distributed at fixed s_2, than at fixed s_1.

Thus, Figure 8 of [25] informs on the deviation from a homogeneous correlation structure of the data.

Then, it is incorrect to use a stationary kernel to parametrise the covariance $\mathbf{\Sigma}_3$, that informs on the covariance between velocity matrices generated at different values of $\boldsymbol{S}$. Our implementation of the dual-layered model tackles this shortcoming of the stationary model. This explains the noted trend of higher 95% HPDs on most parameters learnt using the stationary GP model, compared to the dual-layered model, as observed in comparison between results obtained from runs undertaken with the two models - compare Figure 5.4 to Figure 5.7, and note the comparison in the traces as displayed in Figure 5.7. Indeed, we also note the bigger difference when we compare the learning of $q_1 = 1/\ell_1$ over $q_2 = 1/\ell_2$, in runs that use the stationary model, as distinguished from the non-stationary model. After all, the discontinuity across S_1 is discussed above, to be higher than across S_2.

5.5.2 Effect of varying lookback times, i.e. length of historical data

To check for the effect of the lookback time value $N_{lookback}$, in Figure 5.8, we present traces of the parameters that are learnt in runs undertaken within the dual-layered non-parametric model, at $N_{lookback}$ of 50 and 100. We can compare these traces to those obtained in runs performed under the dual-layered model, with $N_{lookback} = 200$, as displayed in Figure 5.7.

An interesting trend noted in the parameter traces presented in Figure 5.8 for $N_{lookback} = 50, 100$ – and to a lesser extent for $N_{lookback} = 200$, in the results in black in Figure 5.7 – is the roughly periodic nature of these traces. This (rough) periodic fluctuation is more marked for kernel hyperparameters along the direction of S_1 in the input space, than along the direction of S_2.

In Section 4.6.3 we noted that the correlation matrix of each GP in the inner layer, is a symmetric circulant matrix. Then

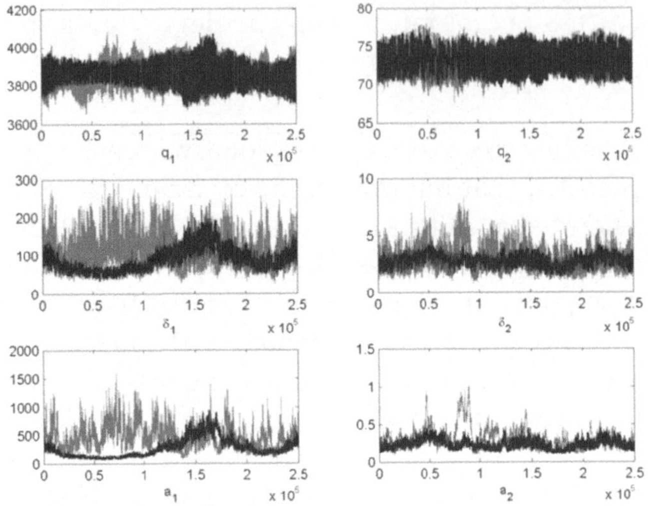

FIGURE 5.8
Comparison of traces of unknown smoothness parameters of $\mathbf{\Sigma}_3$ and hyperparameters of GPs invoked to model these parameters, obtained in runs performed with training data $\mathbf{D}$ and $N_{lookback} = 50$ (in red) and $N_{lookback} = 100$ (in black).

the operation of this correlation matrix, on the vector of output values that are realised at each of the design points relevant here, results in a circular convolution. The length scale ℓ_c that is predicted as the output of function $\eta_c(\cdot)$ – which in turn is learnt using the c-th inner GP – is contributed to by such an operation. Therefore, ℓ_c predicted at different values of the iteration index T, will display rough periodicity that will vary with $N_{lookback}$.

In this application, $c = 1, 2$. This explains the rough periodicity in the traces of ℓ_1 and ℓ_2. (Rough) periodicity in values of the output of $\eta_c(\cdot)$, at inputs (i.e. iteration indices) separated by an interval τ, implies that the correlation between such outputs will also exhibit (similarly rough) periodicity. Additionally, the correlation between the output variables realised at time-step values that are τ iterations apart, decreases as τ increases. Here τ is driven by $N_{lookback}$ s.t. higher is the latter, higher is the period τ. We recall that the correlation function of any (proven to be) stationary inner layer GP is parametrised with an SQE kernel, s.t. the amplitude and length scale conspire to yield the correlation – which we have noted, manifests the rough periodicity discussed above. In this learning scheme, the amplitude and length scale hyperparameters individually manifest the periodicity, with a period of τ time-steps.

Indeed an important question is, what is the "best" $N_{lookback}$, given our data. Such question is itself of relevance, and something similar is discussed intensively under distributed lag models, often within Econometrics [3]. From the point of view of a polynomial (of order $N_{lookback}$) model for the lag operator – that transfers information from the past $N_{lookback}$ realisations of a Stochastic Process to the current iteration – the shape of the trace will be dictated by parameters of this model. If this polynomial admits complex roots, then coefficients of the relevant lag terms will behave like a damped sine function of T. For a different value of $N_{lookback}$, such a pronounced oscillatory trend might not be equally apparent.

Table 5.3 below, and Table 5.1 include results on the learning of each parameter of every considered model, and on the undertaken parameter prediction – tabulated as 95% HPD credible regions. In addition, Table 5.2 presents the effect of inference made with 3 different values of the lookback time.

We notice that the reciprocal of the correlation length scale (q_1) is a couple of orders of magnitude higher than q_2; correlation between values of the sampled function $\boldsymbol{\xi}(\cdot)$, at two different S_1 values (at the same s_2), then wanes more quickly than correlation between sampled functions computed at same s_1 and different S_2 values. Here S_1 is interpreted as the radial coordinate of the observer's location in the Galaxy and S_2 is the observer's angular coordinate. Then it appears that the velocities measured by observers at different radial coordinates, but at the same angle, are correlated over shorter radial-length scales, than velocities mea-

TABLE 5.1
95% HPD credible regions on each learnt parameter, from the stationary GP model

Parameters	using only training data	sampling from posterior predictive	sampling from joint
q_1	[3492.1,4198.1]		[3573.2,4220.8]
q_2	[68.92,76.88]		[68.37,77.33]
$\sigma_{11}^{(1)}$	[0.9837,1.0380]		[0.9797,1.0338]
ρ	[-0.0653,-0.0275]		[-0.0798,-0.0261]
$\sigma_{22}^{(1)}$	[0.3747,0.4234]		[0.3703,0.4237]
s_1	-	[1.8212,2.1532]	[1.8038,2.1960]
s_2	-	[0.0421,1.2052]	[0.0157,1.2172]

TABLE 5.2
95% HPD credible regions on each learnt parameter, from the dual-layered model

Parameters	$N_{lookback} = 200$	$N_{lookback} = 100$	$N_{lookback} = 50$
q_1	[3740.96, 3917.32]	[3710.4, 4011.66]	[3650.92, 4033.51]
q_2	[70.34, 75.70]	[70.42, 76.43]	[68.94, 76.22]
a_1	[78.67, 124.02]	[43.82, 167.35]	[48.27, 219.37]
a_2	[1.88, 3.03]	[2.12, 3.57]	[1.64, 6.16]
d_1	[155.64, 301.65]	[78.47, 521.67]	[123.42, 828.37]
d_2	[0.10, 0.15]	[0.12, 0.46]	[0.10, 0.52]
σ_{31}	[0.97, 1.02]	[0.97, 1.03]	[0.98, 1.02]
σ_{32}	[0.37, 0.41]	[0.37, 0.41]	[0.38, 0.41]
ρ	[-0.076, -0.031]	[-0.073, -0.03]	[-0.075, -0.032]
s_1	[1.83, 2.16]	[1.77, 2.22]	[1.76, 2.24]
s_2	[0.138, 1.15]	[0.112, 1.16]	[0.071, 1.15]

sured by observers at the same radial coordinate, but different angles.

This is understood to be due to the astro-dynamical influences of the Galactic features included by [25] in the simulation that generates the training data that we use here. This simulation incorporates the joint dynamical effect of the Galactic spiral arms and the elongated Galactic bar (made of stars) that rotate at different frequencies (as per the astronomical model responsible for the generation of our training data), pivoted at the centre of the Galaxy. An effect of this joint handiwork of the bar and the spiral arms is to generate distinctive stellar velocity distributions at different radial coordinates, (i.e. along the S_1 direction), at the same angle (s_2). On the other hand, the stellar velocity distributions are more similar at different S_2 values, at the same s_1. This pattern is borne by the work by [22], in which the radial and angular variation of the standard deviations of these bivariate velocity distributions are plotted. Within the framework of this study, it is understandable why correlation length scales are shorter along the S_1 direction, than along S_2.

Furthermore, for the correlation parameter ρ, physics suggests that the correlation will be zero among the two components of a velocity vector. These two components are after all, components of the velocity vector in a 2-dimensional orthogonal basis. However, results indicate a small (negative) correlation between the two components of the stellar velocity vector.

5.5.3 Predicting $s^{(test)}$

Figure 5.3, displays histogram-representations of marginal posterior probability densities of the solar location coordinates $s_1^{(test)}$, $s_2^{(test)}$; q_1^* and q_2^* that get updated once the test data is added to augment the training data, and parameters σ_{11}^{1*}, σ_{22}^{1*} and ρ^*. 95% HPD credible regions computed on each parameter in this inference scheme, are displayed in Table 5.1. These figures display these parameters in the stationary GP model. When the dual-layered model is used, histogram-representations of the marginals of the aforementioned parameters, are displayed in Figure 5.4.

Prediction of $\boldsymbol{s}^{(test)}$ using the dual-layered models gives rise to similar results as when the stationary GP models are used, (see Figure 5.4 that compares the marginals of the solar location parameters sampled from the joint of all unknowns, given all data, in dual-layered models, against those obtained when stationary GP models are used).

The marginal distributions of $s_1^{(test)}$ indicates that the marginal is unimodal and converges well, with modes at about 2 in model units. Distribution of $s_2^{(test)}$ on the other hand is quite strongly skewed towards values of $s_2^{(test)} \lesssim 1$ radians, i.e. $s_2^{(test)} \lesssim 57$ degrees, though the probability mass in this marginal density falls sharply after about 0.4 radians, i.e. about 23 degrees. These values tally quite well with previous work [29]. In that work, using the training data used in this work, (constructed using the astronomical model *sp3bar3_18* discussed by [29]), the marginal distribution of $s_1^{(test)}$ was learnt to be bimodal, with modes at about 1.85 and 2, in model units. The distribution of $s_2^{(test)}$ found by [29] is however more constricted, with a sharp mode at about 0.32 radians, i.e. about 20 degrees. We can notice a mode at about this value, but unlike in the results of [29], the probability mass is not declining to low values beyond about 15 degrees.

5.5.4 Astronomical implications

The radial coordinate of the observer in the Milky Way, i.e. the solar radial location, is dealt with in model units, but will need to be scaled to real galactic unit of distance, which is kilo parsec (kpc). Now, from independent astronomical work, the radial location of the Sun is set as 8 kpc. Then our learnt value of $S_1^{(test)}$ is to be scaled to 8 kpc, which gives 1 model unit of length to be $m := \left(\dfrac{8\text{kpc}}{\text{learnt value of } S_1^{(test)}}\right)$. Our main interest in learning the solar location, is to learn frequency Ω_{bar} with which the Galactic bar is rotating, pivoted at the galactic centre. Here $\Omega_{bar} = \dfrac{v_0}{1 \text{ model unit of length}} = \dfrac{v_0}{m}$, where $v_0 = 220$ km/s (see [25, 13] for details). Here, S_2 is the angular distance between the

Sun-Galactic centre line, and long axis of the bar, and informs on the angular location of the Galactic bar (see Table 5.3).

TABLE 5.3
95% HPD on each Galactic feature parameter learnt from the solar location coordinates learnt using the two predictive inference schemes listed above and as reported in a past paper for the same training and test data.

	95% HPD on Ω_{bar} (km/s/kpc)	angular distance of bar to Sun (°)
from posterior predictive	$[48.11, 57.73]$	$[4.53, 43.62]$
from joint posterior	$[48.25, 57.244]$	$[2.25, 46.80]$
from Chakrabarty et. al 2015	$[46.75, 62.98]$	$[17.60, 79.90]$

Table 5.3 displays the galactic feature parameters derived from the learnt solar location parameters, under the stationary GP model, using sampling from the joint posterior probability of all parameters given all data, and from the posterior predictive of the solar location coordinates given test data and GP parameters already learnt from training data alone. Derived galactic feature parameters are: bar rotational frequency Ω_{bar} in the real astronomical units (km/s/kpc), and angular distance between the bar and the Sun, in degrees. The table includes results from [29].

5.6 Model checking

One way to check for the model and results, given the data at hand, is to generate data from the learnt model, and then compare this generated data with the observed data. Now, the model that we learn, is essentially the tensor-variate GP ($\mathcal{GP}_{non-stat}$) that is used to model the functional relationship $\boldsymbol{\xi}(\cdot)$ between the observable $\boldsymbol{V}$ and the input-space parameter $\boldsymbol{S}$. By saying that we want to generate new data, we imply the prediction of the value of $\boldsymbol{V}$ at a chosen $\boldsymbol{s}$, given the learnt model of $\mathcal{GP}_{non-stat}$. This prediction of new datum on $\boldsymbol{V}$, is fundamentally different from the inverse prediction of the value $\boldsymbol{s}^{(test)}$ of the input-space parameter $\boldsymbol{S}$ that we have undertaken – as discussed above – where the sought $\boldsymbol{s}^{(test)}$ is the value of $\boldsymbol{S}$ at which test data $\boldsymbol{v}^{(test)}$ on $\boldsymbol{V}$ is recorded. However, we can predict the mean and variance of $\boldsymbol{v}^{(test)}$ at a given test input. We undertake such prediction of $\boldsymbol{v}^{(test)}$ values at a given test input, to illustrate the undertaken checking of the model and the results, in the context of the astronomical application that we discuss above.

In this model checking exercise, the unknowns are chosen as certain elements of the cuboidally-shaped data – we will pretend that we know all elements of this dataset, except 100 of its elements. Here we recall that this dataset comprises 216 number of 50×2-dimensional stellar velocity matrices generated by astronomical simulation, at chosen design points $\boldsymbol{s}_1, \ldots, \boldsymbol{s}_{216}$, where these 216 matrices comprise the 3rd-order tensor $\mathbf{D}_V := \{\boldsymbol{v}_1, \vdots \boldsymbol{v}_2, \vdots \ldots, \vdots \boldsymbol{v}_{216}\}$. In the first attempt to model checking, we generate all elements of the p-th stellar velocity matrix $\boldsymbol{v}_p$, (that is generated at the known design point $\boldsymbol{s}_p$), i.e. generate values of $50 \times 2 = 100$ unknown elements of matrix $\boldsymbol{v}_p$. We refer to these unknown elements of $\boldsymbol{v}_p$ as $v_{11}^{(p)}, v_{12}^{(p)}, v_{21}^{(p)}, \ldots, v_{50,2}^{(p)}$, where $v_{i,j}^{(p)}$ is the value of the variable $V_{i,j}^{(p)}$, for $i = 1, \ldots, 50$, and $j = 1, 2$. So, we will learn these values using the 3rd-ordered tensor without the p-th slice, which is the dataset referred to as $\mathbf{D}_V^{(-p)} := \{\boldsymbol{v}_1, \vdots \boldsymbol{v}_2, \vdots \ldots, \vdots \boldsymbol{v}_{p-1}, \vdots \boldsymbol{v}_{p+1}, \vdots \boldsymbol{v}_{216}\}$.

Given data $\mathbf{D}_V^{(-p)}$, the joint posterior predictive of the 100 unknowns, at the learnt modal values $p_1^{(mode)}, p_2^{(mode)}, \sigma_{11}^{(1,mode)}, \sigma_{22}^{(1,mode)}$, $\rho^{(mode)}$ is

$$\pi\left(v_{11}^{(p)}, v_{12}^{(p)}, v_{21}^{(p)}, \ldots, v_{50,2}^{(p)} | \mathbf{D}_V^{(-p)}\right) \propto$$

$$\mathcal{TN}_{2\times 50\times 216}(\hat{\boldsymbol{M}}, \boldsymbol{\Sigma}_1^{(mode)}, \hat{\boldsymbol{\Sigma}}_2, \boldsymbol{\Sigma}_3^{(mode)}),$$

where,

–the 3rd-ordered tensor-valued data that enters the parametric form of the 3rd-ordered tensor Normal density in the RHS, has

elements of its p-th slice, (out of a total of 216 slices), held as unknowns. All other elements of this $2 \times 50 \times 216$-dimensional tensor are known;
$-\boldsymbol{\Sigma}_1^{(mode)}$ is the learnt modal value of the 2×2-dimensional covariance matrix $\boldsymbol{\Sigma}_1$ s.t. its $1,1$-th element is $\sigma_{11}^{(1,mode)}$, $2,2$-th element is $\sigma_{22}^{(1,mode)}$, $1,2$-th element is $\rho^{(mode)}\sqrt{\sigma_{22}^{(1,mode)}\sigma_{11}^{(1,mode)}}$, and the $2,1$-th element is equal to the $1,2$-th element (as this is a covariance matrix);
$-\boldsymbol{\Sigma}_3^{(mode)}$ is the learnt modal value of the 216×216-dimensional covariance matrix $\boldsymbol{\Sigma}_3$, s.t. its ij-th element is
$\exp\left[-(\boldsymbol{s}_i - \boldsymbol{s}_j)^T \boldsymbol{Q}^{(mode)}(\boldsymbol{s}_i - \boldsymbol{s}_j)\right]$, with the non-zero elements of the diagonal 2×2-dimensional $\boldsymbol{Q}^{(mode)}$-matrix given by $q_1^{(mode)}$ and $q_2^{(mode)}$. Here, $\boldsymbol{s}_i$ is the i-th design point, and is known $\forall i = 1, \ldots, 216$.

Then the closed-form expectation and variance of the parameter $V_{i,j}^{(p)}$ are known, $\forall i = 1, \ldots, 50$ and $\forall j = 1, 2$.

For $p = 200$, the predicted mean $\bar{v}_{i,j}^{(200)}$ of $V_{i,j}^{(200)}$, for $i = 1, \ldots, 50$ and $j = 1, 2$, then constitutes the 200-th stellar velocity matrix $\boldsymbol{v}_{200}$. $\bar{v}_{i,j}^{(200)}$ is plotted against the true values of the 100 elements of the $\boldsymbol{v}_{200}$ matrix, in the left panel of Figure 5.9. Thus, each point in this plot is a pair: (empirically observed value of $v_{i,j}^{(200)}, \bar{v}_{i,j}^{(200)}$), and there are $50 \times 2 = 100$ points in this plot. The points are found to lie around the straight line with slope 1. In other words, the values of the elements in the p-th (=200-th) slice of the training data that we learn using our model, are approximately equal to the empirically observed values of these elements. This is a corroboration of our models and results.

We attempt a similar prediction of elements of the training data for other values of p, namely for $p = 190, \ldots, 210$. The learnt values of elements of $\boldsymbol{v}_p$, for each p, is plotted against the empirically observed elements of $\boldsymbol{v}_p$. We have superimposed results for all 20 values of p in the same plot, resulting in the middle panel of Figure 5.9. Again, the values predicted for all 20 slices, are found

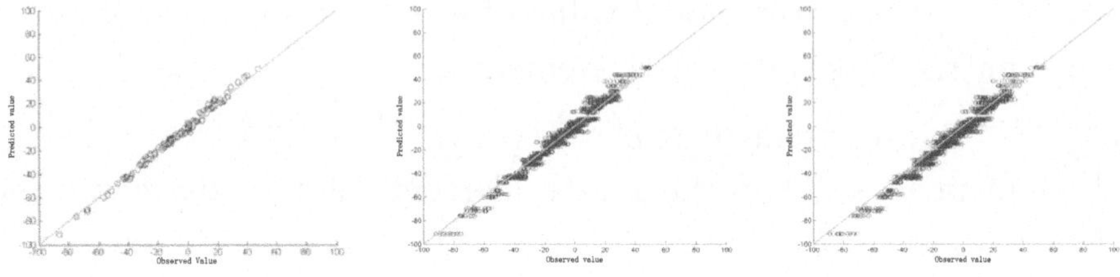

FIGURE 5.9

Left: Comparison of the observed and predicted values of elements of the p-th 50×2-dimensional stellar velocity matrix $\boldsymbol{v}_p$, where 216 such matrices constitute the training data $\mathbf{D}_V$ (on the 2-dimensional velocity vectors of 50 stellar neighbours of the Sun) that is generated by astronomical simulations. The predicted values are the closed-form expected values of the parameters obtained from the dataset that is the 3rd-order tensor $\mathbf{D}_V$ in which only elements of its p-th slice are unknown. For this prediction, we use a chosen summary of the learnt values of the parameters of $\mathcal{GP}_{non-stat}$. This a chosen summary is the mode, of the marginal posterior density of each such learnt GP parameter. In this undertaking, p=200. Equality of the observed and predicted values of the elements of $\boldsymbol{v}_p$ is indicated by the point lying on the straight line with unit slope; the predicted values are found to lie close to this line. *Middle:* Depicts a similar comparison, as displayed in the left panel, but for 20 distinct values of p, namely for $p = 190, 191, \ldots, 210$. *Right:* Depicts the same comparison of observed and predicted values of elements of 20 slices $\boldsymbol{v}_{190}, \ldots, \boldsymbol{v}_{210}$, but this time, the chosen summary of a GP parameter is the mean of its learnt values. Thus, this model-checking exercise checks for the used models and results obtained (given the data at hand) at the mean of the respective posterior.

to be close to the empirical observations, as betrayed by the points lying close to the straight line of unit slope.

Lastly, we wanted to ensure that the encouraging results from our model checking exercise are robust to changes in the posterior summary of the learnt GP parameters. So, we switch to using the mean of the parameter marginal posterior from the posterior mode that we were using before, and carry out the same exercise of predicting elements of slices $\boldsymbol{v}_{190}, \ldots, \boldsymbol{v}_{210}$. Results are displayed in the right panel of Figure 5.9. Again, very encouraging corroboration of our used models and results (of learning the GP parameters) is noted. Indeed, in such model checking exercises, encouraging match between the predictions and the empirical observations lends confidence in the used models and results obtained therefrom, given the data at hand – such models and results are the inputs to this exercise. However, if lack of compatibility is noted in such a model checking exercise, between empirical observations and predictions, then it implies that either the used modelling is wrong, and/or the results obtained therefrom given the data are wrong. However, the model checking exercise that we undertake, vindicates our models and results, given the data at hand.

5.7 (Inverse) learning of the input – relevance in the real world

Learning of the input – rather than the available/measurable output – is of relevance to multiple applications in real life. For example, a physician might be interested in predicting the risk for a new patient to develop a certain disease, where such a risk variable is conceptualised to be influenced by N parameters that include, (say), lifestyle; health; physiological; and demographic variables – some of which can be binary; while others are categorical at multiple levels; and others still are real-valued. For example, lifestyle parameters such as smoking status and co-morbidities can be binary; progression of a co-morbidity can be categorical at four to five levels; while (standardised) dosages of current medication, age and BMI are real-valued variables. Measurements on the risk parameter and each of the N predictor variables are available for only a small number of patients – such as for about 20 patients – before the risk prediction of the new patient is sought. Such a small to a medium-sized training set is typical in rare and terminal diseases, for instance, within Oncology [24]. (Collating data from across too many other institutions – or even too many other patient cohorts – might not be the right thing to do, since the local effects might explain the available data). While it is useless to invoke DNNs in such applications, it is infeasible to use this N dimensional vector-valued variable as the input to the sought GP-modelled function. This is because, the parametrisation of the $N \times N$-dimensional inter-predictor covariance matrix with a non-stationary kernel will demand the learning of N kernel hyperparameters. Then for $N \geq 30$, such learning is difficult, and perhaps not feasible.

One way out is then to learn the function that takes the scalar-valued risk score as the input, and spews out the N-dimensional vector-valued output. Indeed, if we recall from Section 3.2, it was advocated that we use the lower dimensional (of the two) associated variables to be the input to the sought function. Then we learn the vector-valued function that outputs the N-dimensional vector of predictors, and takes the scalar-valued disease risk score as its input, leading to the need to learn only a single kernel hyperparameter at each of the two stages of the dual-layered learning strategy. However, the desired identification of the input risk score for the new patient – whose predictor variables are known – will then need to be undertaken, and such cannot be done in a closed-form way. We then have a clear motivation for performing the (inverse) learning of the input for this new patient.

Along similar lines, we can also motivate the need for identification of one/multiple factor(s) that is/are responsible for a noted value of an econometric or financial marker. Again, the (inverse) learning of the relevant input variable(s) will become relevant then. Indeed, the GP-driven supervised learning strategy that we have discussed in detail in the previous chapter, holds encouraging flexibility to permit the closed-form prediction of the output at test inputs, as well as the with-uncertainty learning of the inputs at which a new output has been realised.

5.8 Summary

This chapter discussed a real-world application of the dual-staged learning method to learn a high-dimensional (matrix-valued) nonlinear and inhomogeneously-correlated function, given a training set, the data on the output within which, is cuboidal in shape, i.e. is 3-dimensional. The ultimate aim of this learning is to learn the inputs at which a new or test output (matrix) has been realised. Indeed, we are not predicting the sought inputs, but we learn the inputs at which the new data points are realised. Such learning is either possible simultaneously with the learning of the function that represents the relationship between the input and the output, or after said functional learning, using the learnt function.

The chapter also discussed a way to check if the model and the results were correct in the available data, and this was undertaken by producing data using the models that were learnt, and subsequently comparing such generated data with the empirically-observed data. If such comparison is encouraging, the technique corroborates if the model + results are correct in the data. However, if said comparison is poor, it is difficult to figure out if the models were incorrect in the first place, and/or the results were erroneously computed.

6

A self-assembled prior on correlation matrices

6.1 Introduction

In the previous chapters, we have discussed the pursuit of a generally high-dimensional inter-variable relationship that is manifest in a function that is underlined by an inhomogeneous correlation structure in general. We have motivated a Stochastic Process-driven modelling of the sought function, i.e. suggested modelling the sought function as a random function that is a sample function of a process, parameters of which we learn, given the available data. Such data has been small to medium in size in our considerations – as is often the case in real-world applications – and such data is typically noted to bear inhomogeneities in its correlation, where we have identified "inhomogeneously correlated" data, as distinguished from data that is not such. Our go-to Stochastic Process has been a Gaussian Process (GP), given that a GP imposes only one constraint on any finite set of sample functions that are drawn from it, while also being easy to generalise to higher dimensions.

The need to address inhomogeneities in the correlation of the sought function led to considering the sought function as a sample function of a GP, each hyperparameter of the covariance function of which is an unknown (random) function of a scalar-valued variable T, s.t. at every new value of T, a new realisation of this unknown (random) function is generated. In fact, this unknown (random) function is treated as a sample function drawn from a distinct GP - that is one of the "inner" GPs that are nested within the mother (non-stationary) "outer" GP. Then, following on from

DOI: 10.1201/9781003296331-6

the above description of the variable T, at every new value of T, a new sample function is drawn from the relevant inner GP.

The above consideration applies to each hyperparameter of the covariance structure of the mother GP. Our discussions included a proof of the stationarity of the inner GPs, that then led to the sufficiency of including only two layers of GPs. Thus, there is no need to model hyperparameters of the covariance function of each of the inner GPs as an unknown function of some other variable. Importantly, the variable T is offered as the time step in any iterative inference that is undertaken to learn the sought function. Indeed, at each new time step of the undertaken iterative inference, the correlation function of each GP is updated. This draws us into the question of evolution of correlation functions – with the input called T.

This chapter is dedicated to such considerations. We will discuss evolution of the correlation function, and in particular, will be interested in the possibility of using knowledge about such evolution towards enriching the learning of the relationship between a pair of variables. Let us start with basics.

6.2 Basics

It is well known that the sum of squares of n *iid* standard Normal variables is the chi-squared distribution with degrees of freedom (dof) n. In other words, if $Y_i \sim \mathcal{N}(0, 1)$, $\sum\limits_{i=1}^{n} Y_i^2 \sim \chi_n^2$. [If the n variables are instead Normally distributed with variance σ^2 each, then $\sum\limits_{i=1}^{n} Y_i^2$ is distributed according to the scaled chi-squared distribution with dof n, i.e. as $\sigma^2\chi_n^2$, [104]]. When this result is extended to n *iid* Normally distributed vector-valued variables $\boldsymbol{Y}_1, \ldots, \boldsymbol{Y}_n$, the sum of the outer product $\boldsymbol{Y}_i\boldsymbol{Y}_i^T$, over all i $(= 1 \ldots, n)$, is Wishart distributed, for $n \geq p$, where $\boldsymbol{Y}_i$ is p-dimensional, [92, 41, 133, 15]. Let us clarify.

Definition 6.1 *Let an n-dimensional vector-valued random vari-*

able $\boldsymbol{W} \in \mathcal{W} \subseteq \mathbb{R}^n$ *be s.t.* $\boldsymbol{W} = (Y_1, \ldots, Y_n)^T$ *and* $\mathbb{E}[Y_i] = 0$, $\forall i = 1, \ldots, n$. *Here* $Y_1, Y_2, \ldots, Y_n$ *are iid.*

We consider p *realisations of* Y_i *at* p *respective values of an input* $\boldsymbol{X}$, *s.t. the output realised at* $\boldsymbol{X} = \boldsymbol{x}_q$ *is denoted* $Y_q^{(i)}$, $\forall q = 1, \ldots, p$, *for* $i = 1, \ldots, n$.

We define the p-dimensional vector

$$\boldsymbol{Y}_i = (Y_1^{(i)}, \ldots, Y_p^{(i)})^T \text{ for } i = 1, \ldots, n.$$

and the n-dimensional vector

$$\boldsymbol{W}_q = (Y_q^{(1)}, \ldots, Y_q^{(n)})^T \text{ for } q = 1, \ldots, p.$$

Then $\boldsymbol{W}_q$ *is the vector-valued output realised at* $\boldsymbol{X} = \boldsymbol{x}_q$.

Then joint density of the p outputs $Y_1^{(i)}, Y_2^{(i)}, \ldots, Y_p^{(i)}$ *- realised respectively at* $\boldsymbol{X} = \boldsymbol{x}_1, \ldots, \boldsymbol{X} = \boldsymbol{x}_p$ *- is*

$$\mathcal{MN}(\underline{0}, \mathbf{V}),$$

where $\mathbf{V}$ *is* $p \times p$*-dimensional. This holds* $\forall i = 1, \ldots, n$.

Definition 6.2 *The sum of the outer product* $\boldsymbol{Y}_i \boldsymbol{Y}_i^T$, *over* $i = 1, \ldots, n$, *is a* $p \times p$*-dimensional matrix-valued random variable* $\mathbf{S}$ *that is distributed according to the Wishart distribution with parameters that are: the dimensionality parameter* p*; the degrees of freedom parameter* n*; and the scale matrix parameter* $\mathbf{V}$. *This is denoted as:*

$$\mathbf{S} = \sum_{i=1}^{n} \boldsymbol{Y}_i \boldsymbol{Y}_i^T \sim \mathcal{W}_p(n, \mathbf{V}).$$

Then the random matrix-valued variable $\mathbf{S}$ is symmetric; positive semi-definite; and it is Wishart distributed. For an invertible $\mathbf{V}$, $\mathbf{S}$ is invertible as long as $n \geq p$ [130].

We realise that for the value $\boldsymbol{y}_i = (y_1^{(i)}, \ldots, y_p^{(i)})^T$ of $\boldsymbol{Y}_i$, the $p \times p$-dimensional matrix $\boldsymbol{y}_i \boldsymbol{y}_i^T$ is s.t. its q, r-th element is $y_q^{(i)} y_r^{(i)}$, $\forall q, r \in \{1, 2, \ldots, p\}$. Then

$$\mathbf{S} = [s_{q,r}] = \left[\sum_{i=1}^{n} y_q^{(i)} y_r^{(i)} \right] \equiv [corr(W_q, W_r)],$$

where

— the random variable W_q has realisations $y_q^{(1)}, y_q^{(2)}, \ldots, y_q^{(n)}$, $\forall q = 1, \ldots, p$, and

— W_r attains values $y_r^{(1)}, y_r^{(2)}, \ldots, y_r^{(n)}$, $\forall r = 1, \ldots, p$.

This is so, since $Y_q^{(i)}$ is standardised, $\forall i = 1, \ldots, n; q = 1, \ldots, p$.

So we see that the $p \times p$-dimensional $\mathbf{S}$ defines the inter-realisation correlation of $\boldsymbol{Y}$, p realisations of which are considered.

6.3 Connection to supervised learning of inter-variable relationship

Above we have considered Y_i to be a random variable s.t. its p realisations are jointly multivariate (p-variate) Normal, with a null vector for mean, and $\mathbf{V}$ for the covariance matrix. To connect this to the supervised learning context that is relevant to this book, let us consider Y_i to be associated with another random variable $\boldsymbol{X} \in \mathcal{X} \subseteq \mathbb{R}^d$, s.t. we can express their state of relationship as $Y_i = f_i(\boldsymbol{x})$, at $\boldsymbol{X} = \boldsymbol{x}$, where the function $f_i(\cdot)$ being unknown, is modelled as a function-valued random variable that is treated as a sample function of a GP – which we choose out of other options, given the advantages of using a GP. Then the $n \times p$-dimensional matrix comprising the n outputs, each of which is realised at p inputs, is equivalently the matrix of sample functions $\tilde{f}_1(\cdot), \ldots, \tilde{f}_n(\cdot)$, each of which is computed at the p inputs – or rather, the p given design points – $\boldsymbol{x}_1, \boldsymbol{x}_2, \ldots, \boldsymbol{x}_p$. This is depicted as follows:

$\tilde{f}_1(\boldsymbol{x}_1)$	$\tilde{f}_2(\boldsymbol{x}_1)$	$\ldots$	$\tilde{f}_n(\boldsymbol{x}_1)$
$\tilde{f}_1(\boldsymbol{x}_2)$	$\tilde{f}_2(\boldsymbol{x}_2)$	$\ldots$	$\tilde{f}_n(\boldsymbol{x}_2)$
$\vdots$	$\vdots$	$\ldots$	$\vdots$
$\vdots$	$\vdots$	$\ldots$	$\vdots$
$\tilde{f}_1(\boldsymbol{x}_p)$	$\tilde{f}_2(\boldsymbol{x}_p)$	$\ldots$	$\tilde{f}_n(\boldsymbol{x}_p)$

Here $\tilde{f}_i(\cdot)$ is a sample function drawn from the GP that we denote $\mathcal{GP}_i$, $i = 1, \ldots, n$. Then what we have so far is a $p \times n$-dimensional matrix of random variables that is each an output of a function modelled as a sample function of the corresponding GP, realised at the p design points $\boldsymbol{x}_1, \boldsymbol{x}_2, \ldots, \boldsymbol{x}_p$. Then we know the distribution of the $p \times p$-dimensional inter-realisation correlation matrix $\mathbf{S}$ of the vector-valued outputs that are realised at the p different design points, given sample functions drawn from the specified GPs: $\mathcal{GP}_1, \ldots, \mathcal{GP}_n$.

Knowledge of the distribution of such a correlation matrix, could then contribute towards the prior probability density of the matrix that comprises correlations between outputs realised across the p design points, obtained using n different invoked GPs that are somehow relevant at the current time, when we pursue the modelling of the relation between the input and the output. Such a correlation is of crucial importance when we undertake supervised learning of inter-variable relationships represented as functions. To ensure tractability of the multivariate Normal likelihood. (or the more high dimensional version of the same), that we need to compute – in order to learn the covariance function of the GPs that are invoked to learn the relation between an output and the input $\boldsymbol{X}$ – the inter-realisation correlation matrix $\mathbf{S}$ needs to be invertible. For the Wishart distributed $\mathbf{S}$, this is ensured if $n \geq p$. Thus, we need to formulate our learning s.t. it includes a robust implementation of the Wishart prior on this correlation matrix. Below, we present a protocol, within the context of iterative inference on relevant parameters, as is used during attempted supervised learning of inter-variable relationships.

6.3.1 How to invoke a Wishart prior on a correlation matrix, within supervised learning

When learning the function that represents the relation between two variables, how do we use a Wishart prior on any/multiple correlation matrix/matrices that is/are relevant to said learning? One way to frame the learning strategy to include such a Wishart-distributed correlation matrix, is by treating the p realisations of Y_i as values computed of a sample function – that is drawn out of $\mathcal{GP}_i$ – at $\boldsymbol{X} = \boldsymbol{x}_1, \ldots, \boldsymbol{x}_p$, ($\forall i = 1, \ldots, n$), where n sample functions are drawn out of respective GPs *in n consecutive time steps of the iterative inference that is undertaken.*

At every time step of the inferential strategy, we look back to values of the hyperparameters of the kernel – with these kernel hyperparameters collated into the vector $\boldsymbol{\Theta} \in \mathcal{H} \subseteq \mathbb{R}^h$. To clarify, these hyperparameters are of the kernel that parametrises the correlation function of the invoked GP. We collate values of $\boldsymbol{\Theta}$ that were current in the previous $n-1$ iterations, as well as the hyperparameters that are proposed in the current iteration. Then in the t-th iteration, this generates the "looked-back" sample $\{\boldsymbol{\theta}_t^{(\star)}, \boldsymbol{\theta}_{t-1}, \boldsymbol{\theta}_{t-2}, \ldots, \boldsymbol{\theta}_{t-(n-1)}\}$ of the kernel hyperparameters. Here, $\boldsymbol{\theta}_t^{(\star)}$ is the value of the hyperparameter vector $\boldsymbol{\Theta}$, proposed in the current, i.e. the t-th iteration, while $\boldsymbol{\theta}_i$ is the value that was accepted as the value of $\boldsymbol{\Theta}$, in the i-th iteration in the past, where $i = t-1, t-2, \ldots, t-(n-1)$.

Definition 6.3 *We define an "average" inter-realisation correlation matrix – or a scale matrix that represents the correlation structure of outputs realised across the p design points – at the t-th iteration, using the mean* $\bar{\boldsymbol{\theta}}_t$ *of the kernel hyperparameters that were collated in the looked-back sample* $\{\boldsymbol{\theta}_t^{(\star)}, \boldsymbol{\theta}_{t-1}, \boldsymbol{\theta}_{t-2}, \ldots, \boldsymbol{\theta}_{t-(n-1)}\}$, *i.e.*

$$\bar{\boldsymbol{\theta}}_t := \frac{\sum\limits_{i=1}^{n-1} \boldsymbol{\theta}_{t-i} + \boldsymbol{\theta}_t^{(\star)}}{n}.$$

Thus, the scale matrix at the t-th iteration is represented as:

$$\mathbf{V}_t := [k(\boldsymbol{x}_q, \boldsymbol{x}_r; \bar{\boldsymbol{\theta}}_t)],$$

where $k(\boldsymbol{x}_q, \boldsymbol{x}_r; \boldsymbol{\theta})$ *is the kernel that parametrises* $corr[W_q, W_r]$ *at hyperparameter* $\boldsymbol{\theta}$, *where in the available data, the variable* W_q *attains values of* $y_q^{(t-(n-1))}, \ldots, y_q^{(t-1)}, y_q^{t\star}$, *and* W_r *attains values of* $y_r^{(t-(n-1))}, \ldots, y_r^{(t-1)}, y_r^{t\star}$, *for* $q, r \in \{1, \ldots, p\}$.

Then, in the t-th iteration, the inter-realisation correlation matrix $\mathbf{S}_t$ is kernel parametrised as:

$$\mathbf{S}_t = [k(\boldsymbol{x}_q, \boldsymbol{x}_r; \boldsymbol{\theta}_t^{(\star)})],$$

where $\boldsymbol{\theta}_t^{(\star)}$ is the vector of kernel hyperparameters that are current in the t-th iteration. Then such a corelation matrix would be Wishart distributed, with parameters p, n and $\mathbf{V}_t$. However, we need to scrutinise such a claim.

The assumption that is made in constructing a scale matrix at the t-th time step of the undertaken inference is that $\mathbf{V}_t$ is the constant scale matrix that is the covariance matrix of the (multivariate Normal) joint distribution of the n outputs, each realised at the p design points, in the $t-k$-th time step, $\forall k = 0, \ldots, n-1$. This is likely to be true if the undertaken inferential strategy has attained equilibrium during the considered "lookback window", which comprises time steps from the current iteration to the $n-1$-th previous iteration. However, in general, $\mathbf{V}_t$ is not a constant with respect to time step T. Then it may be argued that we will need to implement this construct – of collating kernel hyperparameters within the lookback window and computing the scale matrix using the sample average of these collated hyperparameters – only after the inferential strategy has equilibriated. Let the time step index at which the inference equilibriates be n_0; for example, if inference is undertaken with MCMC, burnin occurs by the n_0-th iteration. Then it may be argued that the joint of outputs realised at the design points, at all time steps within the lookback window is multivariate Normal, with a null vector for mean and the same correlation matrix $\boldsymbol{V}_t$, $\forall t \geq n_0$. Here we recall that an output realised at a given value of the input $\boldsymbol{X}$, is the value of a sample function drawn from a GP, computed at this input value.

However, even for $t \geq n_0$, it maybe a step too far to suggest that the inter-realisation correlation matrix generated by collating outputs relevant to the past few time steps of the inference, is Wishart distributed. This would owe to the non-independent nature of the random variables that are the outputs realised at $\boldsymbol{x}_1, \ldots, \boldsymbol{x}_p$ at the t-th time step, and outputs realised similarly at the $t+1$-th time step, $\forall t = 1, \ldots, N_{iter}$. This in turn is due to the generally non-zero correlation amongst $\boldsymbol{\theta}_t$ and $\boldsymbol{\theta}_{t+1}$ that parametrise the correlation function of the GPs relevant to the t-th and $t+1$-th time steps, sample functions of which computed at the design points, define the outputs in question. So in the construct that is discussed above, the inter-realisation correlation matrix $\mathbf{S}_t$ is not Wishart distributed, $\forall t \geq n_0$, with $t \leq N_{iter}$ for an inferential chain that includes N_{iter} iterations.

However, our pursuit is of a prior probability density on the hyperparameters that describe the kernel that parametrises inter-output correlation at $T = t$ – or a function of such hyperparameters, as manifest in the inter-output correlation matrix $\mathbf{S}_t$. In the context of this search for a prior, $W_p(\mathbf{V}_t, n)$ could be used for the prior, for $t \geq n_0$, (and $t \leq N_{iter}$). Indeed, since we are looking for the prior when attempting to learn $f(\cdot)$ – where $Y = f(\boldsymbol{x})$ – we need not adhere to the qualification of n_0 as the equilibration time step of the undertaken inference. It suffices to use $n_0 \geq n$, for an n-step wide lookback window.

In the following subsection, we consider an application to a real world dataset in which the above construct is employed to formulate a Wishart prior density on the inter-realisation correlation matrix in the t-th iterations, for all iterations that are post the n_0-th one, within the (MCMC-based) Bayesian inference that we undertake.

6.3.2 Learning a multivariate function with the Tetouan city data – using a Wishart-distributed correlation matrix

We revisit the Tetouan city data that we have seen in earlier chapters. In particular, we consider the problem of learning the relationship between energy that is consumed in all three zones of the city, associated with the same four metereological parameters as we have seen before in Exercise 2 in Section 3.7 earlier. Such learning is to permit prediction of the energy consumed at test values of the 4-dimensional input vector. We denote the random variable that represents the energy consumed over all zones of the city to be $Y \in \mathcal{Y} \subset \mathbb{R}$, while the inputs are: temperature (denoted X_1); humidity (denoted X_2); general diffuse flows (denoted X_4); and diffuse flows (X_5), s.t. the input variable is

$\boldsymbol{X} = (X_1, X_2, X_4, X_5)^T \in \mathcal{X} \subset \mathbb{R}^4$. We express the state of relationship between Y and $\boldsymbol{X}$ as: $Y = f(\boldsymbol{x})$, where $f(\cdot)$ is a random function. As before, we model this random function as a sample function of a GP, i.e. $f(\cdot) \sim \mathcal{GP}(\mu(\cdot), K(\cdot, \cdot))$, with $\mu(\cdot)$ and $K(\cdot, \cdot)$ the mean function and covariance function of $\mathcal{GP}$ respectively.

The training data $\mathbf{D} := \{(\boldsymbol{x}_i, y_i)\}_{i=1}^M$ is obtained by randomly sampling $M = 203$ data points from the publicly-available dataset at `https://archive.ics.uci.edu/ml/datasets/Power+consumption+of+Tetouan+city`. This is the same dataset that was used in Chapter 3 to accomplish Exercises 1,2 ad 3. Then using the $f(\cdot)$ that is learnt with data $\mathbf{D}$, we predict the energy consumed at each of 50 test values of the 4-dimensional input vector. Again, the test inputs are the same as used in Chapter 3.

Recalling the modelling strategy from Section 3.7.2 of Chapter 3, modelling the sought function $f(\cdot)$ with $\mathcal{GP}$ induces a multivariate-Normal likelihood: $\mathcal{MN}_M(\underset{\sim}{0}, \boldsymbol{\Sigma})$, where standardisation of outputs results in $\boldsymbol{\Sigma}$ being treated as the correlation matrix, i.e. $\boldsymbol{\Sigma} = [corr(Y_i, Y_j)]$, for $i, j = 1, \ldots, M$, where $Y_i = f(\boldsymbol{x}_i)$ for the random function $f : \mathcal{X} \longrightarrow \mathcal{Y}$. This also motivates us invoking a zero-mean $\mathcal{GP}$ to model $f(\cdot)$, i.e. the mean vector of the induced (multivariate Normal) likelihood is a null vector.

We parametrise the correlation matrix $\boldsymbol{\Sigma}$ with the kernel function $k(\cdot, \cdot; \boldsymbol{\theta})$ that bears hyperparameters collated into the vector $\boldsymbol{\theta}$, i.e. $\boldsymbol{\Sigma} = [corr(Y_i, Y_j)] = [k(\boldsymbol{x}_i, \boldsymbol{x}_j; \boldsymbol{\theta})]$. Here we are using a stationary kernel – with hyperparameters that are (unknown) constants – to parametrise the correlation matrix $\boldsymbol{\Sigma}$. So our learning exercise reduces to the learning of each component of this vector $\boldsymbol{\theta}$ of hyperparameters, given data $\mathbf{D}$.

In fact, we employ a simple SQE kernel, with four hyperparameters that are the four length scale hyperparameters $\ell_1, \ell_2, \ell_4, \ell_5$. In other words, in the chosen kernel, $\boldsymbol{\theta} = (\ell_1, \ell_2, \ell_4, \ell_5)^T$. So in the undertaken application that uses an SQE kernel, we define the kernel-parametrised $\boldsymbol{\Sigma}$ correlation function as:

$$\boldsymbol{\Sigma} = [k(\boldsymbol{x}_i, \boldsymbol{x}_j; \ell_1, \ell_2, \ell_4, \ell_5)] = [\exp(-(\boldsymbol{x}_i - \boldsymbol{x}_j)^T \boldsymbol{Q}(\boldsymbol{x}_i - \boldsymbol{x}_j)/2)],$$

where the 4×4-dimensonal square diagonal matrix $\boldsymbol{Q}$ is:

$$\boldsymbol{Q} = [\delta_{c,c'}/(\ell_c)^2], \quad c, c' = 1, 2, 4, 5,$$

with $\delta_{c,c'}$ the delta function that is 0 for $c \neq c'$, and 1 for $c = c'$. In this application, $i, j = 1, \ldots, M$, with M=203.

Then in the t-th iteration, in which the c-th proposed length scale hyperparameter is denoted $\ell_c^{(\star,t)}$ (for $c = 1, 2, 4, 5$), the correlation matrix computed at the proposed hyperparameters is

$$\boldsymbol{\Sigma}_t^{(\star)} = [\exp(-(\boldsymbol{x}_i - \boldsymbol{x}_j)^T \boldsymbol{Q}_t^{(\star)} (\boldsymbol{x}_i - \boldsymbol{x}_j)/2)],$$

where

$$\boldsymbol{Q}_t^{(\star)} = [\delta_{c,c'}/(\ell_c{}^{(\star,t)})^2], \quad c, c' = 1, 2, 4, 5,$$

Again, if the current value of the c-th hyperparameter in the t-th iteration is $\ell_c^{(t-1)}$ – before this c-th hyperparameter is updated in this iteration – the correlation matrix computed at such current values of the hyperparameters, is similarly computed, and denoted as $\boldsymbol{\Sigma}^{(t-1)}$.

Having learnt these four length scale hyperparameters, we can then perform closed-form prediction of the mean and variance of the random variable that is the output realised at one of the N_{test} number of considered test inputs.

So the above consideration, is a repeat of what we have already seen in the learning + prediction exercise that was presented in Section 3.7.2. In the current undertaking, the learning is helped by imposing a Wishart prior on the correlation matrix $\boldsymbol{\Sigma}_t$ that is generated at the t-th iteration of the undertaken inference strategy, using the values of $\ell_1, \ell_2, \ell_4, \ell_5$ that are proposed in this iteration.

At the t-th iteration, we collate the length scale hyperparameters that are proposed in this iteration, and that were accepted at the end of the $t-1$-th, the $t-2$-th, $\ldots$, $t-(n-1)$-th iterations of the undertaken inference, $\forall t \geq n_0$, where we choose n_0. In our experiments, we have chosen different n (with $n < n_0$) and n_0 values – including n of about 500 and n_0 of about 501 – while ensuring that $n > M$. Let the sample of collated length scale hyperparameters be $\{(\ell_1^{(t-n+1)}, \ell_2^{(t-n+1)}, \ell_4^{(t-n+1)}, \ell_5^{(t-n+1)})^T, \ldots, (\ell_1^{(t-1)}, \ell_2^{(t-1)}, \ell_4^{(t-1)}, \ell_5^{(t-1)})^T, (\ell_1^{(\star,t)}, , \ell_2^{(\star,t)}, \ell_4^{(\star,t)}, \ell_5^{(\star,t)})^T\}$. Then the mean $\bar{\ell}_c^{(t)}$ in this sample – collated at the t-th iteration – of values of ℓ_c (for $c = 1, 2, 4, 5$), is used to compute the i, j-th element of the scale matrix at the t-th iteration; $i, j = 1, \ldots, M$. Thus, for the sample mean

$$\bar{\ell_c}^{(t)} = (\ell_c^{(t-n+1)} + \ldots + \ell_c^{(t-1)} + \ell_c^{(\star,t)})/n,$$

the scale matrix at this iteration is

$$\mathbf{V}_t = [\exp(-(\boldsymbol{x}_i - \boldsymbol{x}_j)^T \bar{\boldsymbol{Q}}_t (\boldsymbol{x}_i - \boldsymbol{x}_j)/2)],$$

where

$$\bar{\boldsymbol{Q}}_t = [\delta_{c,c'}/(\bar{\ell_c}^{(t)})^2], \quad c, c' = 1, 2, 4, 5,$$

Then in this t-th iteration, the prior on the correlation matrix is the Wishart density with parameters M, n and scale matrix $\mathbf{V}_t$. This prior probability density computed at the proposed values of the length scale parameters is

$$\pi_0(\mathbf{\Sigma}_t^{(\star)}) = \frac{C|\mathbf{\Sigma}_t^{(\star)}|^{(n-M-1)/2}}{|\mathbf{V}_t|^{(M-1)/2}} \exp\left(-\frac{trace(\mathbf{V}_t^{-1}\mathbf{\Sigma}_t^{(\star)})}{2}\right),$$

where C is a constant.

6.3.2.1 Algorithm for implementing the Wishart-prior included learning

Within the context of MCMC-based inference that is undertaken in the learning + prediction with data $\mathbf{D}$, the algorithm for implementation of this Wishart prior-distributed inter-realisation correlation matrix is the following.

1. Standardise $y_1, \ldots, y_M$ in the available training data $\mathbf{D} = \{(\boldsymbol{x}_i, y_i)\}_{i=1}^M$, with the mean and standard deviation of the sample $\{y_1, y_2, \ldots, y_n\}$.
2. Before the MCMC starts, choose seed values for the length scale hyperparameters $\ell_1, \ell_2, \ell_4, \ell_5$.
3. In the t-th iteration, propose the value $\ell_c^{(\star,t)}$ of hyperparameter ℓ_c, from a Normal proposal density $\mathcal{N}(\ell_c^{(t-1)}, \sigma_c^2)$, for $c = 1, 2, 4, 5$. Compute the correlation matrix $\mathbf{\Sigma}_t^{(\star)}$ as parametrised by an SQE kernel [105], with parameters that are the recently-proposed length scale hyperparameters, i.e. $\ell_1^{(\star,t)}, \ell_2^{(\star,t)}, \ell_4^{(\star,t)}, \ell_5^{(\star,t)}$.
4. If $t \leq n_0 + n$, compute the multivariate Normal likelihood of the hyperparameters in data $\mathbf{D}$, as delineated in Equation 3.3. Compute Normal priors on ℓ_c, at the proposed

value $\ell_c^{(\star,t)}$, for $c = 1, 2, 3, 5$. Choose prior mean and prior variance experimentally. The undertaken empirical illustration used the seed value of ℓ_c as the prior mean and employed the same (large) prior variance values as used in undertaking Exercise 2 in Chapter 3.

5. On the other hand, if $t > n_0 + n$, compute the likelihood as stated in the last point, but this time, collate values of the hyperparameter ℓ_c that were current in the past $n - 1$ iterations, as well as its value proposed in this t-th iteration, into a sample. Compute the mean $\overline{\ell}_c^{(t)}$ of ℓ_c in this sample. Do the above $\forall c = 1, 2, 4, 5$. Compute the scale matrix $\mathbf{V}_t$ as the $M \times M$-dimensional correlation matrix that is kernel-parametrised using an SQE kernel that bears the hyperparameters $\overline{\ell}_1^{(t)}, \overline{\ell}_2^{(t)}, \overline{\ell}_3^{(t)}, \overline{\ell}_5^{(t)}$. Compute the Wishart prior $W_M(n, \mathbf{V}_t)$ upto a global scale, at the correlation matrix $\mathbf{\Sigma}_t^{(\star)}$ that is computed in this t-th iteration with the proposed hyperparameters. Compute the posterior of the hyperparameters given data $\mathbf{D}$, at the proposed hyperparameters.

6. Compute the acceptance ratio as the ratio of the posterior of the hyperparameters given $\mathbf{D}$ at the proposed, to that at the current values of the hyperparameters [107]. If the acceptance ratio equals or exceeds the value of a $Uniform[0, 1]$ random number, accept the proposed value of ℓ_c as the current value; else, $\ell_c^{(t)} = \ell_c^{(t-1)}$, for $c = 1, 2, 4, 5$.

7. Once the MCMC chain is finished, predict the mean and variance of the output that is realised at a test input $\boldsymbol{x}_k^{(test)}$, $\forall k = 1, \ldots, N_{test}$; ($N_{test} = 50$).

6.3.2.2 Results of using Wishart-prior on inter-realisation correlation given Tetouan city data

Figure 6.1 represents the traces of the logarithm of the prior; of logarithm of the likelihood; and of the four length scale hyperparameters that we learn. The left panel of Figure 6.2 represents the uncertainty-included predictions (in black) overplotted on the true values of the output (energy consumed), at the index k of the test point, for $k = 1, \ldots, 50$, while in the right panel, the predicted values of the consumed energy is plotted against the first component of the test input, namely temperature. These figures are compared to Figure 3.6 which was generated in a learning + prediction task that was identical in all respects to what is discussed here, except that here we include a Wishart prior on the correlation matrix $\mathbf{\Sigma}$ of the multivariate Normal likelihood that is induced by the GP that is invoked to model the function that represents the relationship between energy consumed in Tetouan city and four metereological parameters.

6.3.3 Wishart prior on correlation matrix gives better predictions

The superior predictions obtained on the usage of the Wishart prior on $\mathbf{\Sigma}$ – as distinguished from when vague normal priors are used – bears testament to the comparatively higher information content in this prior structure. It is highly desirable that even in generic applications that are not blessed by relevant prior information on the unknowns (that we seek to learn given the data), we design application-specific priors with high information content. Such situations are in fact more common in the real-world than we would like, and information-rich priors are more acutely missed in applications that call for GP-based learning of non-linear, inhomogeneously-correlated functions, (perhaps given small to medium-sized multivariate datasets). Examples of such applications include biological and social sciences in which parametric models for multivariate inter-variable functional relationships are found to be too simplistic to explain observations.

Ideally, we would like the prior on the crucially-important correlation structure of the invoked GP, to be generic and as informative as possible, so that a high level of accuracy of predictions is attained, while results are not sensitive to the exact parametrisation of the prior. Instead, a universal prior structure is desired, s.t. the implementation of the prior on the cor-

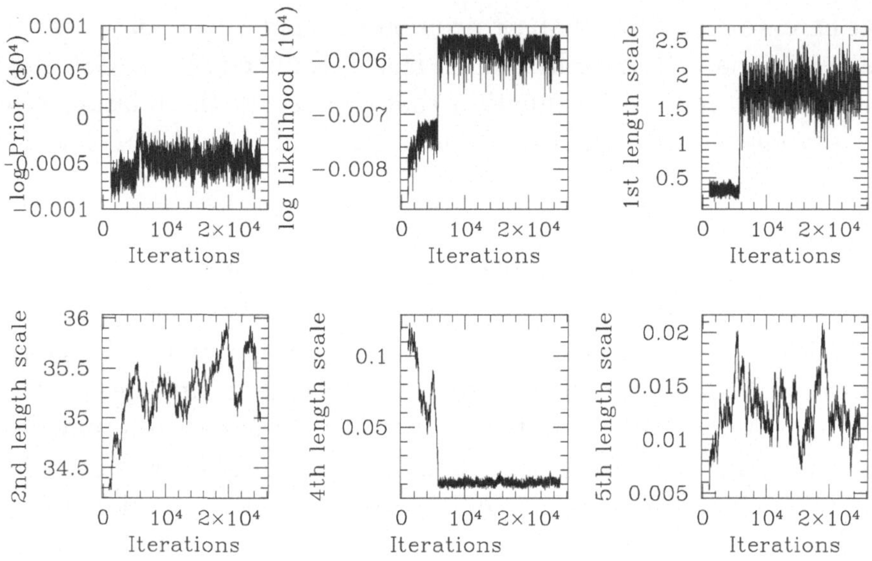

FIGURE 6.1
Figure showing traces of the logarithm of the prior (*top left*); of log likelihood (*top middle*); and of length scale hyperparameters $\ell_1, \ell_2, \ell_4, \ell_5$ in the other four panels – in a run undertaken to learn the GP-generated function that takes the 4-dimensional vector $(X_1, X_2, X_4, X_5)^T$ of metereological parameters as its input, and outputs the energy consumption in Tetouan city. In this learning, the invoked GP is stationary, while a Wishart prior is used on the correlation matrix that informs on correlation between pairs of outputs realised at input pairs.

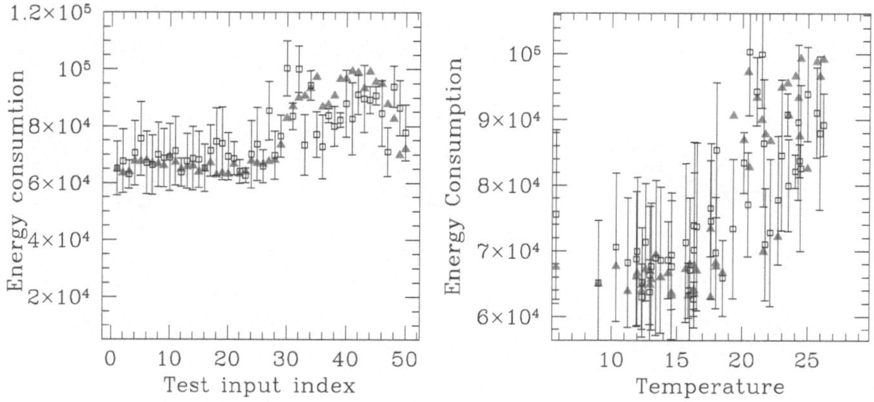

FIGURE 6.2
Figure showing predictions of energy consumed at test inputs, computed from the learning, results of which are presented in Figure 6.1. The predictions are plotted in black against the index of the test input vector (in the *left panel*), True values of the consumed energy are shown in green (or grey in the black and white version). In the *right panel*, same predictions are plotted against the test values of X_1, i.e. temperature. This figure is the same as Figure 3.6 that presented results of using the same test on training data, except in that run, a Wishart prior was not used on the correlation matrix of the multivariate Normal likelihood.

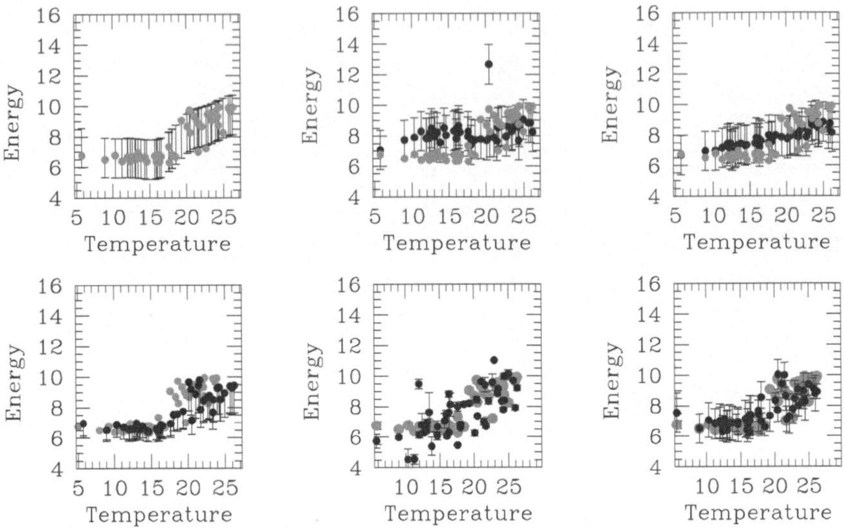

FIGURE 6.3
Predictions on the energy consumption plotted against the input X_1 (temperature), from various runs. Predicted values are in black while true values are in grey – or green, in the coloured version. The *top panels* represent the predictions from learning the energy-temperature relation using the fully non-parametric kernel within a dual-layered learning strategy (*top left*); the non-stationary version of this kernel (*top middle*); the blended kernel (*top right*). The top panels are a reproduction of Figure 4.7. On the *bottom left* is shown the results of predictions following the learning of the univariate energy-temperature relation using a stationary kernel. The other two panels in the bottom row depict results of learning of the multivariate function that represents the relation between energy consumed and a 4-dimensional input vector of meteorological parameters, including temperature. The *bottom right* panel plots the predicted consumed energy against test temperatures when a Wishart prior is used on the correlation matrix of the multivariate Normal likelihood, while the results shown in the *bottom middle* panel are from a learning exercise that did not invoke any informative prior.

relation is a standard protocol relevant acoss disciplines, fed by the data at hand and the learning relevant to the undertaken application. It is this wishlist that the Wishart-driven prior on the inter-realisation correlation matrix successfully relates to.

Indeed, as we say above, for an ideal prior, the protocol of implementation is the same across applications, and yet, the information in the prior is obtained from the learning strategy itself. In our advanced prior strategy, monitoring of prior values across iterations informs on the relevance of the correlation scales that are being worked with. Thus, if the scale matrix – that is computed as an "average" correlation over the n iterations of the undertaken inference – is very different from the correlation manifest in the GP that is currently proposed, the (exponential term in the) Wishart density will bear the signature of such a (wide) difference. Given that our Wishart-prior formulation emerges out of the undertaken learning itself, we refer to it as "self assembled".

6.3.3.1 Discussing the empirical illustration using the Tetouan city data

In the application using the publicly-available Tetouan city data, we find that once the Wishart priors are used, the predictions on energy consumption across indices of the test inputs, tally very well with the true values of the consumed energy. In fact, while these predictions are better than when learning was made using a stationary GP, the predictions are as good as when a dual-layered fully non-parametric kernel model (that is also equivalently non-stationary) was used to make the learning. Thus, the implementation of the learning is easy, while the predictions are good. Figure 6.3 represents the predictions on consumed energy that are made at a given set of test temperatures:

1. using a fully non-parametric kernel within the dual-layered GP model (discussed in Section 4.6.3) that is employed to perform the learning of the univariate function that is the relation between temperature and energy consumption;

2. using the non-stationary equivalent kernel that is discussed in Section 4.8.5 invoked to learn the univariate function;
3. using the Blended model that is discussed in Section 4.9.1 to again learn the univariate function;
4. using a stationary SQE kernel and vague Normal priors, employed within a univariate function learning;
5. using a similar stationary SQE kernel and vague Normal priors in the learning of a multivariate function;
6. using a stationary SQE kernel and a Wishart-distributed prior to learn the function that represents the relation between energy consumption and four meteorological parameters.

The inter-comparison of the results suggest proficiency of the usage of the Wishart prior-included model. In fact, the models that are manifestly non-stationary, do not come across as any better than the model that uses this prior – and this is in spite of the fact that the non-stationary models learn a univariate function, while the Wishart prior-included model is used to learn a multivariate function that takes a 4-dimensional variable as its input. Additionally, the Wishart prior-included model is easier to implement than the non-stationary models.

6.4 Temporal evolution of correlation matrix

As [131] have stated, the process $\{\mathbf{S}_t\}_{t\geq 0}$ is a Generalised Wishart Process (GWP), where $\mathbf{S}_t \in \mathbb{R}^{p\times p}$ distributed as $\mathcal{W}_p(n, \mathbf{V}_t)$, with the time-dependent scale matrix $\mathbf{V}_t \in \mathbb{R}^{p\times p}$ the time-dependent parameter of the Wishart marginal. Here n and p are constant integers. $\mathbf{S}_t$ can be interpreted as the inter-realisation correlation matrix of np random variables at time t.

To contextualise this framework to the GP-based supervised learning that we discuss in the book, at time t, each of these np

random variables is realised – as the output of one of n random functions, computed at one of p inputs, where any of these n random functions is modelled as a sample function of one of n GPs, that differ from each other within n iterations of the pursued inference. Indeed, a fresh sample function is drawn from the relevant GP, when the output is sought at one of the p inputs. If t is the iteration index marking the equilibrium phase of the undertaken inference, then the n GPs vary – in their covariance functions – only slightly. Thus, when using MCMC for inference, if t is the index of a post-burnin iteration, the GPs in question are distinguished by the slight differences amongst the hyperparameters of the kernel – that parametrises the correlation function of any such GP. [131] considered this framework to be dynamic, s.t. the considered random functions are varying with time. Similarly, in our GP-based supervised learning setup, the covariance function of the invoked GPs change with time. In our setup, the Wishart marginal of $\mathbf{S}_t$ – $\forall t \geq 0$ – is useful only for $n \geq p$.

To summarise, in the GP-modelled function learning that we undertake, the inter-realisation correlation matrix is varying with the discrete-valued variable T that is the index of the iteration of the inference strategy. This correlation matrix is varying with t, and the scale matrix as per our definition above, varies with time. While the scale matrix is likely to be similar at all post-burnin iterations, i.e. $\forall t \geq N_{burnin}$ - where burnin of the MCMC chain run to undertake inference, occurs at the N_{burnin}-th iteration – for $t < N_{burnin}$, the scale matrix that parametrises the Wishart prior *pdf* on the correlation matrix, is expected to vary from one iteration to another. This detail notwithstanding, the collection of correlation matrices collated over iterations during which the Wishart prior is employed, is a GWP with distribution at time t given by the $\mathcal{W}_p(n, \mathbf{V}_t)$ density.

[71] go beyond the usage of Wishart priors on correlation matrices, to report a general dynamical modelling of said matrices. [83] present a way for the dynamic estimation of the covariance relevant to brain connectivity estimation, using Wishart processes. [93] advance an approximation of the posterior of a deep Wishart process. [59] discuss inference strategies for Wishart (and inverse

Wishart) processes, for the dynamical modelling of heteroskedastic covariance matrix of a multivariate time series.

[131] offer the closed-form step-one-ahead forecasting of the output variable at the subsequent time point $t+1$, where data is available on the variable till time t. Such forecasting banks on knowledge of evolution of the correlation matrix, where we can specify the distribution of all correlation matrices relevant to times till the t-th time point. Thus, any such step-one-ahead forecasting employs past trends in the evolution of the correlation matrix, in the forecasting. However, it often happens in real-world nonlinear dynamical systems, that evolution in the future does not necessarily follow evolution in the past. So it is desirable to identify independent constraints that link the future with the past. Leaving such discussions aside, we next address the possibility of borrowing Wishart distributed-correlation within the dual-staged GPs-based learning strategy.

6.5 Wishart prior on inter-component correlation matrix in high-dimensional learning

Another context in which the Wishart prior is useful, is when one is attempting the learning of a high-dimensional function $\boldsymbol{f}(\cdot)$, and we seek an informative prior on a correlation matrix that informs on the correlation between the pairs of variables $Y_i^{(a)}$ and $Y_j^{(a)}$ that are output at input values of $\boldsymbol{x}_i$ and $\boldsymbol{x}_j$ respectively. Here $\boldsymbol{Y}_i = \boldsymbol{f}(\boldsymbol{x}_i)$, and $Y_i^{(a)}$ is a component of the tensor-valued $\boldsymbol{Y}$.

To simplify the discussions, let us consider $\boldsymbol{Y}$ to be a vector-valued variable s.t. $\boldsymbol{Y} \in \mathcal{Y} \subset \mathbb{R}^p$. As has been the go-to technique throughout the book, the exercise of learning the p-dimensional vector-valued function $\boldsymbol{f}(\cdot)$ is undertaken by modelling this function with a (vector-variate) GP. The resulting likelihood of model parameters in the (standardised) training data $\{(\boldsymbol{x}_i, \boldsymbol{y}_i)\}_{i=1}^N$ is a matrix Normal density, with

— a mean matrix;

— an $N \times N$-dimensional matrix $\mathbf{\Sigma}_1$ that infoms on correlation between pairs of outputs realised at a corresponding pair of values of $\boldsymbol{X}$. So we refere to $\mathbf{\Sigma}_1$ as the inter-realisation correlation matrix. Here, such an inter-realisation correlation is kernel parametrised with kernel $K(\cdot, \cdot; \theta_1, \ldots, \theta_h)]$ that bears the h hyperparameters $\theta_1, \ldots, \theta_h$;

— a $p \times p$-dimensional matrix $\mathbf{\Sigma}_2$ that informs on the correlation between pair of components of $\boldsymbol{Y}$. So we refer to $\mathbf{\Sigma}_2$ as the inter-component correlation matrix.

Then for standardised data on the outputs, $\mathbf{\Sigma}_1 = [corr(\boldsymbol{Y}_q, \boldsymbol{Y}_r)] = [K(\boldsymbol{x}_q, \boldsymbol{x}_r; \theta_1, \ldots, \theta_h)]$, $(q, r = 1, \ldots, N)$.

However, there exists no effective "input" variable, in terms of which, the inter-component correlation matrix $\mathbf{\Sigma}_2 = [corr(Y_a, Y_{a'})]$ can be kernel parametrised, $(a, a' = 1, \ldots, p)$.

However we can resort to the direct learning of the $p(p-1)/2$ distinct elements of $\mathbf{\Sigma}_2$, where such elements populate the upper (or lower) triangle of this matrix above (or below) the diagonal, by placing a highly informative Wishart prior on $\mathbf{\Sigma}_2$. Such direct learning is in principle possible, if for example, MCMC is employed to undertake inference.

At the t-th iteration of the undertaken inference, the inter-component correlation matrix is

$$\mathbf{\Sigma}_2(t) \sim \mathcal{W}_p(\mathbf{V}(t), t_0),$$

where $t_0 > p$ and at this t-th iteration, the scale matrix $\mathbf{V}(t)$ could be the empirically estimated inter-component matrix, computed using t_0 number of realisations of $\boldsymbol{Y}$ as recorded during the $t-1$-th, $t-2$-th, ..., $t-t_0$-th iterations of the inference that is undertaken to learn $\boldsymbol{f}(\cdot)$, (where $\boldsymbol{Y} = \boldsymbol{f}(\boldsymbol{x})$). Such a scale matrix would be t-dependent, and so we have denoted it using the notation $\mathbf{V}(t)$. Then the prior on $\mathbf{\Sigma}_2$ is

$$\pi_0(\mathbf{\Sigma}_2(t)) = C|\mathbf{\Sigma}_2(t)|^{(t_0-p-1)/2} \exp\left(-\frac{trace(\mathbf{V}(t)^{-1}\mathbf{\Sigma}_2(t))}{2}\right),$$

where C is a constant independent of $\mathbf{\Sigma}_2(t)$ and of t.

For a judiciously chosen t_0, such a prior would definitely enhance the information content of the model and help with the learning. However, the learning of the $p(p-1)/2$ components of $\mathbf{\Sigma}_2$ is computationally burdensome.

6.6 Summary

Real-world learning endeavours are often not blessed by the availability of strong priors, and yet, the learner wishes for an informative prior that would encourage accuracy of predictions made with their learnt function – that represents the relation between a pair of variables. This chapter presents a pan-disciplinary prior that is informed by the learning exercise itself, and is ubiquitous in form s,t, it does not need fine tuning when a new training set is considered. Such a prior structure stems from the Wishart-distributed inter-realisation correlation matrix, any element of which is given by the correlation between the pair of outputs of (random) functions that are computed at respective design points. At an iteration of the iterative inference that is being used to undertake the learning, any element of the scale matrix parameter of this prior density is a correlation, given by averaging over the previous n number of realisations of this correlation, as the learning proceeds. Including the prior is demonstrated to promote accuracy of predictions, over cases when other weak priors were invoked – even compared to when a fully non-parametric and non-stationary kernel is used. We also briefly discussed the possibility of using a Wishart prior on the inter-component correlation matrix, in the learning of high-dimensional functions.

Bibliography

[1] Robert J. Adler. An introduction to continuity, extrema, and related topics for general gaussian processes. *Lecture Notes-Monograph Series*, 12:i–155, 1990.

[2] Samer Albahra, Tom Gorbett, Scott Robertson, Giana D'Aleo, Sushasree Vasudevan Suseel Kumar, Samuel Ockunzzi, Daniel Lallo, Bo Hu, and Hooman H. Rashidi. Artificial intelligence and machine learning overview in pathology & laboratory medicine: A general review of data preprocessing and basic supervised concepts. *Seminars in Diagnostic Pathology*, 40(2):71–87, 2023.

[3] Shirley Almon. The distributed lag between capital appropriations and expenditures. *Econometrica*, 33(1):178–196, 1965.

[4] Matías Altamirano and Felipe Tobar. Nonstationary multi-output Gaussian processes via harmonizable spectral mixtures. In *International Conference on Artificial Intelligence and Statistics*, pages 3204–3218. PMLR, 2022.

[5] S.F. Arnold. *The Theory of Linear Models and Multivariate Analysis.* Probability and Statistics Series. Wiley, 1981.

[6] Ishana Attri, Lalit Kumar Awasthi, Teek Parval Sharma, and Priyanka Rathee. A review of deep learning techniques used in agriculture. *Ecological Informatics*, 77:102217, 2023.

[7] Coryn A. L. Bailer-Jones. *Practical Bayesian Inference: A Primer for Physical Scientists.* Cambridge University Press, 2017.

[8] D. Barber. *Bayesian Reasoning and Machine Learning.* Cambridge University Press, 2012.

[9] Johnathan M Bardsley and Tiangang Cui. A Metropolis-Hastings-within-Gibbs sampler for nonlinear hierarchical-Bayesian inverse problems. *2017 MATRIX Annals*, pages 3–12, 2019.

[10] T. Beckers. *Gaussian Process Based Modeling and Control with Guarantees.* Technische Universität München, 2020.

[11] J. Belk. Non-measurable sets. One of the notes on advanced analysis topics that a course on Measure Thoery was based on, `https://e.math.cornell.edu/people/belk/measuretheory/NonMeasurableSets.pdf`.

[12] Yu K Belyaev. Continuity and hölder's conditions for sample functions of stationary gaussian processes. In *Proc. 4th Berkeley Symp. Math. Statist. and Prob*, volume 2, pages 22–33, 1961.

[13] J. Binney and M. Merrifield. *Galactic Astronomy.* Princeton Series in Astrophysics. Princeton University Press, 1998.

[14] George D Birkhoff. Proof of the ergodic theorem. *Proceedings of the National Academy of Sciences*, 17(12):656–660, 1931.

[15] Adrian N Bishop, Pierre Del Moral, Angèle Niclas, et al. An introduction to Wishart matrix moments. *Foundations and Trends® in Machine Learning*, 11(2):97–218, 2018.

[16] G.E.P. Box and G.C. Tiao. *Bayesian Inference in Statistical Analysis.* Wiley Classics Library. Wiley, 2011.

[17] Ronald N. Bracewell. *Fourier and Other Mathematical Transforms*, pages 1149–1175. John Wiley & Sons, Ltd, 2009.

[18] Steve Brooks, Andrew Gelman, Galin Jones, and Xiao-Li Meng. *Handbook of Markov Chain Monte Carlo.* CRC press, 2011.

[19] Stamatis Cambanis. On some continuity and differentiability properties of paths of gaussian processes. *Journal of Multivariate Analysis*, 3(4):420–434, 1973.

[20] David Milton Cargill. The axiom of choice and the paradoxes of the sphere. OpenBU, `https://open.bu.edu/handle/2144/7825`, 1961.

[21] George Casella, Christian P. Robert, and Martin T. Wells. Generalized accept-reject sampling schemes. *Lecture Notes-Monograph Series*, 45:342–347, 2004.

[22] D. Chakrabarty. Patterns in the outer parts of galactic discs. *Monthly Notices of the Royal Astronomical Society*, 352(2):427–439, 08 2004.

[23] D. Chakrabarty. *Learning in the Absence of Training Data.* Springer International Publishing, 2023.

[24] D. Chakrabarty, K. Wang, G. Roy, A. Bhojgaria, C. Zhang, J. Pavlu, and J. Chakrabartty. Constructing training set using distance between learnt graphical models of time series data on patient physiology, to predict disease scores. *PloS one*, 18(10):e0292404, 2023.

[25] Dalia Chakrabarty. Phase space structure in the solar neighbourhood. *Astronomy & Astrophysics*, 467(1):145–162, 2007.

[26] Dalia Chakrabarty. Automated learning of gravitational mass of elliptical galaxies. *Journal of the Franklin Institute*, 360(3):1635–1671, 2023.

[27] Dalia Chakrabarty. *Bespoke Learning in Static Systems: Application to Learning Sub-surface Material Density Function*, pages 153–188. Springer International Publishing, Cham, 2023.

[28] Dalia Chakrabarty. *Learning the Temporally-Evolving Evolution-Driving Function of a Dynamical System, to Forecast Future States: Forecasting New COVID19 Infection*

Numbers, pages 23–100. Springer International Publishing, Cham, 2023.

[29] Dalia Chakrabarty, Munmun Biswas, and Sourabh Bhattacharya. Bayesian nonparametric estimation of milky way parameters using matrix-variate data, in a new gaussian process based method. *Electronic Journal of Statistics*, 9:1378–1403, 2015.

[30] Dalia Chakrabarty, Nare Gabrielyan, Fabio Rigat, Richard Beanland, and Shashi Paul. Bayesian estimation of density via multiple sequential inversions of two-dimensional images with application to electron microscopy. *Technometrics*, 57(2):217–233, 2015.

[31] Patrick Cheridito, Hideyuki Kawaguchi, and Makoto Maejima. Fractional Ornstein-Uhlenbeck processes. *Electronic Journal of Probability*, 8:1–14, 2003.

[32] Siddhartha Chib and Edward Greenberg. Understanding the metropolis-hastings algorithm. *The American Statistician*, 49(4):327–335, 1995.

[33] Hikmet Can Cubukcu, Deniz Ilhan Topcu, and Sedef Yenice. Machine learning-based clinical decision support using laboratory data. *Clinical Chemistry and Laboratory Medicine (CCLM)*, 62(5):793–823, 2024.

[34] M. Cui and D.Y. Zhang. Artificial intelligence and computational pathology. *Laboratory Investigation*, 101(4):412–422, 2021.

[35] Petros Dellaportas and Gareth O. Roberts. *An Introduction to MCMC*, pages 1–41. Springer New York, New York, NY, 2003.

[36] M.F. Dixon, I. Halperin, and P. Bilokon. *Machine Learning in Finance: From Theory to Practice*. Springer International Publishing, 2020.

[37] F. Donado, R. E. Moctezuma, L. Lopez-Flores, M. Medina-Noyola, and J. L. Arauz-Lara. Brownian motion in non-equilibrium systems and the Ornstein-Uhlenbeck stochastic process. *Scientific Reports*, 7(12614), 2017.

[38] J.L. Doob. Probability in function space. An address delivered on August 22, 1946, `https://www.ams.org/journals/bull/1947-53-01/S0002-9904-1947-08728-0/S0002-9904-1947-08728-0.pdf`.

[39] David Draper. *Bayesian Multilevel Analysis and MCMC*, pages 77–139. Springer New York, New York, NY, 2008.

[40] M.N. Dvivedi. *The Mandukyopanishad: With Gaudapada's Karikas and the Bhashya of Sankara ; Translated Into English by Manilal N. Dvivedi*. Asian Humanities Press, 2003.

[41] Morris L. Eaton. Chapter 8: The Wishart distribution. In *IMS Lecture Notes Monogr. Series*, pages 302–333, 2007.

[42] Bradley Efron and Trevor Hastie. *Objective Bayes Inference and MCMC*, page 233–263. Institute of Mathematical Statistics Monographs. Cambridge University Press, 2016.

[43] D.F. Elliott. *Handbook of Digital Signal Processing: Engineeing Applicatiions*. Academic Press, 1987.

[44] William R Emerson. The Hausdorff Paradox for general group actions. *Journal of Functional Analysis*, 32(2):213–227, 1979.

[45] Marek Fisz. Characterization of sample functions of stochastic processes by some absolute probabilities. In *Proceedings of the Fourth Berkeley Symposium on Mathematical Statistics and Probability*, volume 2, pages 143–151. Univ of California Press, 1961.

[46] G. E. Forsythe. Generation and testing of 1,217,370 "random" binary digits of the SWAC (abstract). *Bulletin of the American Mathematical Society*, 57(4):304, 1951.

[47] J.F.L. Gall. *Measure Theory, Probability, and Stochastic Processes.* Graduate Texts in Mathematics. Springer International Publishing, 2023.

[48] P.R. Gamble and J. Blackwell. *Knowledge Management: A State of the Art Guide.* Kogan Page Series. Kogan Page, 2001.

[49] D. Gamerman. *Markov Chain Monte Carlo: Stochastic Simulation for Bayesian Inference.* Chapman & Hall/CRC Texts in Statistical Science. Taylor & Francis, 1997.

[50] Alan E. Gelfand. Gibbs sampling. *Journal of the American Statistical Association*, 95(452):1300–1304, 2000.

[51] A. Gelman, J.B. Carlin, H.S. Stern, and D.B. Rubin. *Bayesian Data Analysis, Second Edition.* Chapman & Hall/CRC Texts in Statistical Science. Taylor & Francis, 2003.

[52] Christian Gianoglio, Edoardo Ragusa, Rodolfo Zunino, and Paolo Gastaldo. Efficient digital implementation of n-mode tensor-matrix multiplication. In *2021 IEEE 3rd International Conference on Artificial Intelligence Circuits and Systems (AICAS)*, pages 1–4, 2021.

[53] Josiah Willard Gibbs. *Elementary principles in statistical mechanics: developed with especial reference to the rational foundations of thermodynamics.* C. Scribner's sons, 1902.

[54] W. R. Gilks, N. G. Best, and K. K. C. Tan. Adaptive rejection metropolis sampling within gibbs sampling. *Journal of the Royal Statistical Society. Series C (Applied Statistics)*, 44(4):455–472, 1995.

[55] G. Grimmett and D. Stirzaker. *Probability and Random Processes.* Probability and Random Processes. OUP Oxford, 2001.

[56] A.K. Gupta and D.K. Nagar. *Matrix Variate Distributions.* Monographs and Surveys in Pure and Applied Mathematics. Taylor & Francis, 1999.

[57] M.E. Gurtin. *An Introduction to Continuum Mechanics.* ISSN. Elsevier Science, 1982.

[58] W. Keith Hastings. Monte Carlo sampling methods using Markov Chains and their applications. *Biometrika*, 1970.

[59] Creighton Heaukulani and Mark van der Wilk. Scalable bayesian dynamic covariance modeling with variational wishart and inverse wishart processes. *Advances in Neural Information Processing Systems*, 32, 2019.

[60] Markus Heinonen, Henrik Mannerström, Juho Rousu, Samuel Kaski, and Harri Lähdesmäki. Non-stationary Gaussian Process regression with Hamiltonian Monte Carlo. In Arthur Gretton and Christian C. Robert, editors, *Proceedings of the 19th International Conference on Artificial Intelligence and Statistics*, volume 51 of *Proceedings of Machine Learning Research*, pages 732–740. PMLR, 2016.

[61] T. Hey, K. Butler, S. Jackson, and J. Thiyagalingam. Machine learning and big scientific data. *Philosophical Transactions of the Royal Society A: Mathematical, Physical and Engineering Sciences*, 378(2166):20190054, 2020.

[62] P.G. Hoel, S.C. Port, and C.J. Stone. *Introduction to Stochastic Processes.* Waveland Press, 1986.

[63] R.A. Horn and C.R. Johnson. *Matrix Analysis.* Cambridge University Press, 2012.

[64] Dionissios T Hristopulos. Non-separable covariance kernels for spatiotemporal gaussian processes based on a hybrid spectral method and the harmonic oscillator. *IEEE Transactions on Information Theory*, 2023.

[65] A. Huizing. The value of a rose: Rising above objectivism and subjectivism. All Sprouts Content 197, `http://aisel.aisnet.org/sprouts\_all/197`, 2008.

[66] Lundberg I., Brand J.E., and Jeon N. Researcher reasoning meets computational capacity: Machine learning for social science. *Social Science Research*, PMID: 36334925(1), 2022.

[67] G.R. Iversen. *Bayesian Statistical Inference.* Number no. 43 in Bayesian Statistical Inference. SAGE Publications, 1984.

[68] G. James, D. Witten, T. Hastie, and R. Tibshirani. *An Introduction to Statistical Learning: with Applications in R.* Springer Texts in Statistics. Springer New York, 2013.

[69] P.W. Jones and P. Smith. *Stochastic Processes: An Introduction, Second Edition.* Chapman & Hall/CRC Texts in Statistical Science. Taylor & Francis, 2010.

[70] K.R. Koch. *Introduction to Bayesian Statistics.* Springer Berlin Heidelberg, 2007.

[71] Shiwei Lan, Andrew Holbrook, Gabriel A Elias, Norbert J Fortin, Hernando Ombao, and Babak Shahbaba. Flexible bayesian dynamic modeling of correlation and covariance matrices. *Bayesian analysis*, 15(4):1199–1228, 2020.

[72] Krzysztof Latuszyński, Gareth O. Roberts, and Jeffrey S. Rosenthal. Adaptive Gibbs samplers and related MCMC methods. *The Annals of Applied Probability*, 23(1):66 – 98, 2013.

[73] Neal Lawrence. *MCMC using Hamiltonian dynamics*, page 113–162. Handbook of Markov Chain Monte Carlo, eds. S. Brooks, A. Gelman, G. Jones and X. L. Meng. CRC Press, 2011.

[74] P.M. Lee. *Bayesian Statistics: An Introduction.* Wiley, 2012.

[75] D. Lemm, G.F. von Rudorff, and O.A. von Lilienfeld. Machine learning based energy-free structure predictions of molecules, transition states, and solids. *Nature Communications*, 12:4468, 2021.

[76] H. Levine and Y. Tu. Machine learning meets physics: A two-way street. *PNAS*, 121(27):e2403580121, 2024.

[77] Finn Lindgren, Håvard Rue, and Johan Lindström. An explicit link between Gaussian fields and Gaussian Markov random fields: the stochastic partial differential equation approach. *Journal of the Royal Statistical Society: Series B (Statistical Methodology)*, 73(4):423–498, 2011.

[78] Clayton Littlejohn. *Objectivism and subjectivism in epistemology*, page 142–160. Mitova, Veli (Ed.). Cambridge University Press, 2018.

[79] D.J.C. MacKay. *Information Theory, Inference and Learning Algorithms.* Cambridge University Press, 2003.

[80] Ameur M. Manceur and Pierre Dutilleul. Maximum likelihood estimation for the tensor normal distribution: Algorithm, minimum sample size, and empirical bias and dispersion. *J. Comput. Appl. Math.*, 239:37–49, 2013.

[81] D.L. Mayer and R.O. Collier. *Bayesian Statistics.* Ninth annual Phi Delta Kappa symposium on educational research, `https://books.google.co.uk/books?id=frmVvwEACAAJ`. F. E. Peacock, 1970.

[82] Arman Melkumyan and Fabio Ramos. Multi-kernel Gaussian processes. In *Twenty-second international joint conference on artificial intelligence*, 2011.

[83] Rui Meng, Fan Yang, and Won Hwa Kim. Dynamic covariance estimation via predictive Wishart process with an application on brain connectivity estimation. *Computational Statistics & Data Analysis*, 185:107763, 2023.

[84] Nicholas Metropolis, Arianna W Rosenbluth, Marshall N Rosenbluth, Augusta H Teller, and Edward Teller. Equation of state calculations by fast computing machines. *The journal of chemical physics*, 21(6):1087–1092, 1953.

[85] Calvin C. Moore. Ergodic theorem, ergodic theory, and statistical mechanics. *Proceedings of the National Academy of Sciences*, 112(7):1907–1911, 2015.

[86] Sendhil Mullainathan and Jann Spiess. Machine learning: An applied econometric approach. *Journal of Economic Perspectives*, 31(2):87–106, 2017.

[87] Peter Mörters and Yuval Peres. *Brownian Motion.* Cambridge Series in Statistical and Probabilistic Mathematics. Cambridge University Press, 2010.

[88] Radford M. Neal. Slice sampling. *The Annals of Statistics*, 31(3):705 – 767, 2003.

[89] J v Neumann. Proof of the quasi-ergodic hypothesis. *Proceedings of the National Academy of Sciences*, 18(1):70–82, 1932.

[90] Maximilian P. Niroomand, Luke Dicks, Edward O. Pyzer-Knapp, and David J. Wales. Insights into machine learning models from chemical physics: an energy landscapes approach (el for ml). *Digital Discovery*, 3(4):637–648, 2024.

[91] Marcus M. Noack, Hengrui Luo, and Mark D. Risser. A unifying perspective on non-stationary kernels for deeper Gaussian processes. *APL Machine Learning*, 2(1):010902, 2024.

[92] Steven W Nydick. The Wishart and Inverse Wishart distributions. *Electronic Journal of Statistics*, 6(1-19), 2012.

[93] Sebastian Ober and Laurence Aitchison. A variational approximate posterior for the deep Wishart process. *Advances in Neural Information Processing Systems*, 34:6567–6579, 2021.

[94] A. O'Hagan and M.G. Kendall. *Kendall's Advanced Theory of Statistics: Bayesian inference. Volume 2B.* Number v. 2, pt. 2 in Kendall's Advanced Theory of Statistics, 6th Ed. Edward Arnold, 1994.

[95] Christopher Paciorek and Mark Schervish. Nonstationary covariance functions for gaussian process regression. *Advances in neural information processing systems*, 16, 2003.

[96] Christopher J Paciorek and Mark J Schervish. Spatial modelling using a new class of nonstationary covariance functions. *Environmetrics: The official journal of the International Environmetrics Society*, 17(5):483–506, 2006.

[97] Omiros Papaspiliopoulos and Gareth Roberts. Stability of the Gibbs sampler for Bayesian hierarchical models. *The Annals of Statistics*, 36(1):95 – 117, 2008.

[98] G.A. Pavliotis. *Stochastic Processes and Applications: Diffusion Processes, the Fokker-Planck and Langevin Equations*. Texts in Applied Mathematics. Springer New York, 2014.

[99] Ludovico Piccinato. *Bayesian Statistical Inference: An Overview*, pages 51–63. Springer International Publishing, Cham, 2014.

[100] Christian Plagemann, Kristian Kersting, and Wolfram Burgard. Nonstationary gaussian process regression using point estimates of local smoothness. In Walter Daelemans, Bart Goethals, and Katharina Morik, editors, *Machine Learning and Knowledge Discovery in Databases*, pages 204–219, Berlin, Heidelberg, 2008. Springer Berlin Heidelberg.

[101] N.G. Polson and G.C. Tiao. *Bayesian Inference*. Number 2 in Bayesian Inference. E. Elgar Publishing Company, 1995.

[102] Emilio Porcu, Reinhard Furrer, and Douglas Nychka. 30 years of space-time covariance functions. *WIREs Computational Statistics*, 13(2):e1512, 2021.

[103] Naveed Rabbani, Grace Y.E. Kim, Carlos J. Suarez, and Jonathan H. Chen. Applications of machine learning in routine laboratory medicine: Current state and future directions. *Clinical Biochemistry*, 103:1–7, 2022.

[104] Kandethody M. Ramachandran and Chris P. Tsokos. Chapter 4 - sampling distributions. In Kandethody M. Ramachandran and Chris P. Tsokos, editors, *Mathematical Statistics with Applications in R (Third Edition)*, pages 147–177. Academic Press, third edition edition, 2021.

[105] C.E. Rasmussen and C.K.I. Williams. *Gaussian Processes for Machine Learning*. Adaptive Computation and Machine Learning series. MIT Press, 2005.

[106] Sami Remes, Markus Heinonen, and Samuel Kaski. Nonstationary spectral kernels. *Advances in neural information processing systems*, 30, 2017.

[107] C. Robert and G. Casella. *Monte Carlo Statistical Methods*. Springer Texts in Statistics. Springer New York, 2005.

[108] G. O. Roberts and O. Stramer. Langevin diffusions and metropolis-hastings algorithms. *Methodology and computing in applied probability*, 4:337–357, 2002.

[109] Gareth O. Roberts and Jeffrey S. Rosenthal. Harris recurrence of Metropolis-within-Gibbs and trans-dimensional Markov Chains. *The Annals of Applied Probability*, 16(4):2123–2139, 2006.

[110] L.C.G. Rogers and D. Williams. *Diffusions, Markov Processes, and Martingales: Volume 1, Foundations*. Cambridge Mathematical Library. Cambridge University Press, 2000.

[111] Oscar Rojo and Héctor Rojo. Some results on symmetric circulant matrices and on symmetric centrosymmetric matrices. *Linear Algebra and its Applications*, 392(1):211–233, November 2004.

[112] J.S. Rosenthal. *A First Look at Rigorous Probability Theory*. G - Reference,Information and Interdisciplinary Subjects Series. World Scientific, 2000.

[113] Abdul Rahim Salam and Abdelaaziz El Hibaoui. Comparison of machine learning algorithms for the power consumption prediction : Case study of Tetouan city. *2018 6th International Renewable and Sustainable Energy Conference (IRSEC)*, pages 1–5, 2018.

[114] Paul D Sampson and Peter Guttorp. Nonparametric estimation of nonstationary spatial covariance structure. *Journal of the American Statistical Association*, 87(417):108–119, 1992.

[115] T.J. Santner, B.J. Williams, and W.I. Notz. *The Design and Analysis of Computer Experiments.* Springer Series in Statistics. Springer New York, 2019.

[116] Jonathan Schmidt, Mário R. G. Marques, Silvana Botti, and Miguel A. L. Marques. Recent advances and applications of machine learning in solid-state materials science. *npj Computational Mathematics*, 5:83, August 2019.

[117] Douglas Scott and Ali Frolop. Deeper learning in astronomy. *arXiv preprint arXiv:2403.19937*, 2024.

[118] Chris Sherlock, Paul Fearnhead, and Gareth O. Roberts. The Random Walk Metropolis: Linking Theory and Practice Through a Case Study. *Statistical Science*, 25(2):172 – 190, 2010.

[119] X. Si, X. Wu, and Z. et al Li. An all-in-one seismic phase picking, location, and association network for multi-task multi-station earthquake monitoring. *Commun Earth Environ*, 5:22, 2024.

[120] Fabio Sigrist, Hans R. Künsch, and Werner A. Stahel. Stochastic Partial Differential Equation Based Modelling of Large Space–Time Data Sets. *Journal of the Royal Statistical Society. Series B*, 77(1):3–33, 2015.

[121] Robert M. Solovay. A model of set-theory in which every set of reals is Lebesgue measurable. *Annals of Mathematics*, 92(1):1–56, 1970.

[122] D.W. Stroock. *An Introduction to Markov Processes*. Graduate Texts in Mathematics. Springer Berlin Heidelberg, 2013.

[123] T. Tao. *An Introduction to Measure Theory*. Graduate Studies in Mathematics. American Mathematical Society, 2021.

[124] R.J. Thierauf. *Knowledge Management Systems for Business*. Quorum Books, 1999.

[125] Ricardo Carrizo Vergara, Denis Allard, and Nicolas Desassis. A general framework for SPDE-based stationary random fields. *Bernoulli*, 28(1):1 – 32, 2022.

[126] Stan Wagon. *The Banach-Tarski Paradox*. Encyclopedia of Mathematics and its Applications. Cambridge University Press, 1985.

[127] Kangrui Wang, Oliver Hamelijnck, Theodoros Damoulas, and Mark Steel. Non-separable non-stationary random fields. In *International Conference on Machine Learning*, pages 9887–9897. PMLR, 2020.

[128] L. Wasserman. *All of Statistics: A Concise Course in Statistical Inference*. Springer Texts in Statistics. Springer, 2003.

[129] Peter Whittle. Stochastic processes in several dimensions. *Bulletin of the International Statistical Institute*, 40:974–994, 1963.

[130] Andrew Gordon Wilson and Zoubin Ghahramani. Generalised Wishart processes. In Fábio Gagliardi Cozman and Avi Pfeffer, editors, *UAI 2011, Proceedings of the Twenty-Seventh Conference on Uncertainty in Artificial Intelligence, Barcelona, Spain, July 14-17, 2011*, pages 736–744. AUAI Press, 2011.

[131] Andrew Gordon Wilson and Zoubin Ghahramani. Modelling input varying correlations between multiple responses. In Peter A. Flach, Tijl De Bie, and Nello Cristianini, editors, *Machine Learning and Knowledge Discovery in Databases*,

pages 858–861, Berlin, Heidelberg, 2012. Springer Berlin Heidelberg.

[132] Changqing Xu and Kaijie Xu. Random tensors and their normal distributions. *arXiv preprint arXiv:1908.01131*, 2019.

[133] Zhiyong Zhang. A note on Wishart and Inverse Wishart priors for covariance matrix. *Journal of Behavioral Data Science*, 1(2):119–126, 2021.

Index